Structure and Function
of Intrinsically Disordered Proteins

Structure and Function
of Intrinsically Disordered Proteins

Peter Tompa

CRC Press
Taylor & Francis Group
Boca Raton London New York

CRC Press is an imprint of the
Taylor & Francis Group an **informa** business
A CHAPMAN & HALL BOOK

Chapman & Hall/CRC
Taylor & Francis Group
6000 Broken Sound Parkway NW, Suite 300
Boca Raton, FL 33487-2742

Printed in the United States of America on acid-free paper
10 9 8 7 6 5 4 3 2 1

International Standard Book Number: 978-1-4200-7892-3 (Hardback)

Library of Congress Cataloging-in-Publication Data

Tompa, Peter.
Structure and function of intrinsically disordered proteins / Peter Tompa.
p. ; cm.
Includes bibliographical references and index.
ISBN 978-1-4200-7892-3 (hardcover : alk. paper)
1. Proteins--Pathophysiology. 2. Proteins--Structure-activity relationships. 3. Proteins--Metabolism--Disorders. I. Title.
[DNLM: 1. Protein Conformation. 2. Protein Denaturation. 3. Protein Folding. 4. Structure-Activity Relationship. QU 55.9 T662s 2009]

RC632.P7T66 2009
612.3'98--dc22 2009011360

Visit the Taylor & Francis Web site at
http://www.taylorandfrancis.com

and the CRC Press Web site at
http://www.crcpress.com

Dedication

I wish to give special thanks to my mother, who is a mathematician, and my father, who is a physicist. From them, I inherited a love for science and a spiritual foundation without which this book would not exist. Ironically, the most profound message I took from them was not about science, but poetry. They instilled in me a respect for the power of language that has been a driving force for this book. I also thank my wife, Csilla, and our daughters, Rozi and Bori, for their love, patience, and encouragement during the endless days of writing.

Peter Tompa

Dedication

I wish to give special thanks to my mother, who is a mathematician, and my father, who is a physicist. From them, I inherited a love for science and a spiritual foundation without which this book would not exist. Ironically, the most profound message I took from them was not about science, but poetry. They instilled in me a respect for the power of language that has been a driving force for this book. I also thank my wife, Csilla, and our daughters, Rozi and Bori, for their love, patience, and encouragement during the endless days of writing.

Peter Tompa

Dedication

Contents

Foreword

Professor Sir Alan Fersht

It is now half a century since the first crystal structure of a protein (myoglobin) was published, soon to be followed by a series of high-resolution structures. For most of this time, we have admired the beautiful structures of proteins comprised of well-packed helices, sheets linked by turns. A well-folded, albeit dynamic, structure was thought to be the hallmark of protein function. This view was also built on the previous half century where such ideas as lock-and-key specificity of enzymes and complementarity of antibody to antigen structure were guiding principles. The subsequent 50,000 or so structures that are deposited in the Protein Data Bank (PDB) have provided the foundation for our understanding of how enzymes, receptors, transporters, and structural proteins function. Accordingly, it came as a shock to discover that many proteins or regions of proteins are not ordered, but intrinsically disordered. Like all paradigm shifts, the existence of intrinsically disordered or unstructured proteins (IDPs or IUPs), was not immediately accepted and there is still skepticism in some quarters. But, it now seems that ordered proteins and domains cover only about half of the sequence space in various proteomes. Protein disorder reaches high proportions in higher eukaryotes, and is intimately linked with the functions of signaling and regulation, also often causally linked with debilitating diseases such as cancer and neurodegenerative disorders. Peter Tompa's fine comprehensive overview of this rapidly advancing field, *Structure and Function of Intrinsically Disordered Proteins*, is of timely importance, both for its documentation and for emphasizing the importance of IDPs in biology and in protein science.

Peter Tompa addresses the structure, function, and evolution of IDPs at a variety of levels. After a short introduction to the history of the recognition of the phenomenon, he provides insight into the physical principles of protein structure and describes in detail the biophysical techniques applicable for the characterization of IDPs. The book also highlights bioinformatics and proteomic techniques applied for their large-scale discovery and characterization. Detailed description of the structural ensemble of disordered proteins leads to chapters focusing on the functional insight gained by recognizing disorder, such as the functional classification of IDPs, the extension of the structure-function paradigm, and the involvement of structural disorder in disease. This book demonstrates Peter Tompa's considerable command of the field, providing appropriate examples and ample details in every respect. Its coverage of even the latest developments in the field is impressive, and the author manages to strike a good balance between detail and concept to lead the reader through this novel field. Thus, it can be recommended to a wide audience, including researchers actively pursuing IDP research, informed professionals interested in this novel concept, and undergraduate to graduate level students who take on studies in protein science, biochemistry, or molecular biology. Since results of the field of protein disorder also shed new light on the etiology of many well-known diseases, I can also recommend this book to physicians and biomedical researchers, who must better understand the role of structural disorder in human diseases in their efforts to develop novel remedies against them.

Alan Fersht
Cambridge, 2009

Preface

Throughout my career, I have shared the view of many colleagues—that the structure of proteins must relate to their function, as clearly witnessed by tens of thousands of actual structures solved and deposited in various databases. The turning point in my research came when I came across the idea of *disorder* in proteins, and I became interested in understanding these rare exceptions to the rule. The introduction of this simple idea has unleashed a wealth of information on proteins that defy the structure-function paradigm. It turns out that these proteins are rather prevalent and play important regulatory and signaling roles.

Due to the breathtaking pace at which novel information is generated these days, it is both easy and difficult to write a book on this subject. It is easy because so many observations have been made that the subject lends itself to extensive coverage. It is difficult because the concepts are in continuous flux due to the constant outpouring of information. At least one thing is clear: The structural disorder of proteins deserves to be surveyed comprehensively. This book is my attempt to help reach this ambitious goal. I hope it will inspire future work in the field.

About the Author

Peter Tompa, Ph.D., graduated from the University of Budapest (Eötvös Loránd University [ELTE]) in 1984, and has been working since then at the Institute of Enzymology, Biological Research Center of the Hungarian Academy of Sciences in Budapest, Hungary. His basic disciplines are protein science and enzymology, and his research activity has included studies on interactions of soluble enzymes, proteins within the cytoskeleton, and the calcium-activated intracellular protease, calpain. Having studied two intrinsically disordered proteins (IDPs)—microtubule-associated protein 2 (MAP2) and calpastatin—in 2000, Dr. Tompa's interest turned to studying proteins without a well-defined structure, which has been his major activity ever since.

Dr. Tompa played an influential role in initiating studies on IDPs and has been instrumental in the subsequent rapid expansion of the field. He has contributed basic discoveries to the field, such as the role of transient structural elements in the recognition function of IDPs, their possible chaperone functions, the evolution of IDPs by repeat expansion, the role of the structural malleability of IDPs in their promiscuous activity (i.e., moonlighting), and the persistence of structural disorder in the partner-bound state (fuzziness). He studied the molecular principles of the interactions of IDPs, tackled the problems of determining their concentrations, developed a novel algorithm for the sequence-based prediction of disorder, and created the characterization of a novel 2-D electrophoresis technique for their rapid experimental identification. These novel methods enabled the characterization of the evolutionary advance and prevalence in eukaryotic proteomes of protein disorder. Dr. Tompa has also contributed several influential reviews and several book chapters. In all, he has contributed 30 papers to the field, including some basic ones toward developing the concept of disorder.

Dr. Tompa has taught courses on protein disorder at the ELTE University in Budapest, Hungary, and at the Weizmann Institute of Science in Rehovot, Israel. In 2007, he was invited to the East China University of Science and Technology (ECUST, Shanghai) to lecture on IDPs. He was an invited speaker at the first international meeting of the field—the IDP subgroup meeting at the Annual Meeting of the American Biophysical Society, March 3–7, 2007, in Baltimore, Maryland. He organized the IDP section at the European Biophysical Society Annual Meeting, July 14–18, 2007, in London, England. He was also the main organizer of the first conference of the field, the European Molecular Biology Organization (EMBO) workshop "Intrinsically Unfolded Proteins: From Structure to Function," May 20–24, 2007, in Budapest, Hungary (http://embo-iup.enzim.hu).

About the Author

Acknowledgments

This book could not have been written without my colleagues Bianka Agoston, Denes Kovacs, Attila Farkas, Veronika Csizmok, Zoltan Bozoki, Eszter Hazy, Agnes Tantos, Lajos Kalmar, Hedi Hegyi, Istvan Simon, Andras Perczel, Robert Kiss, Kalman Tompa, and Zsuzsa Dosztanyi, who helped with the figures and gave their advice and comments on the text. I am also indebted to my editor, Luna Han, for her inspiration and assistance throughout the various phases of writing.

Abbreviations and Acronyms

2DE: two-dimensional electrophoresis
4E-BP: 4E-binding protein
ACF: autocorrelation function
AChase: acetylcholinesterase
ACTR: activator for thyroid hormone and retinoid receptors
AD: Alzheimer's disease
AFM: atomic force microscopy
ANS: 1-anilino-8-naphthalene-sulfonic acid
APC: adenomatous polyposis coli
APC/C: anaphase-promoting complex/cyclosome
APP: amyloid precursor protein
ATF: architectural transcription factor
AU: analytical ultracentrifugation
BLAST: Basic Local Alignment Search Tool
BP: biological process
BSA: bovine serum albumin
BSE: bovine spongiform encephalopathy
BSP: bone sialoprotein
CaM: calmodulin
CaMBT: calmodulin-binding target
CBD: cellulose binding domain
CD: circular dichroism
Cdk: cyclin-dependent kinase
CDP: conserved disorder prediction
CFTR: cystic fibrosis transmembrane conductance regulator
CH: charge-hydropahty (plot)
CJD: Creutzfeldt–Jakob disease
CKI: Cip/Kip Cdk inhibitor
CPEB: cytoplasmic polyadenylation element binding protein
CREB: cyclic-AMP response element-binding protein
CSI: chemical shift index
CTD: C-terminal domain
CVD: cardiovascular disease
cytD: cytoplasmic domain
DBD: DNA binding domain

D-box: destruction-box
Df31: decondensation factor 31
DHFR: dihydrofolate reductase
DHN: dehydrin
DHPR: dihydropyridine receptor
DLS: dynamic light scattering
DP: dual personality
DSC: differential scanning calorimetry
DSSP: dictionary of protein secondary structure
ECM: extracellular matrix
EFP: EWS fusion protein
eIF4E: eukaryotic translation initiation factor 4E
eIF4F: eukaryotic translation initiation factor 4F
eIF4G1: eukaryotic translation initiation factor 4G1
ELM: eukaryotic linear motif
EM: electron microscopy
EOM: ensemble optimization method
EPR: electron paramagnetic resonance
ERD: early responsive to dehydration
ESR: electron spin resonance
EWS: Ewing's sarcoma
FCS: fluorescence correlation spectroscopy
FID: free induction decay
FITC: fluorescein-isothiocyanate
FMRP: fragile X mental retardation protein
FnBPA: fibronectin binding protein(A)
FRET: fluorescence resonance energy transfer (also Forster resonance energy transfer)
FTIR: Fourier-transform infrared spectroscopy
GARP: glutamic acid-rich protein
GBD: GTPase-binding domain
GF: gel filtration (chromatography)
GFP: green fluorescent protein
Gnd-HCl: guanidine-hydrochloride
GO: gene ontology
HAT: histone acetyltransferase
HCAP: human cancer-associated proteins
HD: Huntington's disease
HMG: high-mobility group
HMGA: high-mobility group protein A
hnRNPA1: heteronuclear ribonucleoprotein A1
HSQC: heteronuclear single quantum coherence
HTS: high-throughput screening
HTT: Huntingtin
HXMS: hydrogen/deuterium exchange mass-spectrometry
HCA: hydrophobic cluster analysis
I2: inhibitor-2

IDP: intrinsically disordered protein
IDR: intrinsically disordered region
IFSU: intrinsically folded structural unit
ILK: integrin-linked kinase
ITC: isothermal titration calorimetry
IULD: intrinsically unstructured linker domain
IUP: intrinsically unstructured protein
KID: kinase inhibitory domain (in p27) or kinase-inducible domain (in CREB)
KIX: KID-binding domain
LEA: late-embryogenesis abundant
LEF: lymphocyte enhancer binding factor
LH: linker helix
LM: linear motif
MAP2: microtubule-associated protein 2
MAPK: mitogen-activated protein kinase
MBP: myelin basic protein
MD: molecular dynamics
MDM2: murine-double minute 2
MeCP2: methyl CpG-binding protein 2
MF: molecular function
MG: molten globule
MLA: machine-learning algorithm
MMP-9: matrix metalloproteinase 9
MoRE: molecular recognition element
MoRF: molecular recognition feature
MS: mass spectrometry
MSCRAMM: microbial surface components recognizing adhesive matrix molecules
MT: microtubule
MTBR: microtubule-binding region/repeat
M_w: molecular mass
NACP: non-Aβ component of Alzheimer's disease amyloid plaques
NATA: N-acetyl tryptophane amide
NCBD: nuclear coactivator binding domain
NLS: nuclear localization signal
NMR: nuclear magnetic resonance
NN: neural network
NOE: nuclear Overhauser effect
NPC: nuclear pore complex
NQO1: NAD(P)H quinine oxidoreductase 1
NRS: non-restricted site
NTD: N-terminal domain
NU: natively unfolded
Nup: nucleoporin
ODC: ornithine decarboxylase
OPN: osteopontin
PAGE: polyacrylamide gel electrophoresis

PCA: perchloro-acetic acid
PCNA: proliferating cell nuclear antigen
PDB: Protein Data Bank
PEST regions: regions enriched in Pro, Glu, Ser, and Thr
PEVK: Pro, Glu, Val, Lys-rich region
PFG: pulsed-field gradient
PG-SLED: pulse gradient stimulated echo longitudinal encode-decode
PHF: paired helical filament
PIC: pre-initiation complex
PKA: protein kinase A
PMG: pre-molten globule
PONDR®: predictor of natural disordered regions
PP1: protein phosphatase 1
PPII: polyproline II helix
PRE: paramagnetic resonance enhancement
PRG: proline-rich glycoprotein
ProTa: prothymosin alpha
PrP: prion protein
PRP: proline-rich (glyco)protein
PRR: proline-rich region
PSD: post-synaptic density
PSE: preformed structural element
PTB: phospho-tyrosine-binding domain
PTM: post-translational modification
PTTG: pituitary tumor transforming gene
PVA: potato virus A
RD: regulatory domain
RDC: residual dipolar coupling
REM: reflection electron microscopy
RNAP II: RNA polymerase II
RNase: ribonuclease
ROA: Raman optical activity
ROC: receiver operating curve
RP: ribosomal protein
RS: restricted site
RyR: ryanodine receptor
SANS: small-angle neutron scattering
SARA: Smad-anchor for receptor activation
SAS: small-angle scattering
SAXS: small-angle X-ray scattering
SBD: Smad-binding domain
SCS: secondary chemical shift
SDS: sodium dodecyl sulfate
SDSL: site-directed spin-labeling
SE: sedimentation equilibrium
SEC: size-exclusion chromatography

SIBLING: small integrin-binding ligand, N-linked glycoprotein
Sir3p: silent information regulator 3 protein
SLiM: short linear motif
SLBP: stem-loop binding protein
SV: sedimentation velocity
SVM: support vector machine
TAD: trans-activator domain
TAP-tag: tandem affinity purification tag
TBD: tubulin-binding domain
TBP: TATA-box binding protein
TCA: trichloro-acetic acid
Tcf: T-cell factor
TEM: transmission electron microscopy
TFE: trifluoroethanol
TGF-β: transforming growth factor beta
TMAO: trimethylamine N-oxide
TS: transition state
TSE: transmissible spongiform encephalopathy
TTR: transthyretin
Tβ4: thymosin β4
UV: ultraviolet
VNTR: variable number tandem repeat
WASP: Wiskott–Aldrich syndrome protein
WH1: WASP homology domain 1
WH2: WASP homology domain 2
Y2H: yeast two-hybrid

Principles of Protein Structure and Function

1

The principles of protein structure surveyed in this chapter have been established mostly by studying globular proteins. The structure of a globular protein can be described by the coordinates of all its atoms, but this information is often too complex to interpret in terms of function. Thus, scientists have devised a hierarchical vocabulary that can describe different levels of structure from the sequence of amino acids to the spatial arrangement of subunits. Further levels of complexity, such as post-translational modifications, the process of acquiring the 3-D structure (folding), and the description of the unfolded state resembling intrinsically disordered proteins (IDPs), also pertain to the comprehensive structural description of proteins. It has long been thought that the description of (ordered) proteins by these concepts provides a universal key to understanding protein function, a notion termed the classical structure-function paradigm. This book is devoted to demonstrating how this knowledge can be extended to understand how IDPs function.

1.1 PHYSICAL FORCES THAT SHAPE PROTEIN STRUCTURE

Because structural biology has its roots in studying globular (ordered) proteins, the classical concepts of protein structure are better suited for the description of ordered than disordered proteins. Usually, four hierarchical levels are distinguished, such as primary structure (sequence of amino acids in the polypeptide chain), secondary structure (local, often repetitive structural elements [i.e., α-helix, β-strand, turn and coil]), tertiary structure (the fold in space of the entire polypeptide chain, also meaning the spatial arrangement of its secondary structural elements), and quaternary structure (stoichiometry and spatial arrangement of subunits in a multi-subunit protein). These basic structural principles are covered in many textbooks (Garrett and Grisham 2007; Stryer 1995) and serve as a starting point for the description of the "structure" of IDPs (Chapter 10). Apparently, the physical principles governing the structural organization of the two classes of proteins are the same; only the balance between various components and dynamics of the emerging structure differ.

It is generally held that the three-dimensional (3-D) structure of a protein is encoded in the succession (sequence) of its amino acid building blocks, as governed by an intricate interplay of physical forces between its atoms. The fundamental interaction is the covalent bond, which defines the connectivity of atoms. Because nonbonded atoms cannot penetrate each other, at a very short-range repulsion is the decisive type of nonbonding interaction within the protein, which precludes a large number of structural states (Rose et al. 2006) and limits the universe of available 3-D folds of the protein. Repulsion is indiscriminate; thus fine details of the structure of a protein are decided by specific weak attractive forces.

The most widespread attractive/repulsive interaction between atoms or segments of the protein occurs between partially separated charges, characterized by a dipole moment. Different types of this interaction, collectively termed the van der Waals interaction, may occur between permanent dipoles, between a permanent dipole and an induced dipole, and between two induced dipoles. Atoms may also have unit charges: several amino acids have either acidic or basic character (i.e., they possess a net electric charge at neutral pH) (Figure 1.1). Charges of opposite sign attract each other, whereas charges of the same sign repulse each other, with the strength of their interaction described by Coulomb's law:

$$F = k\frac{q_1 \times q_2}{r^2} \tag{1.1}$$

where q_1 and q_2 are the charges at each atom separated by a distance r, and constant k depends on the dielectric constant of the medium. The attractive interaction of two opposite charges is called a salt bridge.

A physical interaction of special importance in protein structure is the hydrogen bond (H-bond), which is the attraction between two electronegative atoms mediated by a hydrogen atom covalently linked to one of them (donor). Attraction between the partial positive charge of the hydrogen atom and the partial negative charge of the adjacent electronegative atom (acceptor) results in the formation of a partial covalent bond. H-bonds typically form between carbonyl oxygens and amide hydrogens of the backbone, and represent the major stabilizing force of repetitive (or turn) secondary structural elements.

The structural fate of proteins also depends on the interactions of their residues with solvent water. In general, polar residues interact favorably, whereas apolar residues interact unfavorably with water. The relative tendency of amino acids to interact with water is expressed in terms of hydrophobicity or hydropathy, numerically expressed in scales such as the "Kyte–Doolittle" (Kyte and Doolittle 1982) and "Sweet–Eisenberg" (Sweet and Eisenberg 1983) scales (see Figure 1.1). The importance of this interaction stems from the fact that the vicinity of an apolar/hydrophobic residue limits the conformational freedom of water molecules. Thus, the release of such water molecules is highly favorable and provides for the hydrophobic effect, which drives protein folding (see Section 1.6).

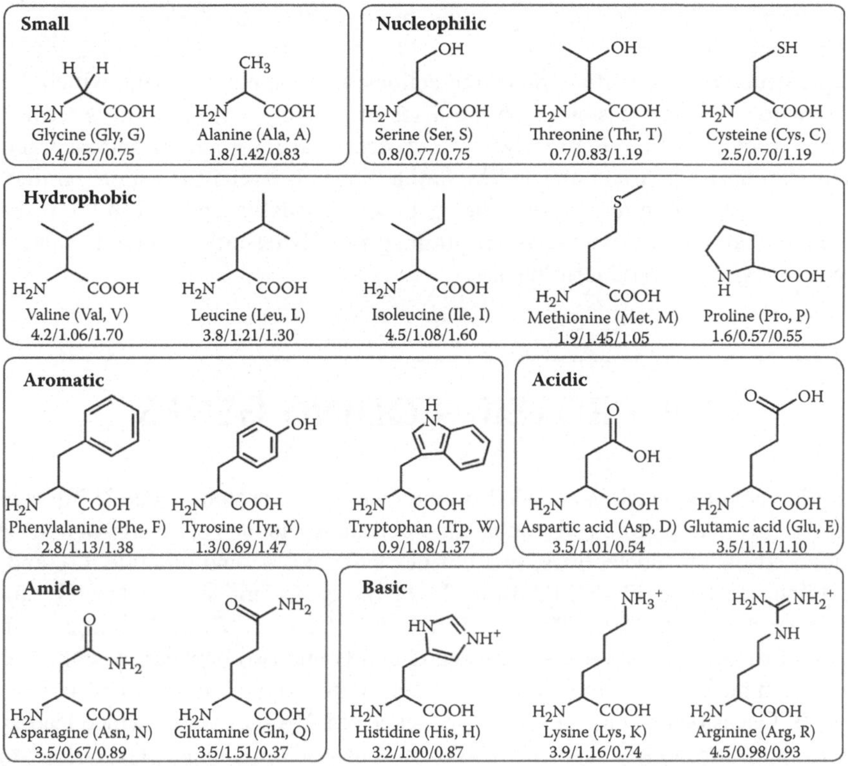

FIGURE 1.1 Basic features of the 20 amino acids of proteins. The structure and basic physicochemical features of amino acids (shown by their standard three-letter and one-letter codes). The three numbers below the name represent their tendency to interact with water (hydropathy or hydrophobicity, as given by the Kyte–Doolittle scale [data from Kyte and Doolittle 1982]), and preference to be found in secondary structural elements α-helix and β-sheet [data from Chou and Fasman 1978].

1.2 PRIMARY STRUCTURE: AMINO ACID SEQUENCE

The general structure of the 20 amino acids that make up proteins is H_2N-CH(R)-COOH (i.e., they contain a uniform α-amino-carboxylic acid element, which makes up the backbone of the protein and a variable "R" side-chain) (Figure 1.1). The only exception is proline, which is an imino acid, with its side-chain bonded to its amino nitrogen. Amino acids can exist in two different enantiomeric forms, L and D, because their α-carbon is surrounded by four different substituents. Proteins in the living world exclusively use the L form. The side-chain of amino acids endows them with unique physicochemical characteristics (e.g. acidic, basic, small hydrophilic, nucleophilic, and large hydrophobic) (Figure 1.1), which provides proteins the chemical toolkit to build a unique structure and mediate their interactions with the environment, which defines their function. The succession of amino acids in the polypeptide chain is termed the primary structure, which is usually the first structural information obtained about a protein. Amino acids are connected by amide bonds between their alpha amino and carboxylic groups (Figure 1.2), and the sequence is usually rendered in the direction from the first amino acid with a free α-amino group (N-terminus) to the last one, with a free carboxylic group (C-terminus).

1.3 PROTEIN-CODING GENES

The amino acid sequence is encoded by the gene of the protein. Genetic information is almost universally stored in double-stranded deoxyribonucleic acid (DNA), which is composed of four nucleotides, two purines (adenine, A, and guanine, G), and two pyrimidines (cytosine, C, and thymine, T; this latter nucleotide is replaced by uracil, U, in ribonucleic acid, RNA). The two strands of DNA are structurally complementary, because of three hydrogen bonds between C and G and two between A with T. Thus, in DNA synthesis during replication of a cell, both original strands can be used as templates. Occasionally, changes in the sequence of DNA may occur, which lead to differences in the amino acid sequence of the protein. Such mutations play a basic role in evolution by generating variants of the protein.

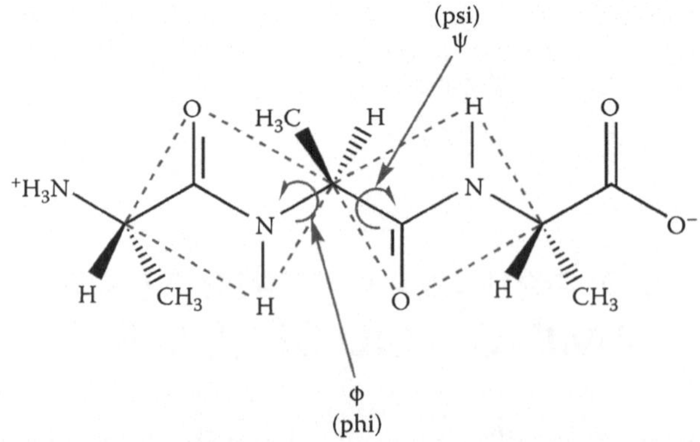

FIGURE 1.2 Local structure and dihedral angle around the α-carbon in a polypeptide chain. A useful descriptor of local conformation of a polypeptide chain is the pair of dihedral angles of the rotation of two planar peptide bonds around the alpha carbon. The example shown is an Ala_3 tripeptide: The four atoms of the peptide bonds on either side of Cα are found within a plane, the position of which is described by two torsion angles, Φ (defined as C′-N-Cα-C′) and Ψ (defined as N-Cα-C′-N).

The basic units of genetic information are genes, which, at the first approximation, correspond to segments of DNA that encode for a protein. The information is first transcribed, in which a messenger RNA (mRNA) molecule is synthesized and then translated by the ribosome to give rise to a protein molecule. The sequence of amino acids within the polypeptide chain is defined by the succession of codons (nucleotide triplets) within the gene. Because there are four types of nucleotides in DNA, $3^4 = 64$ different codons exist. Four of these signal for the initiation (start codon, AUG, also encoding for Met) and termination (stop codons, UAA, UAG, and UGA) of protein synthesis; thus 61 actually encode for amino acids. Several amino acids have more than one corresponding codon (i.e., the genetic code is redundant in this sense).

The nucleotide sequence of the gene and the amino acid sequence of the polypeptide chain of the protein are colinear (i.e., codons read from the 5′ end, defined by the 5′ hydroxyl group of ribose units within the backbone of DNA) toward the 3′ end of the gene correspond to amino acids, starting from the N-terminus toward the C-terminus within the protein. The sequence is also determined by covalent changes in mRNA, because its intervening regions (introns) are removed, and the rest (exons) are joined together in a process termed *splicing*, to yield mature mRNA. It is estimated that in about one-half to two-thirds of eukaryotic genes, splicing can occur in more than one way (Blencowe 2006; Huang et al. 2005; Kim, Magen, and Ast 2007), and such alternative splicing generates variants of the same protein.

1.4 POST-TRANSLATIONAL MODIFICATIONS OF AMINO ACIDS

The polypeptide chain after synthesis may function without further chemical modification, but in many cases it may undergo additional post-translational chemical modifications, which either extend the range of chemical functionalities of amino acids or change the function of the protein for the purposes of regulation.

The most frequent modification is the formation of disulfide bonds between thiol groups of Cys residues, which is intimate to the stabilization of 3-D structure. Disulfide bonds can form spontaneously, but their formation can also be assisted by specific enzymes known as protein disulfide isomerases. The tendency of cysteines to spontaneously form disulfide bonds is probably one of the major reasons why IDPs have a low level of this amino acid.

Another common modification of side-chains is the enzymatic phosphorylation of Ser, Thr, and Tyr residues. This modification is carried out by protein kinases, and it can be reversed by protein phosphatases. Reversible phosphorylation often causes changes in function, and thus it is used very extensively for regulatory purposes. Specificity of recognition by the kinase comes from the primary sequence flanking the site of phosphorylation.

Proteins may also be glycosylated on their amine or hydroxyl groups. Such modification adds single or multiple branched carbohydrate moieties to the polypeptide chain (i.e., N-linked glycans are attached to the amide nitrogen of Asn, and O-linked glycans are attached to the hydroxy oxygen of Ser or Thr side chains), which may increase solubility, lengthen the biological lifetime of the protein, or modify its interactions with other constituents of the cell. Glycosylated proteins are often involved in highly specific cell–cell contacts or interactions between the cell and the extracellular matrix.

There are many other less-frequent but important regulatory post-translational modifications, such as acetylation, myristoylation, methylation, sulfonylation, and nitrosylation. A special modification is targeted at the backbone of the protein: Enzymes of proteolytic activity may cleave off segments of the polypeptide chain, with the remaining fragment(s) having an activity different from the intact protein. Such limited proteolysis may be carried out by the proteasome, a large multi-protein complex primarily involved in protein degradation (see Chapter 8, Section 8.3.1) (Liu et al. 2003).

The enzymatic modification of proteins may occur at a few specific residues only, but proteins may also undergo spontaneous chemical modifications in the absence of modifying enzymes at their chemically labile residues. For example, Cys and Met residues may be oxidized, Asn and Gln residues may undergo spontaneous deamidation, or Ser residues may be glycosylated by glucose. Such modifications usually have severe functional consequences.

1.5 HIERARCHICAL DESCRIPTION OF STRUCTURE

Concurrent to synthesis, the polypeptide chain folds up into a unique 3-D structure that is required for its function. It is generally agreed that the primary structure determines the 3-D structure attained (Anfinsen 1973), which can be described by a set of coordinates of the equilibrium position of all the atoms of the protein. The coordinates of thousands of atoms, however, do not provide a sense of comprehensibility of the structure and function of the protein. To this end, a scheme of simplified hierarchical description of the structure at increasing levels of complexity is applied.

1.5.1 Secondary Structure

The local spatial organization of the polypeptide chain is termed *secondary structure* and is usually thought of as repetitive conformational elements. An elegant description of structure at this level has been suggested by Ramachandran in the form of a plot of dihedral (torsion) angles around the α-carbon atom of amino acids (Ramachandran, Ramakrishnan, and Sasisekharan 1963). There are many different possible local arrangements of adjacent residues, but most of them can be described as belonging to one of four major classes.

The four atoms of the peptide bond (–N(H)–C(=O)–) connecting adjacent amino acids in the sequence occupy a planar orientation due to the partial double-bond character of the central N–C bond. Thus, local structure around the α-carbon of an amino acid can be adequately described by two torsion angles, Φ (phi) and Ψ (psi), corresponding to the rotation of the two adjoining amide plains around the bond connecting them to the Cα (Figure 1.2). The Ramachandran plot (Figure 1.3) describes local conformation of a polypeptide chain by the Φ,Ψ pairs. Large parts of the plot correspond to disallowed conformations, which do not occur in actual proteins. Different amino acids have different propensities to occur in different regions of the plot (i.e., in different secondary structural elements) (Figure 1.1).

Residues in α-helices typically adopt backbone Φ,Ψ dihedral angles around –60°, –45° (Figure 1.3). The resulting structure is repetitive, in which the polypeptide chain takes turns so that the carbonyl oxygen of each peptide bond is H-bonded to the amide hydrogen of the fourth peptide bond in the chain. The helix has 3.6 residues per turn, with the H-bonds lying almost parallel with its axis (Figure 1.4A). Often, the distribution of residues in the sequence creates an α-helix with sides of distinct physicochemical character (i.e., an amphipathic helix, which has a hydrophobic/apolar and a hydrophilic/polar face).

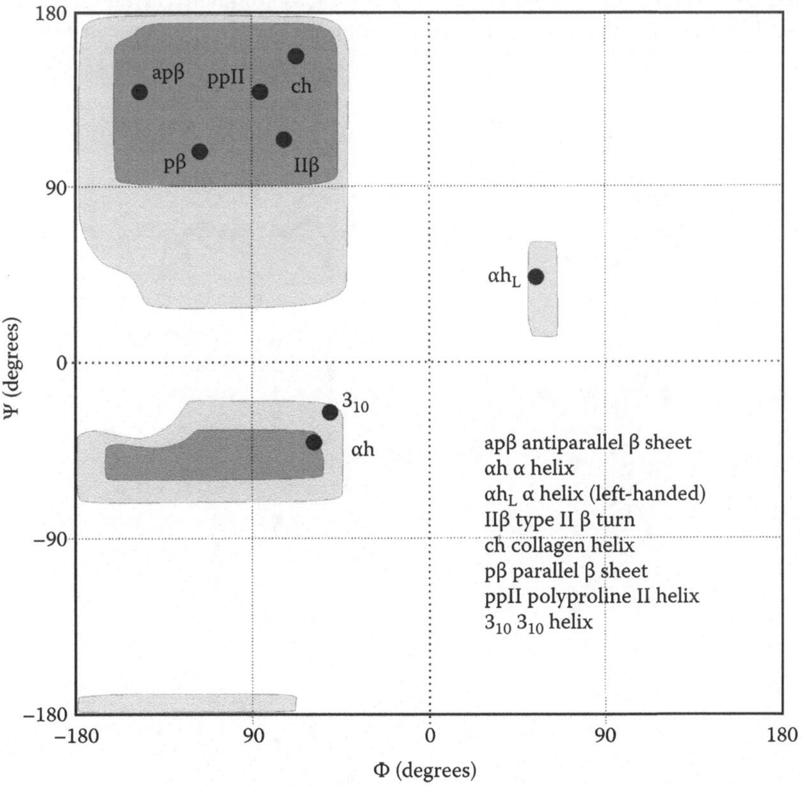

FIGURE 1.3 Ramachandran plot: peptide bond dihedral angles in proteins. A Ramachandran plot of major preferred (dark gray) and allowed (light gray) Φ, Ψ angle pairs in proteins, with the position of repetitive secondary structures marked. Most of the area (white) on the plot corresponds to disallowed conformations.

Another basic building block of proteins is the β-sheet, constituted by an extended helical structure, the β-strand. The energetically preferred dihedral angles of the β-strand are around –135°, 135° (Figure 1.3). The polypeptide chain in a β-strand is almost fully extended, having 2.0 residues per turn, and it cannot make intrachain H-bonds (Figure 1.4B). Thus, a β-strand is only stable when it is part of a sheet, in which adjacent strands are connected by interchain H-bonds. The two basic arrangements are parallel and antiparallel β-sheets. A special case of β-sheets is the amyloid, a pathological state of proteins when individual molecules form a fiber of practically infinite length. The amyloid, in the structural sense, is a β-sheet composed of strands, which can be extended indefinitely (see Chapter 15, Section 15.5.3).

To form a globular structure, the direction of the polypeptide chain has to reverse, which is usually attained by a secondary structural element called the turn. The turn has several variants, depending on which amino acids provide it critical stabilizing interactions. The most common arrangement is the β-turn (Figure 1.3), which is frequently used for reverting the chain to form a β-hairpin of two adjacent antiparallel β-strands. A β-turn is characterized by H-bonds in which the donor and acceptor residues are separated by three residues (i → i+3 H-bonding); less-frequent variants are the γ-turn (i → i+2 H-bonding) and α-turn (i → i+4 H-bonding). Due to the variety of types

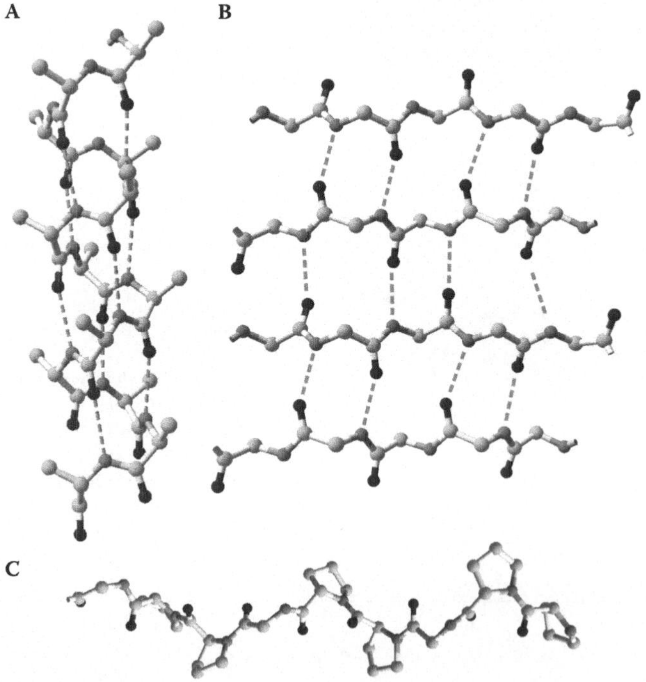

FIGURE 1.4 Typical secondary structural elements of proteins. Local conformation of the polypeptide chain in a protein often assumes repetitive conformations, such as α-helix (A, an oligo-Ala segment), β-sheet (B, shown here to be composed of antiparallel strands, TOP7 structure, pdb 1qys), or PPII helix (C, collagen model peptide, pdb 2d3f).

of turns and the structural heterogeneity of individual residues in them, they occupy different regions in the Ramachandran plot.

Left-handed PPII helix conformation has not been recognized for a long time as an individual secondary structural element, but comprehensive studies of ordered (Adzhubei and Sternberg 1993) and intrinsically disordered (Syme et al. 2002) proteins provided evidence for the frequent occurrence of this structural element. PPII is the most fully extended secondary structural state of the polypeptide chain, with about three residues per turn (Figure 1.4C). The polypeptide chain in PPII conformation derives its stability mostly from H-bonds made with water molecules. Its location (–75°, 150°) on the Ramachandran plot (Figure 1.3) partially overlaps with the β-strand region.

There are segments of proteins which cannot be described by the repetition of any of the previously described structural states, but their local conformation varies from residue to residue. These regions are called coils, and at the extreme the whole protein may be constituted of such segments (i.e., loopy proteins) (Liu, Tan, and Rost 2002). When (part of) a protein fluctuates among many alternative conformations, without a discernible preference for any of the foregoing secondary structural states, it is termed a *random coil.* Although a fully structureless state probably does not exist even under highly denaturing conditions (Kohn et al. 2004; Shortle 1996), this expression very frequently occurs in the literature.

Secondary structural elements in actual proteins are usually not confined to the very narrow range of Φ,Ψ angles defined, which introduces some uncertainty into structural annotation of residues. This situation may be treated by applying a more thorough set of definitions, as suggested in the dictionary of protein secondary structure (DSSP) approach (Kabsch and Sander 1983). The DSSP scheme is founded not on angles but on the presence/absence of H-bonds, defined by a threshold value –0.5 kcal/mole of interaction calculated from partial charges and interatomic distances. Two elementary H-bond types are defined, and a *turn* occurs when there is a H-bond between C=O of residue (i) and NH of residue (i + n), where n = 3, 4, or 5, whereas a *bridge* is defined between two (parallal or antiparallel) stretches of tripeptides if the actual residues (i and j) that form a H-bond are more remote in sequence than in the case of the turn. A minimal helix is then defined as two consecutive n-turns, whereas a longer helix is described as overlapping minimal helices (an α-helix, for example, is described as repeating 4-turns). A *ladder* is defined as a set of consecutive bridges of identical type, whereas a *sheet* is defined as one or more ladders connected by shared residues. The extraction of such patterns from structures is easily automated.

1.5.2 Tertiary Structure

Strictly speaking, the tertiary structure of a protein is the folding of its polypeptide chain in 3-D space, described by the coordinates of all its atoms. Because this description is difficult to interpret, usually a simplified description of the topology of its secondary structural elements is used instead.

In the case of certain elongated fibrous proteins, the structure is built up of a simple secondary structural element. The long fiber of α-keratin, for example, is composed of a structural unit of two long α-helices twisted around each other (termed a two-stranded coiled coil). Fibroin and β-keratin found in silk fibers are composed of stacked antiparallel β-sheets. Collagen is a special type of helical structure made up of three helices wound up around each other, each in a conformation close to the PPII helix.

Structures of globular proteins are more complex, because their polypeptide chain folds up into a compact globule. Their interior is usually filled by tightly packed hydrophobic residues, with very few cavities and water mostly excluded. Their packing density is usually on the order of 0.72–0.77, close to that of contacting solid spheres. Most of the polar side chains point outward and interact with solvent water.

Globular proteins represent an enormous variety of individual structures, but considering the arrangement of secondary structural elements they usually fall into one of four broad structural classes (Figure 1.5): antiparallel α-helix, parallel or mixed β-sheet, antiparallel β-sheet, and small metal- and disulfide-rich proteins (Garrett and Grisham 2007). Antiparallel α-helix proteins are dominated by α-helices, usually packed in an antiparallel arrangement, with a slight twist of the helix bundle, as exemplified by hemagglutinin (Figure 1.5A). In the second class, structures are arranged around parallel or mixed β-sheets. Because a parallel sheet distributes hydrophobic side chains on both of its sides, neither side can be exposed to the solvent; thus the sheet is typically found within the core of the structure of proteins, such as in the eight-stranded β-barrel of triose-phosphate isomerase (Figure 1.5B). The hydrophobic residues of antiparallel β-sheets are located on just one side of the sheet, which usually exist as one of two sheets juxtaposed, with their opposite faces exposed to the solvent. An example is soybean trypsin inhibitor (Figure 1.5C). Small proteins often do not fit into any of these categories, because their structure is heavily influenced by liganded metals or disulfide bonds, without which their structure is usually unstable. A characteristic example is insulin (Figure 1.5D).

These descriptions of tertiary structures only apply to simple proteins composed of a single globular unit. Real proteins are usually more complex, containing several autonomous structural regions, which are termed *domains* (Copley, Goodstadt, and Ponting 2003; Ponting et al. 2000; Vogel et al. 2004). In these cases, the above descriptions of tertiary structure actually apply to domains, defined by three distinct definitions. The original definition is that of an autonomous structural unit of a protein (Wetlaufer 1973) (i.e., an element that has the same structure whether or not part of the protein). This structural view is used synonymously to the concept of *fold*, which emphasizes the ability of a domain to acquire a well-defined tertiary structure on its own (Han et al. 2007). A domain may also be considered as a segment of the protein that can be recognized in distinct genetic contexts by virtue of sequence similarity, when it is called a *module* (Patthy 1996). Underlying these definitions is the idea that a domain is a *functional unit* of the protein that carries a distinct function on its own (Vogel et al. 2004).

An additional and often underappreciated level of structural complexity stems from the fact that proteins are not static, but rather undergo constant motions. Mobility has two basic types: one is best approximated as harmonic atomic/collective oscillations about the single, most stable equilibrium conformation, and the other is directed motions of whole segments of the protein (i.e., conformational changes that often form part of the function). The atomic vibrations are very fast, occurring on the order of picoseconds, whereas conformational changes may be much slower, taking seconds or even longer.

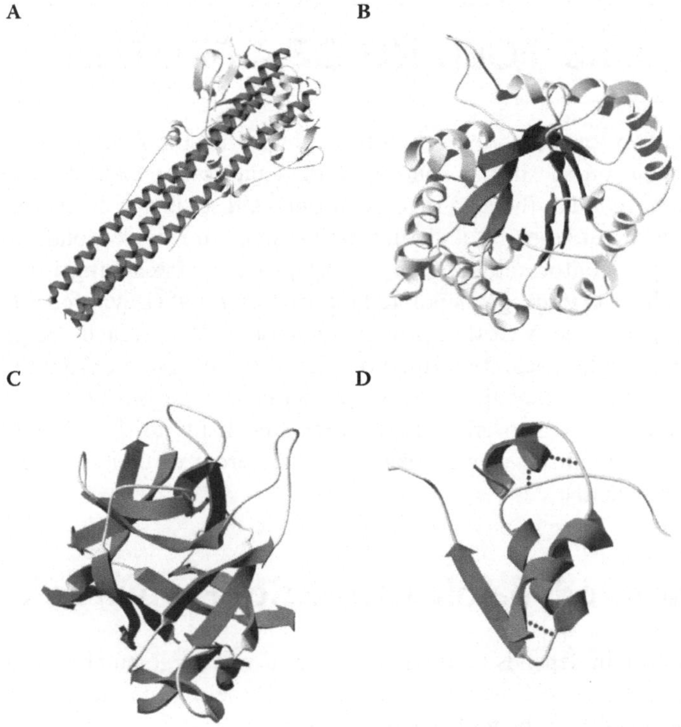

FIGURE 1.5 The four major classes of structures of globular proteins. Ribbon diagrams of globular proteins that represent the four major structural classes. (A) Hemagglutinin (pdb 1htm) belongs to the class of antiparallel α-helix proteins. (B) The eight-stranded β-barrel of triose-phosphate isomerase (TIM, pdb 1r2t) represents the class of parallel or mixed β-sheet proteins. (C) Antiparallel β-sheet structures are exemplified by soybean trypsin inhibitor (pdb 1avu). (D) The structure of small proteins that do not fit into the previous categories is often organized around liganded metals or disulfide bonds, as is the case of insulin (pdb 2zp6).

1.5.3 Quaternary Structure

The native functional state of a protein is often not a single folded polypeptide chain but an assembly of several chains (subunits) in a stable oligomeric species. This quaternary structure can be described by the stoichiometry of subunits, their spatial relations, and eventually the full description of the coordinates of all the atoms of the oligomer. The oligomer may be composed of identical (homomultimer) or different (heteromultimer) subunits, the interaction surfaces of which can be identical (isologous interaction) or different (heterologous interaction). For example, alcohol dehydrogeanse is a symmetric dimer of two identical subunits. Hemoglobin, on the other hand, is composed as a dimer of dimers of two different subunits and has a structure $\alpha_2\beta_2$. More complex cases are tubulin, which is an αβ dimeric protein that polymerizes to form microtubules $(\alpha\beta)_n$, and the closed structure of the coat of tomato bushy stunt virus composed of 180 subunits (Garrett and Grisham 2007).

1.6 FOLDING OF A PROTEIN

In his seminal experiments, Anfinsen has shown that the folded state of a protein does not depend on the initial conditions of denaturation (Anfinsen 1973). These observations have led to the idea that the amino acid sequence determines the native 3-D structure unequivocally, and this native structure corresponds to the global minimum in the conformational space. The process of reaching this state by the polypeptide chain is folding. As pointed out by Levinthal (Levinthal 1969), the conformational space is so vast that proteins cannot fold and reach the global energy minimum by a random search (termed the *Levinthal paradox*). From these two considerations, it follows that the amino acid sequence not only encodes the native, functional state but it also encodes the path that leads to this state. The process of protein folding can be viewed from three different directions: thermodynamic, kinetic, and mechanistic.

1.6.1 Thermodynamic Aspects of Protein Folding

Folding of a protein depends on the interplay of a very large number of weak interactions, including those with water molecules, which determine the difference between the Gibbs free energy of the folded and unfolded states:

$$\Delta G_{total} = \Delta H_{chain} + \Delta H_{solvent} - T\Delta S_{chain} - T\Delta S_{solvent} \tag{1.2}$$

Individual components arise from differences between the unfolded and folded states (Garrett and Grisham 2007). The folded structure is highly ordered; thus $-T\Delta S_{chain}$ is a large positive quantity in the equation. The other terms depend on the nature of amino acid residues in the chain. Apolar groups can better interact with water than with each other; thus ΔH_{chain} is somewhat favorable to the unfolded state. On the other hand, $\Delta H_{solvent}$ is slightly favorable for the folded state, because water molecules can better interact with other water molecules than with exposed apolar side chains. The critical component of the equation, $-T\Delta S_{solvent}$, is large and negative in the presence of apolar groups and strongly favors the folded state, because interaction with apolar groups forces water molecules to become ordered. Usually, this hydrophobic effect drives the burial of apolar residues within the interior of a globule. The net thermodynamic gain of large unfavorable and favorable components, however, is rather small, usually on the order of 10 kcal/mol (40 kJ/mol) for typical globular proteins (Baldwin 2007; Makhatadze and Privalov 1995).

The thermodynamics of folding may also be interpreted in terms of the landscape theory, which approaches folding by the free energy surface of states in the entire conformational space. The underlying assumption is that folding occurs over a funnel-like energy surface (Figure 1.6) that leads from practically all possible starting positions to the global minimum (Dill and Chan 1997). In terms of the terminology of reaction kinetics, this means that there is no single transition state along the folding pathway,

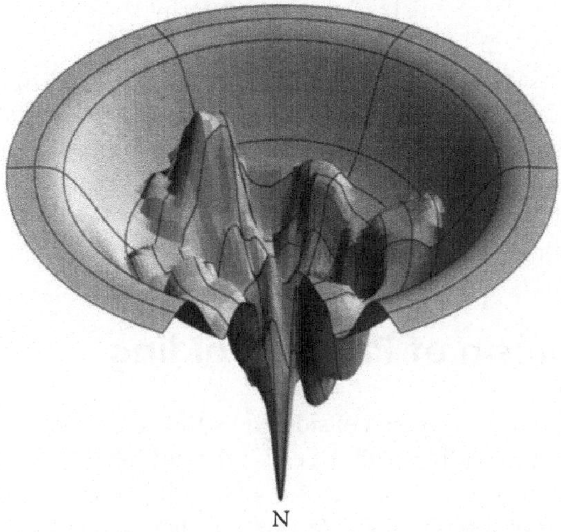

FIGURE 1.6 The folding funnel of proteins. Protein folding usually occurs over a funnel-like surface in the conformational space. The shape of the funnel and a global minimum corresponding to the native (N) state ensure that the protein folds into the same structure from practically any initial denatured/unfolded state. The walls of the funnel are not perfectly smooth, and their ruggedness may occasionally halt folding in local minima (folding traps) of conformational energy. Reproduced with permission from Dill and Chan (1997), *Nat. Struct. Biol.* 4, 10–19. Copyright by Nature Publishing Group.

rather there are several alternatives, and the transition state is only adequately described by a transition-state ensemble. Because the walls of the funnel are not perfectly smooth, folding may occasionally be halted in local minima (folding traps), which manifests itself in both the kinetics and the mechanism of folding.

1.6.2 Kinetic Aspects of Protein Folding

Studies of the kinetics of folding may provide information on the number of folding intermediates and energy barriers and the identity of critical interactions, which may also distinguish between different possible mechanisms. For example, "downhill" or "run-down" folding is a process in which a protein folds without encountering any significant macroscopic free energy barrier. More realistic for most proteins is "two-state" folding (Jackson and Fersht 1991), which assumes folding from the unfolded state (U) to the native state (N) through a single energy barrier:

$$U \leftrightarrow N \tag{1.3}$$

If the protein folds through intermediate state(s) of local energy minima:

$$U \leftrightarrow I \leftrightarrow N \tag{1.4}$$

the structure of these intermediates can be studied by structural techniques (Uversky and Ptytsin 1994). Protein folding may occur on a wide range of timescales. Very fast (run-down) folders, usually small, single-domain proteins, can acquire the native structure in microseconds, whereas it may take minutes or even hours for slow folders to arrive at the native structural state. For these proteins, Pro isomerizations or rearrangements of wrong disulfide bonds can be rate-limiting; thus they must pass through a number of intermediate states that may be considered as "misfolded" with reference to the native conformation (Kim and Baldwin 1990).

1.6.3 Mechanism of Protein Folding

The mechanism of folding may be considered as the atomic-level description of intermediate states along the folding path. Because the unfolded state is a dynamic ensemble, such a description is usually limited to global characteristics and critical features of the transition state of folding. Such descriptions have suggested two seemingly contrasting mechanisms (Figure 1.7). In the "framework" model, the primary event is the stabilization of local secondary structural elements, which then establish long-range tertiary contacts to build up the structure. At the other extreme, initial compaction of the structure driven by hydrophobic interactions, without appreciable formation of secondary structural elements (Daggett and Fersht 2003; Gianni et al. 2003), may occur in a "hydrophobic collapse." Secondary structural elements and tertiary interactions fully form only within the confines of this collapsed state.

These mechanisms have been the subject of a great number of studies, which suggest a possible way of unification. For example, Φ-value analysis of the transition state shows that secondary and tertiary structures often form in parallel as the molecule undergoes a general collapse (Otzen et al. 1994). Good correlation between the decrease in hydrodynamic volume and increase in secondary structure during folding can also be generally observed (Uversky and Fink 2002). In general, the evidence for compact intermediates completely lacking secondary structure and extended states with highly ordered secondary structure is rather limited. The key assumption of the ensuing unified model, "nucleation condensation" (Figure 1.7), is the parallel formation and stabilization of secondary and tertiary structure, in the transition state of which long-range hydrophobic interactions stabilize otherwise weak secondary structural elements (Daggett and Fersht 2003). Deviations in both directions, depending on the strength of the different types of interactions, are feasible.

1.6.4 Folding and Chaperones

Because high macromolecular concentrations in the cell (see Chapter 8, Section 8.1) may cause aggregation of unfolded/partially folded proteins, the process of folding of a protein within the cell is different from that in the test tube. Specialized proteins known as molecular chaperones have evolved to assist the folding of other proteins by several related mechanisms, such as promoting the correct folding of a polypeptide chain,

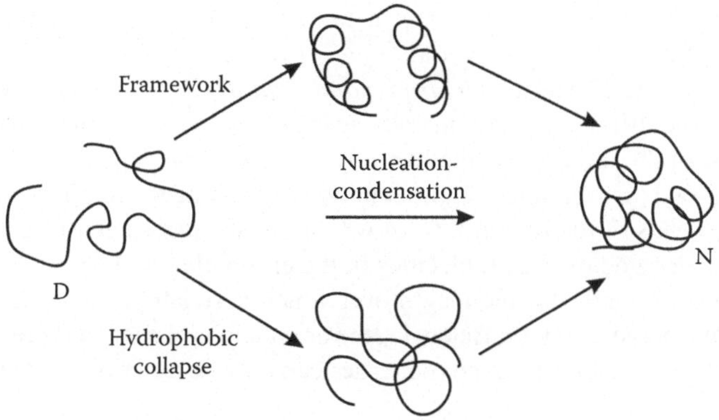

FIGURE 1.7 Mechanisms of protein folding. The scheme of the three possible mechanisms of protein folding. The "framework" model assumes that secondary structural elements form in the open state of the chain, and tertiary contacts are made by these pre-formed elements. The "hydrophobic collapse" model suggests that folding is initiated by the compaction of the polypeptide chain around a hydrophobic core, followed by the formation of secondary structural elements. A combination of the two models, "nucleation-condensation," states that the formation of secondary and tertiary structure occurs in parallel, in a mutually cooperative manner. Reproduced with permission from Daggett and Fersht (2003), *Trends Biochem. Sci.* 28, 18–25. Copyright by Elsevier.

preventing its inappropriate interactions potentially leading to aggregation, and facilitating the assembly of multi-subunit proteins (Csermely 1999; Ellis 2006). Many chaperones have been first described as heat shock proteins (Hsps), because their expression is stimulated by stress conditions such as elevated temperatures. Best-known examples are Hsp90, Hsp70, and Hsp60/Hsp10, known in bacteria as HtpG, DnaK, and GroEL/GroES (see also Chapter 12, Section 12.3.1). These chaperones use ATP energy in their action.

1.7 UNFOLDING OF A PROTEIN: LESSONS FROM POLYMER THEORY

The interest in describing the unfolded/denatured state of proteins has been motivated by the fact that it serves as a reference point for understanding both thermodynamic and mechanistic aspects of folding. Unfolded states are usually generated by denaturing conditions, such as high concentrations of urea (8M) or guanidine hydrochloride (Gnd-HCl, 6M), low pH (2.0), or high temperatures (90–100°C). Structural description of denatured states of globular proteins is in most direct association with describing IDPs (discussed in detail in Chapter 10, Section 10.4). A first approximation of such descriptions is by one of several global hydrodynamic parameters, such as the following:

1. Radius of gyration (R_G, the root mean square distance of atoms from the center of mass, averaged over all molecules and over time)
2. Stokes radius, also termed *hydrodynamic radius* (R_S, R_H, the radius of a hard sphere that diffuses at the same rate as the given molecule; the corresponding volume of the molecule is the hydrodynamic volume, V_H)
3. End-to-end distribution (R_N, the function describing the distribution between the two ends of the protein), from which mean-squared end-to-end distance ($<L^2>$, averaged over all molecules in the ensemble), derives
4. Persistence length (L_P, the length over which correlations in the directions of units of a polymer is lost). Below L_P, the orientations of segments are correlated, whereas for longer pieces the properties can only be described statistically.

These parameters are deeply rooted in polymer theory pioneered by Flory (reviewed in [Flory 1969]) applied to the field of proteins by Tanford (Tanford 1968). The two most frequent models for describing polypeptide chains are the "freely jointed chain" and the "wormlike chain." In the freely jointed chain (Flory 1969), the chain is divided into *N* statistical segments (beads) of size *b*, connected by virtual bonds. The chain performs a random walk, with mean squared distance between units separated by *N* segments, $<R_N^2> = b^2N$, and the radius of gyration:

$$R_G = \frac{1}{\sqrt{6bN^{1/2}}} \tag{1.5}$$

Formally, this description requires that the segments behave independently of each other. To describe chains of finite length and flexibility, the wormlike chain model was developed, into which later refinements introduced the effects of heterogeneity of residues, steric exclusion, and differences in the solvation of side chains. An important parameter of this model is v (second virial coefficient), which describes interactions within the chain. $v < 0$ indicates attraction between segments and a tendency for global collapse, whereas $v > 0$ corresponds to repulsive interactions and indicates an overall swelling of the chain beyond its predicted Gaussian dimensions. $v = 0$ reproduces the ideal walk (i.e., the Gaussian or random-coil behavior). As a function of v, the radius of gyration scales as

$$R_G = R_0 N^{\mu} \tag{1.6}$$

where R_0 is a constant related to persistence length, and μ is the scaling factor, which, depending on v, may take on the value $\mu = 0.33$ (collapsed, spherical molecule in poor solvent), $\mu = 0.5$ (random chain in an ideal solvent, also termed Θ solvent), and $\mu = 0.588$ (an extended volume random coil in a good solvent). $<L^2>$ for unfolded proteins is also expected to scale linearly with chain length ($<L^2> = L_0N$) (Fitzkee and Rose 2004). Tanford provided experimental evidence for the random-coil behavior in the case of globular proteins under highly denaturing conditions by intrinsic viscosity measurements, which yielded $\mu = 0.67$ (Tanford, Kawahara, and Lapanje 1966). In small-angle

X-ray scattering (SAXS) experiments, $\mu = 0.598$ was obtained for denatured proteins (Kohn et al. 2004).

These concepts of polymer theory were adopted by the field of protein folding, from where it arrived to the field of IDPs. Besides the state of random coil, observations of more compact intermediates have led to the concept of molten globule (MG) (Ptytsyn and Uversky 1994) and the somewhat less compact pre-molten globule (PMG) (Uversky and Ptytsin 1994) states. MG is characterized by a large internal flexibility of side chains and backbone, with characteristic hydrodynamic parameters, such as R_G, R_S 1.5–2.0 times larger than that of globular proteins.

1.8 THE LIMITS OF GLOBAL DESCRIPTIONS OF THE UNFOLDED STATE

A simple thought experiment warns that the description of a structural ensemble by a single parameter may be inappropriate and rather misleading, because random-coil statistics is not unique to featureless polymers, even if it successfully predicts the experimentally determined R_G of denatured proteins (Fitzkee and Rose 2004). In this experiment, backbone torsion angles of structures of known globular proteins were varied at random for 8% of the residues while keeping the remaining 92% fixed in their native conformation: The models appear as random coil by the criteria of end-to-end distances and mean R_G. This conclusion can be experimentally corroborated by SAXS of proteins denatured under strongly denaturing conditions (Kohn et al. 2004). These denatured proteins have significant residual structure by spectroscopic studies, but their R_G deviates significantly from the exponent 0.598, a value very close to that expected for excluded volume random coils, in only 2 out of 28 cases. The limitations of describing the denatured state as a random coil are also suggested by the thermodynamic effects of point mutations on folding (Shortle 1996). Many point mutations affect the denatured state, rather than the native state, which suggests residual structure in the unfolded state. Thus, by many distinct observations, the denatured states of proteins have residual structure, which may serve the process of folding. This concept readily penetrated into the field of IDPs.

1.9 DATABASES OF PROTEINS AND PROTEIN STRUCTURES

There are several comprehensive databases of structural and functional information on proteins. The primary and generic database is the Universal Protein Resource, UniProt (http://www.uniprot.org/), which is a comprehensive resource for protein sequence and annotation data. UniProt is built on SwissProt and TrEMBL (http://www.expasy.ch/sprot/), which contain sequence information, functional annotations, and cross-references to many protein resources on organisms, structure, function, interactions, ontology, domains, and other features. The most complete resource of sequence information is in the GenBank at the National Center for Bitotechnology Information (NCBI, http://www.nlm.nih.gov). The database of structures of ordered proteins and nucleic acids is the Protein Data Bank (PDB), which now also appears as the worldwide PDB (http://www.wwpdb.org/). PDB contains the atomic coordinates of more than 50,000 structures solved by X-ray crystallography and nuclear magnetic resonance (NMR) imaging.

The information on sequence and structure in UniProt and PDB requires interpretation in terms of domains (see Section 1.5.2). For example, the Pfam (protein families, http://pfam.sanger.ac.uk/) database (Bateman et al. 2002) is based on hidden Markov models and multiple sequence alignments, emphasizing the evolutionary conservation of protein domains. The SMART (simple modular architecture research tool, http://smart.embl-heidelberg.de/) approach for identifying domains focuses on genetic mobility (Letunic et al. 2002). Other databases employ structural definitions of domains, emphasizing their capacity for autonomous folding. For example, the CATH database (class, architecture, topology, and homologous superfamily, http://www.cathdb.info/) contains a hierarchical classification of protein domain structures studied at four different levels (Orengo et al. 1997), and the SCOP (structural classification of proteins, http://scop.mrc-lmb.cam.ac.uk/scop/) database is based on an evolutionary classification that builds on conserved structural features (Murzin et al. 1995).

1.10 DISPROT: THE DATABASE OF DISORDERED PROTEINS

Comprehensive data on IDPs are found in the DisProt database (http://www.disprot.org/), which contains data on more than 500 proteins, in which experimental evidence of disorder has been provided for about 1,100 segments (Sickmeier et al. 2007). The techniques are dominated by X-ray crystallography, NMR, and circular dichroism (CD). The database can be searched by keywords and sequence similarity, contains links to most other protein databases (UniProt, SwissProt, NCBI, PDB, PubMed), and is annotated by additional functional and structural information. Many intrinsically disordered regions (IDRs) are also found in PDB, as detailed in Chapter 5, Section 5.1.

1.11 THE CLASSICAL STRUCTURE-FUNCTION PARADIGM

At the center of the classical structure-function paradigm is the idea that protein function depends on a well-defined 3-D structure, and detailed description of this structure holds the key to understanding function. The basic idea is that the unique

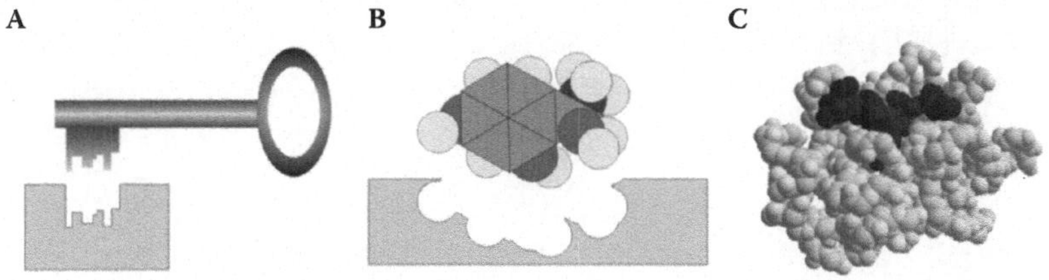

FIGURE 1.8 A well-defined 3-D structure is required for enzyme activity. (A) The classical model of lock-and-key was formulated by Emil Fisher in 1894 to explain stereo-specificity of enzyme catalysis (Fisher 1894). (B) The model assumes that the substrate fits tightly to the binding site on the enzyme as a key into its lock. (C) The perfect fit between the enzyme and its substrate can be mimicked by a tight complex with its inhibitor (trypsin in complex with serpin, pdb 1k90).

spatial pattern of properly placed amino acid residues creates a special physico-chemical microenvironment tailored for the tight and extremely specific binding of ligands, catalysis of chemical reactions, translocation of ions or small molecules, or assembly of specific macromolecular complexes. The success of this paradigm is demonstrated by the structures deposited into the PDB and countless reports on the functional details of enzymes, receptors, transporters, membrane channels, building blocks, mechanochemical proteins, and many other types of proteins (Fersht 1985; Garrett and Grisham 2007; Stryer 1995). The example of enzymes illustrates some of the key points.

Enzymes have a rather well-defined binding pocket for the formation of an enzyme-substrate (ES) complex, as formulated in the classical concept of the lock-and-key hypothesis (Figure 1.8A; see also Chapter 2, Section 2.2.1), which suggested a tight fit between the binding pocket and substrate (Figure 1.8B). The concept is substantiated by the structure of a large number of enzyme-substrate and enzyme-inhibitor complexes (Figure 1.8C). The current theory assumes a perfect fit between the enzyme and the transition state (TS) of the reaction (i.e., that acceleration of the conversion of substrate results from the stabilization of the highest-energy state on the reaction path). Enzymes lower the energy of TS by several possible mechanisms, such as proximity/orientation, the formation of transient covalent bonds, general acid/base catalysis, and complementarity in structure with TS. The residues that directly take part in accelerating the reaction make up the active site. Whereas this model is much more elaborate in its details than the lock-and-key, its basic premise is the perfect positioning of residues, which can only be ensured by a well-defined 3-D structure.

A Brief History of Protein Disorder 2

The classical structure-function paradigm has been the dominant view of proteins for a long time. This chapter provides a historical overview of how observations contradicting this notion have slowly accumulated for decades, eventually leading to the recognition of the possible generality of the phenomenon of structural disorder.

2.1 CAN WE DEFINE DISORDER?

Despite a good deal of structural/functional data that contrast the classical structure-function paradigm (see the DisProt database [Sickmeier et al. 2007]), it is surprisingly difficult to give a definition of structural disorder that applies in a variety of theoretical and experimental situations. Because this is the least we can expect from a book dedicated to this subject, we will show that there are several possible ways of defining disorder, none of which is generally agreed upon.

The classical approach that originates from the first observations is to simply contrast disorder with order, which provides a sort of negative definition but captures the essence of the phenomenon. Ordered proteins can be characterized by a well-defined 3-D structure; therefore disordered proteins are the ones that do not have a well-defined structure. Although it does not clearly define the subject, it does work as an operational definition, and, in many experimental situations, it makes it easy to decide where a protein in this binary classification scheme belongs.

Based on this premise, one might try to give a strict definition, which might be of value for theoreticians but does not offer too much help in actual experimental situations. By definition, structures of ordered proteins reside in a characteristic global minimum in the conformational space. Their atoms fluctuate in time around their equilibrium positions thus defined, and because this applies to all the protein molecules in an ensemble, this permits their structure (i.e., coordinates corresponding to this equilibrium state) to be determined. From this direction, we may simply define disordered proteins as the ones that do not have a single global minimum in the conformational space, but a multiplicity of accessible structural states separated by low energy barriers. The protein constantly fluctuates between these, and function stems from this structural ensemble.

The most severe limitation of this approach is that we have no idea what the actual conformational energy surface of even a single intrinsically disordered protein (IDP) looks like. Even if we had an idea, it would be very difficult to generalize and painstaking to verify if a novel protein conforms to what is considered the rule. From a practical point of view, it seems more useful to return to an operational definition that rests on the result of one or a combination of several experimental protocols.

Actually, the whole field of protein disorder has grown out of observations that the behavior of a protein is in contrast with what a "protein expert" would expect. If a protein can be boiled and still does not precipitate, this behavior can hardly be interpreted in terms of the traditional view of proteins. If it has practically no secondary structural elements by circular dichroism (CD), it is very suspicious behavior. If it is so sensitive to proteolysis that it cannot be purified without fragmentation, we have a good reason to suspect that our protein is not an "ordinary" one. If we see poor resonance dispersion on a nuclear magnetic resonance (NMR) spectrum, there is every reason to assume it is disordered. Of course, its structure might have been spoiled during expression and/or purification, but additional test-tube experiments may provide evidence that the protein is in a state compatible with function. Having accepted such very simple rules of thumb, the field of protein disorder got to a jump start around the year 2000.

2.2 THE HISTORY OF DISORDER

Recognition of the phenomenon of protein disorder has taken many years, especially if we consider how long the central role of proteins in life had been known. The major reason for this long neglect is the early recognition of the importance of protein structure (i.e., protein order) in the function of many proteins (reviewed in [Dunker et al. 2001]). The explanatory power of the classical structure-function paradigm (Chapter 1, Section 1.11) and the range of evidence in support of it have actually put a mental break to recognizing that many proteins were different.

2.2.1 The Legacy of the Lock-and-Key Hypothesis

The victorious history of the structure-function paradigm started with the recognition of stereospecificity in enzyme catalysis, such as the observation that extracts of beer yeast (containing invertase) hydrolyzed α-glucosides but not β-glucosides, whereas emulsin hydrolyzed the latter but not the former. These observations were explained by the "lock-and-key" model (Fisher 1894), which suggested a strict geometric complementarity of the enzyme and substrate (see Chapter 1, Section 1.11 and Figure 1.8). Although not even the fact that enzymes were proteins was clear, the right inference from this model was that an enzyme had to have a well-defined structure. This conclusion was corroborated by later observations that conditions that caused denaturation of proteins (treatment by acid, alkali, or urea) led to the loss of enzyme activity. These denatured proteins could not be crystallized; thus it was concluded (Mirsky and Pauling 1936) that "the characteristic specific properties of native proteins we attribute to their uniquely defined configurations. The denatured protein molecule we consider to be characterized by the absence of a uniquely defined configuration." As a further piece of evidence, Anfinsen observed that ribonuclease (RNase) could be renatured *in vitro* (i.e., it regained activity upon regaining structure) (Anfinsen 1973). The determination by X-ray crystallography of the first protein structures, myoglobin (Kendrew et al. 1958), and hemoglobin (Perutz 1960), followed by the structure of the first enzyme, lysozyme (Blake et al. 1965), have practically precluded any alternative view. These milestones in the history of protein science along with more than 50,000 protein structures in the Protein Data Bank (PDB) have placed it on a firm ground that a highly specific 3-D structure is the absolute and necessary prerequisite of protein function.

2.2.2 Structural Adaptability of Binding Sites

The first exceptions to the static lock-and-key model were provided by observations that the binding site of a protein was not necessarily strictly complementary to its substrate, and that certain enzymes were capable of catalyzing the conversion of several different, although related, substrates. For example, Karush has reported that serum albumin—which is not an enzyme—exhibited a nearly universal capacity for the high-affinity binding of small hydrophobic molecules (Karush 1950). He inferred that the binding site of albumin assumes a large number of configurations in equilibrium, and the best-fitting one becomes stablized upon interacting with a given small molecule. He termed the phenomenon "configurational adaptability."

With respect to enzymes, the lock-and-key theory was criticized on theoretical grounds, because it was realized that it fails to explain the stabilization of the transition state of the reaction. To cope with this problem, Koshland (Koshland 1958) suggested a modification of the model that the substrate does not simply bind to a rigid active site but to an active site that is structurally adapted to the shape that enables it to perform its catalytic function. The active site has the flexibility to accommodate the highest-energy intermediate of conversion (i.e., the "transition state"), which was first put forward by Pauling (Pauling 1948). The underlying configurational adaptability of enzymes was termed "induced fit" by Koshland. Such adaptability in binding was also demonstrated in the case of antigene binding by antibodies achieved by surface regions of high segmental mobility (Westhof et al. 1984). It was suggested that such mobility may make it easier to adjust to an epitope to which the exact geometry of the protein did not fit. For example, the X-ray structure of an antibody, SPE7, in complex with two entirely unrelated antigens (James, Roversi, and Tawfik 2003), demonstrated significant differences in the two complexes.

2.2.3 Polymer Theory and Protein Folding

As outlined in Chapter 1, Section 1.7, many concepts of IDPs came from studies on polymers and the unfolding/folding reactions of globular proteins. Their analogous behavior with some native proteins was slowly recognized.

2.2.4 Caseins Are Different

Perhaps the first documented biophysical study of protein disorder was an optical rotation measurement that compared native casein with native and denatured globular proteins, leading to the conclusion that casein in milk occurs in an unfolded configuration (McMeekin 1952). The importance of this first case comes from the implication that this state is important for the function of the protein (i.e., its digestibility in mother's milk). After this first account on disorder, caseins in the literature have been accepted as structurally "unusual" proteins, considered apparently random coils, because they had little α-helical structure (Creamer, Richardson, and Parry 1981) and had shown no sign of denaturation upon heating (Paulsson and Dejmek 1990). Hydrodynamic measurements also indicated their rather open conformational states: The R_G of β-casein, for example, is much larger than that expected for a globular protein of the same molecular mass (M_W) (Payens and Vreeman 1982).

To account for all these observations, Holt and Sawyer have described the sequence features that are important for maintaining such a structural state (Holt and Sawyer 1993). They observed that fully conserved residues in caseins are extremely rare, and condensation into ordered structures was inhibited by certain conserved features of the primary structure, allowing the protein to maintain an open and mobile conformation. To emphasize that the protein is functional and is not completely featureless despite the observed conformational behavior, they suggested it be termed *rheomorphic* (from the Greek *rheos*, meaning "stream," and *morphe*, meaning "form") instead of *random coil* (Holt and Sawyer 1993). Although this term has not become generally accepted, it represented the first attempt to provide a definition and a functional interpretation of the structural state of an IDP.

2.2.5 If a Protein Does Not Crystallize

The common view that neither the inability to crystallize a protein nor the observation that its segment is missing from an X-ray structure suggests a disordered state has also delayed the recognition of protein disorder. True, ordered proteins often stubbornly resist repeated attempts of crystallization, and missing coordinates might result from a failure to solve the phase problem, crystal defects, or accidental proteolytic removal of the segment during purification. As a matter of fact, disordered segments are often intentionally removed to enable crystallization and are subsequently thought of as "unimportant" for function. Only by the advent of structure solution by NMR has it become compelling that proteins do have disordered segments that often fall within functional regions.

An instructive case of this behavior is that of myelin basic protein (MBP), the major extrinsic protein of the myelin sheath, which underlies membrane compaction at cytoplasmic apposition in the central nervous system. MBP has been extensively studied due to its central role in demyelinating diseases, such as multiple sclerosis. CD experiments have shown MBP to have little, if any, α-helix or β-conformation (Thomas, Weser, and Hempel 1977). In a serious attempt to crystallize and structurally characterize the protein

(Sedzik and Kirschner 1992), it was found that "despite our efforts which included 4,600 different conditions, we were unable to induce crystallization of MBP . . . when it is removed from its native environment in the myelin membrane . . . the protein adopts a random coil conformation and persists as a population of structurally non-identical molecules." Currently, a better definition of disorder is not available. Still, people were reluctant to take notice, although some protein segments with no discernible electron density were already recognized as essential for function (Alber et al. 1983; Bode, Schwager, and Huber 1978; Spolar and Record 1994).

Another protein that resisted crystallization for a long time is microtubule-associated protein 2 (MAP2), which also represents an example of the early observation of protein disorder (Hernandez, Avila, and Andreu 1986). MAP2, the homolog of tau protein involved in Alzheimer's disease (see Chapter 15, Section 15.3.1), is a neuronal microtubule (MT) binding protein, which stabilizes the unstable MT polymer composed of αβ tubulin dimers. This protein was among the first to be recognized as disordered under native, functional conditions. Avila and colleagues showed that heat treatment did not affect its behavior, as assessed by several biophysical techniques (Hernandez et al. 1986). The high frictional ratio, $f/f_0 = 3.7$, in sedimentation equilibrium and gel chromatography suggested that MAP2 was clearly not globular but had either a very elongated shape or an unordered expanded structure. Very little secondary structural content was seen by CD, and this feature was independent of the purification procedure. Overall, MAP2 in solution was described as an "unordered, very flexible and non-compact" protein (Hernandez et al. 1986). Interestingly, a few years later, it was demonstrated by small-angle X-ray scattering (SAXS), CD, and Fourier-transform infrared spectroscopy (FTIR) that tau protein could "behave as if it were denatured, having no compact fold, but a highly extended, random Gaussian polymer, with a minimal content of ordered secondary structure" (Schweers et al. 1994). Because MAP2 and tau are functional analogues but show limited sequence similarity, this similar structural behavior might have focused attention on function of an IDP being maintained in the face of little sequence conservation, an idea the field returned to much later (see Chapter 13, Section 13.4).

It was Paul Sigler, who came closest to generalizing the concept of structural disorder. In a seminal paper on transcription factors (Sigler 1988), Sigler described their DNA-binding domains as structurally well-defined, whereas he noted that mutational and structural studies of their trans-activator domains (TADs) "suggest a disquieting picture of a conformationally ill-defined polypeptide that can function almost irrespective of sequence, provided only that there is a sufficient excess of acidic residues clustered or peppered about." He concluded that eukaryotic transcription relies on nearly shapeless molecules termed "acid blobs" or "negative noodles." Disorder, without using the word, is clearly described as "whereas crystal structures at atomic resolution are crucial to our understanding of … specific molecular interactions, we can imagine many assemblies … whose function requires strong but less precisely defined arrangements than the ones we have seen crystallographically."

2.2.6 The Advent of NMR

The conceptual transition was also anticipated by Dobson (Dobson 1993). He noted that NMR, due to its increasing resolution, shows that terminal segments, loops, and linker regions in otherwise globular proteins actually have such high main-chain mobilities that qualify them as disordered. He suggested that such regions in interleukin-4, GroES, pyruvate dehydrogenase, and eglin c are actually involved in protein–protein recognition, in which disorder provides advantages, such as facilitated spatial search and reduction of binding energy without compromising specificity. These concepts of functional advantages turned out to be critical for developing the paradigm of protein disorder.

Much controversy surrounded the structural state of prothymosin alpha (ProTa), a small acidic protein of 109 amino acids in length. Though its exact function was—and still is—not known, its evolutionary conservation and wide tissue distribution suggested an essential biological role. Gel-filtration experiments suggested an apparent molar mass five times greater than that calculated from the amino acid sequence, whereas sedimentation equilibrium measurements gave the correct molecular mass (Haritos et al. 1989). Proton NMR and CD suggested a disordered chain (Watts et al. 1990), whereas unusual sodium dodecyl sulfate polyacrylamide gel electrophoresis (SDS-PAGE) mobility was interpreted in terms of the protein being a stable dimer (Cordero et al. 1992). The controversy of the physical state of ProTa was eventually settled by a detailed investigation that combined SAXS, dynamic light scattering (DLS), mass spectrometry (MS), and CD (Gast et al. 1995). The results clearly indicated that ProTa is monomeric but adopts a random coil-like conformation with no regular secondary structure. R_S (30.7 Å) and R_G (47.6 Å) are 1.77 and 3.42 times larger than those expected for a compactly folded protein of its length. A cautious note of generalization has been worded by the claim "the finding that a biologically active protein molecule with 109 amino acid residues adopts a random coil conformation under physiological conditions raises the question whether this is a rare or a hitherto-overlooked but widespread phenomenon in the field of macromolecular polypeptides." However, the major impact of this work resulted from its title, which put it in plain terms "Prothymosin Alpha: A Biologically Active Protein with Random Coil Conformation."

A variety of techniques have been used in the conformational analysis of human α-synuclein (Weinreb et al. 1996), also known as the non-Aβ component of Alzheimer's disease amyloid plaque (NACP, see Chapter 15, Section 15.3.1 and Section 15.3.2.1). An elongated shape of the protein was indicated by its much larger R_S and slower sedimentation than a globular protein of similar M_W. CD and FTIR indicated the absence of significant amounts of secondary structure, whereas CD and ultraviolet (UV) spectroscopy suggested the lack of a hydrophobic core. Its conformational properties were unchanged by boiling and were insensitive to denaturants. It was suggested that NACP exists as a mixture of rapidly equilibrating extended conformers and that it probably represents the emerging class of "natively unfolded" proteins. Again, the impact of the study stems from the title "NACP, a Protein Implicated in Alzheimer's Disease and Learning, is Natively Unfolded."

The functional importance of the unfolded state was raised in the case of the bacterial transcription regulator FlgM (Plaxco and Gross 1997). This bacterial protein is an inhibitor of the transcription factor σ^{28}, and its function is to down-regulate the synthesis of flagellar proteins when the assembly of the flagellum is completed. The protein is depleted from cells by being transported through the central channel of flagella, which are not completed and capped yet. Once flagella are ready, FlgM gets trapped in the cell; it inhibits σ^{28} and shuts down expression of flagellar proteins. The argument for its disorder comes from the fact that the channels of flagella are too narrow for FlgM to wriggle through, unless it is in an unfolded state. Although this idea might be challenged, connecting function with disorder of a protein certainly had a significant impact on the field, as signified by the title again: "The Importance of Being Unfolded." This view was supported by NMR studies of the protein (Daughdrill et al. 1997; Daughdrill, Hanely, and Dahlquist 1998).

An important element of the functionality of IDPs was brought up by Wright and colleagues upon studying the cyclin-dependent kinase (Cdk) inhibitor $p21^{Cip1}$, a protein important for the p53-dependent control of cell cycle (Kriwacki et al. 1996). Not only was it shown by proteolytic mapping, CD spectroscopy, and NMR that the binding domain of $p21^{Cip1}$ lacks stable secondary or tertiary structure in the unbound state, it was also demonstrated that the protein adopts a stable conformational state when it binds its partner, Cdk2. It was suggested that the induced folding process enables $p21^{Cip1}$ to bind and inhibit a diverse family of cyclin-Cdk complexes, including Cyclin A-Cdk2, Cyclin E-Cdk2, and Cyclin D-Cdk4. Thus, structural disorder was possibly associated with binding promiscuity, as suggested explicitly in the title "Conformational Disorder Mediates Binding Diversity."

2.3 SO WE HAVE DISORDERED PROTEINS

In all, the concept of disorder appeared rather clearly in many studies; only the generality of this phenomenon was missed, and no attempt was made to generalize function in terms of structural disorder. The increase in the number of examples eventually demanded that the structure-function paradigm be reassessed.

In a series of papers, Dunker and colleagues touched upon several important generalities of structural disorder (Garner et al. 1998; Romero et al. 1998; and Dunker et al. 1998). By collecting ordered and disordered regions identified by either X-ray crystallography, NMR, or CD, neural networks to predict disorder from amino acid sequence could be trained (Garner et al. 1998). This way, it was demonstrated that structural disorder comprises a sequence-dependent category distinct from that of ordered protein structure. By applying the predictor to the Swiss-Prot database (Romero et al. 1998), more than 15,000 proteins were shown to contain disordered regions of at least 40 consecutive amino acids, which suggested that structural disorder is a general phenomenon. Based on theoretical considerations and experimental examples, they suggested that disordered regions might be primarily involved in binding to other molecules (Dunker et al. 1998). Upon binding, the disordered region may become ordered, which uncouples affinity and specificity, thus providing benefits in fine-tuning molecular interaction networks.

Another paper by Wright and Dyson also suggested the generality of "intrinsically unstructured" proteins (Wright and Dyson 1999). In addition, they also argued that these proteins or segments of proteins have important functions. This link between function and the "lack" of structure demanded that the structure-function paradigm be reexamined. Through a variety of examples disordered regions were demonstrated to be frequently found in proteins involved in DNA and RNA binding, transcription, translation, cell-cycle regulation, and membrane fusion, and even in amyloid formation. The examples pointed to the involvement of unstructured proteins in regulatory functions, in which the lack of structure might confer functional advantages.

The analysis of amino acid preferences of "natively unfolded" proteins provided some additional insight into the characteristics of disordered proteins (Uversky, Gillespie, and Fink 2000a). It was demonstrated that these "natively unfolded' proteins are specifically localized within a unique region of the net charge–mean hydrophobicity phase space, which indicated that a combination of low overall hydrophobicity and large net charge is primarily responsible for their inability to fold into well-defined structures. This observation forms the basis of many bioinformatic predictors (Chapter 9) as well as our understanding of the physical principles underlying disorder.

The transition in concept was solidified by many other contributions. The most critical elements of the new view have been the following:

1. The classification of molecular functions of IDPs (Dunker et al. 2002; Tompa 2002, 2005)
2. The development of ever more advanced bioinformatic predictors and their involvement in the critical assessment of techniques for protein structure prediction (CASP) experiment (Bordoli, Kiefer, and Schwedel 2007; Jin and Dunbrack 2005; Melamud and Moult 2003)
3. The development of the first database of protein disorder, DisProt (Sickmeier et al. 2007; Vucetic et al. 2005)
4. The realization that disorder prevails in proteins involved in disease, such as cancer (Iakoucheva et al. 2002)
5. Residue-level description of the local structural preferences of IDPs (Bertoncini et al. 2005; Lee et al. 2000; Mukrasch et al. 2005)
6. Global description of the structural ensemble of IDPs by the combination of NMR, SAXS and molecular dynamics simulations (Dedmon et al. 2005; von Ossowski et al. 2005)
7. The realization of the prevalent binding mechanism of IDPs by short recognition elements (Fuxreiter, Tompa, and Simon 2007; Oldfield et al. 2005b) and the experimental analysis of the mechanism of folding coupled to binding (Sugase, Dyson, and Wright 2007)
8. The suggestion that disorder may enable multiple functions (i.e., moonlighting) of proteins (Tompa et al. 2005)
9. The recognition of "fuzziness" (i.e., that disorder may also prevail in the bound state of disordered proteins) (Tompa and Fuxreiter 2008)
10. The extension of structure-function studies to within the cell (Dedmon et al. 2002; McNulty, Young, and Pielak 2006)

These achievements have been reviewed and discussed in many excellent reviews (Demchenko 2001; Dunker et al. 2002; Dunker et al. 2001; Dyson and Wright 2002a, 2005; Fink 2005; Tompa 2002, 2003a, 2005; Uversky 2002a,b; Uversky, Oldfield, and

Dunker 2005; Wright and Dyson 1999). Their message is clear: The success of the field and rapid progress in diverse directions leads to an ever more complete understanding of the interplay between structure and function of proteins that do not fold into well-defined 3-D structures. Detailed studies of such "intrinsically unstructured" (IUP), "intrinsically disordered" (IDP), or "natively unfolded" (NU) proteins or intrinsically disordered regions (IDR) drive the development of a novel structure-function paradigm that can encompass all distinct structural states of proteins.

3 Indirect Techniques for Recognizing and Characterizing Protein Disorder

This chapter covers techniques that provide the first line of evidence on the unusual structural state of disordered proteins. These simple techniques are usually applicable on full-length proteins, and they are considered indirect, because they do not directly provide structural information but suggest a behavior from which the disordered nature of the protein can be inferred.

3.1 RESISTANCE TO HEAT

The unusual behavior that certain proteins resist boiling temperatures, which cause "ordinary" globular proteins to precipitate, remained a curiosity unexplained for a long time. Such a behavior was described in the case of many intrinsically disordered proteins (IDPs) (see Figure 3.1), such as microtubule-associated protein 2 (MAP2) (Hernandez et al. 1986), calpastatin (Hackel, Konno, and Hinz 2000), α-synuclein (Weinreb et al. 1996), stathmin (Belmont and Mitchison 1996), epsin (Kalthoff et al. 2002), $p21^{Cip1}$ (Kriwacki et al. 1996), protein phosphatase inhibitor 1 (Nimmo and Cohen 1978), 4E-BP1 (Fletcher and Wagner 1998) and 4E-BP3 (Poulin et al. 1998), involucrin (Etoh, Simon, and Green 1986), Df31 (Crevel and Cotterill 1995), inhibitor of PKA (PKIα) (Hauer et al. 1999a; Thomas et al. 1991), the homolog of PPI2 (Glc8) (Tung, Wang, and Chan 1995), group 2 LEA proteins ERD10 and ERD14 (Kovacs et al. 2008), fesselin (Khaymina et al. 2007), caldesmon (Bretscher 1984; Lynch, Riseman, and Bretscher 1987), calreticulin (Kim et al. 2000b), TPPP/p25 (Kovacs et al. 2004), and Pro-rich proteins of rat parotid glands (Muenzer et al. 1979). This behavior is very often exploited in the purification of proteins, to such an extent that it was actually suggested as a general method to purify recombinant IDPs (Kalthoff 2003). The reason for the heat

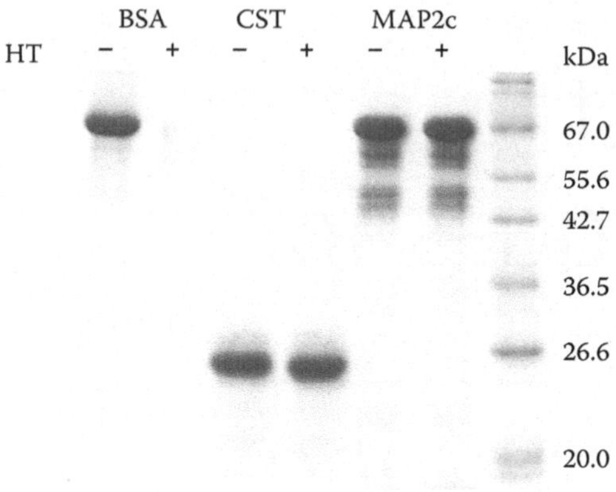

FIGURE 3.1 Heat stability and anomalous SDS-PAGE mobility of IDPs. This SDS-PAGE demonstrates both heat stability and anomalous SDS-PAGE mobility of IDPs. The supernatants of a globular control (BSA) and two IDPs (human calpastatin domain 1, CST, and the juvenile form of microtubule-associated protein 2, MAP2c) were run on the gel without (–) or with (+) heat-treatment at 100°C 10 min and subsequent centrifugation. BSA precipitates, whereas IDPs stay in solution under these conditions. The apparent M_W of the IDPs is much higher than their absolute M_W determined from their sequence: 25 kDa vs. 15 kDa for CST and 67 kDa vs. 50 kDa for MAP2c.

resistance of IDPs resides in their unusual amino acid composition; that is, they are highly charged and have a low content of hydrophobic amino acids (Dunker et al. 2001; Uversky, Gillespie, and Fink 2000a), due to which they do not expose hydrophobic residues that would make them aggregate at elevated temperatures.

The possible generality of the phenomenon was addressed in the study of Kim and colleagues, who characterized the heat stability of proteins in cell extracts and found that 20 and 70 wt% of total proteins are heat-resistant in Jurkat T-cell lysates and human serum, respectively (Kim et al. 2000b). The heat-stable proteins are, in many cases, disordered. The correlation of heat resistance and disorder is also underlined by a few proteomic studies of disorder (see Chapter 7), in which an initial heat treatment was used to enrich cellular extracts for IDPs, as confirmed by subsequent mass spectrometry (MS) identification of proteins (Csizmok et al. 2006; Galea et al. 2006).

Whereas this technique is undoubtedly simple and effective in isolating and identifying potential IDPs, it should never be neglected that although the protein survives boiling, it may suffer irreversible chemical changes during the treatment. At the first approximation, the structural ensemble of IDPs undergoes a reversible change at elevated temperatures and returns to its native conformational state upon returning to ambient temperature. For example, the structure and function of MAP2 prepared with and without heat treatment showed no critical differences (Hernandez et al. 1986). In a similar comparative study, the heat-treated and untreated caldesmon also showed no detectable differences in certain functions (Bretscher 1984; Lynch et al. 1987) but showed significant alterations in CaM-binding (Zhuang, Mabuchi, and Wang 1996). It should also be borne in mind that most proteins are neither fully ordered nor fully disordered but contain ordered and disordered regions at different ratios. Even if such a protein survives heating as a whole, the ordered part may be irreversibly damaged. In addition, even if the conformation of the protein is largely unchanged, several of its residues may undergo chemical conversion that might compromise function. Among many possibilities, deamidation of Gln and Asn and oxidation of Cys and Met residues are the most trivial.

3.2 RESISTANCE TO CHEMICAL DENATURATION

Resistance to chemicals, such as lowering pH by acids that usually causes denaturation of ordered proteins, is another notable feature of IDPs. Unlike globular proteins, IDPs remain soluble under such extreme conditions, as reported for α-synuclein (Uversky 2003), ProTa (Uversky et al. 1999), thymosin-β4 (Tβ4) (Watts et al. 1990), and myelin basic protein (MBP) (Thomas, Weser, and Hempel 1977). Although it is not used as frequently as heat, the generality of this relation is suggested by a comparative analysis (see Chapter 7), in which IDPs were shown to resist treatment with TCA and PCA, which could be used for their enrichment and subsequent proteomic 2DE-MS analysis (Cortese et al. 2005).

The reasons for acid resistance of IDPs also derive from their amino acid composition, although not in such a straightforward manner as their heat resistance. For structured proteins, negatively charged side chains are protonated upon lowering the pH, which leads to charge imbalances disrupting salt bridges and causing aggregation (Dill and Shortle 1991). As noted by Uversky (Uversky 2002a), lowering the pH has an opposite effect on IDPs. These are usually highly charged at neutral pH, whereas decreasing the pH titrates their acidic surface groups and lowers their net charge, thus reducing electrostatic repulsion and shifting their conformational ensemble toward more compact states.

Whereas IDPs are largely insensitive to chemical denaturation, the application of denaturants may also provide information on residual structure in their ensemble. For example, a change in the circular dichroism (CD) spectrum toward the random coil state in the presence of 8M urea may signal the presence of transient structural elements (see Chapter 10, Section 10.2.1).

3.3 UNUSUAL SDS-PAGE MOBILITY

Sodium dodecyl sulfate (SDS) polyacrylamide gel electrophoresis (PAGE) is one of the most frequently used techniques in molecular biology (Laemmli 1970). Its simplicity to perform, ability to visualize proteins and potential contamination, and power to provide accurate information on molecular mass (M_W) have made it indispensable in protein science. Its applicability for M_W determination stems from the fact that proteins are denatured in SDS (i.e., they unfold and acquire a roughly uniform unfolded state). Their charge/mass ratio becomes uniform, because all proteins bind about the same relative amount of the negatively charged SDS (1.4 g/g). Under these conditions, their mobility in an electric field is the same, only to be discriminated by size due to the friction elicited by the gel matrix. By an appropriate calibration, their M_W can be determined from the distance they cover.

IDPs, however, are notorious for an anomalous behavior in SDS-PAGE, because they appear to have larger apparent M_W than the real one deduced from sequence or determined by mass spectrometry (MS) (Figure 3.1). To mention some examples, unusually high M_W was reported for ProTa (20.6/12 kDa, for observed/real M_W [Cordero et al. 1992]), Df31 (31/18.5 kDa [Crevel, Huikeshoven, and Cotterill 2001]), calpastatin (107/77.1 kDa [Takano et al. 1988]), nuclear histone binding N1/N2 protein (110/64.8 kDa [Kleinschmidt et al. 1986]), the cytoplasmic domain of gliotactin (30/23.6 kDa [Zeev-Ben-Mordehai et al. 2003]), XPA (42/31 kDa [Iakoucheva et al. 2001b]), ERD10 (45/29 kDa [Kovacs et al. 2008]), ERD14 (37/20 kDa [Kovacs et al. 2008]), Glc8 (30.5/22.9 kDa [Tung et al. 1995]), dehydrin-like protein VCaB45 from *A. graveolens* (45/16.5 kDa [Heyen et al. 2002]), α-synuclein (19/14.5 kDa [Weinreb et al. 1996]), DARPP32 (32/23.0 kDa [Hemmings et al. 1984]), and caldesmon (150/88.7 kDa [Hayashi et al. 1989]). In general, it was found that the M_W of IDPs is usually overestimated by SDS-PAGE by a factor of 1.2–1.8 (Tompa 2002). The reason again is in their unusual amino acid composition, due to which they tend to bind less SDS and move more slowly through the gel than globular proteins.

3.4 ENHANCED PROTEOLYTIC SENSITIVITY

Proteolytic cleavage of proteins requires two features: the presence of an appropriate recognition sequence (consensus site) for the given protease and the local structural adaptability of the protein. As also discussed in Chapter 12, Section 12.2.2, Fontana and colleagues (Fontana et al. 1986; Fontana et al. 1997a; Fontana et al. 1997b) and Thornton and colleagues (Hubbard, Eisenmenger, and Thornton 1994) demonstrated that local flexibility/disorder of the substrate promotes its proteolysis, because the substrate has to structurally adapt to the active site of the protease along a continuous stretch of about 12 residues. In globular proteins, unfolded/partly folded states and/or locally flexible linkers/loops are the primary targets of proteolytic attack (Fontana et al. 1997a; Fontana et al. 1997b). From the typical free energy of unfolding of a globular protein (on the order of 5–10 kcal/mol), a simple thermodynamic argument suggests that an IDP may be as much as 5–7 orders of magnitude more sensitive to proteolysis (Hubbard, Beynon, and Thornton 1998; Hubbard et al. 1994). Whereas this result rests on serious simplifications, actual observations indicated that IDPs are orders of magnitude more sensitive to proteolysis than ordered proteins (Dunker et al. 2002; Tompa 2002; Uversky 2002a).

In practical terms, it is generally observed that, in the case of IDPs, proteases at very low enzyme:substrate ratios (typically 1:100–1:1000, as compared to 1:10–1:50 in the case of globular proteins) are sufficient to cause rapid degradation. Such experiments provided evidence for the disorder of $p21^{Cip1}$ (Kriwacki et al. 1997), the intracellular domain of Jagged-1 (Popovic et al. 2006), eIF4E (Hershey et al. 1999), lambda N (Greenblatt and Li 1982), Dsp16 (Lisse et al. 1996), XPA (Iakoucheva et al. 2001a), and plant dehydrins ERD10 and 14 (Kovacs et al. 2008), for example. In addition (see also Chapter 12, Section 12.2.2 and Section 12.6.2.3), a general correlation between disorder and proteolytic sensitivity was also observed in the case of calmodulin (CaM)-binding proteins. Often, CaM-dependent enzymes are stimulated by limited proteolytic digestion, which cleaves into their locally disordered CaM-recognition element (CaMBT).

The potential generality of the proteolytic sensitivity of IDPs was also addressed in a proteomic study of IDPs (Galea et al. 2006). As detailed in Chapter 7, Section 7.2, heat treatment and subsequent 2DE-MS identification was used to describe the disordered complement of the proteome of mouse fibroblast cells. By bioinformatic analysis, the proteins identified are significantly enriched for disorder, which was also confirmed by limited proteolysis that affected IDPs much more than ordered proteins.

3.5 LIMITED PROTEOLYSIS AND LOCAL STRUCTURE

Proteolysis under controlled conditions leads to a partial degradation of the substrate, which may be used to probe into the structure of IDPs (see Chapter 12, Section 12.2.2). This approach has provided ample insight into the structural topology of mostly ordered proteins, delineating their flexible segments. It is much less appreciated that at very low protease concentrations IDPs also undergo limited proteolysis, which implies their non-fully random structural organization. In the case of caldesmon (Marston and Redwood 1991), nucleoplasmin (Dingwall et al. 1987), the TAD of GC4N (Hope, Mahadevan, Struhl 1988), CREB KID (Richards et al. 1996), stathmin (Redeker et al. 2000), BRCA1 (Mark et al. 2005), calpastatin (Csizmok et al. 2005), tau (Steiner et al. 1990), and MAP2 (Wille, Mandelkow, and Mandelkow 1992b), the location of preferential cleavage site(s) is not random but correlates with their organization into larger functional and possibly structural segments. An appealing interpretation of these observations is that transient short- and/or long-range structural organization ensures preferential spatial exposure of certain regions in these IDPs.

3.6 DIFFERENTIAL SCANNING CALORIMETRY

Differential scanning calorimetry (DSC) is a thermoanalytic technique that is somewhat neglected in characterizing IDPs. The technique relies on measuring the difference of heat required to increase the temperature of a sample relative to that of a reference, from which heat capacity as a function of temperature can be determined (Privalov 1979, 1982). The technique is particularly sensitive to heat-capacity changes accompanying phase transitions, such as the temperature-induced unfolding of a globular protein. The unfolding event appears as a heat-absorption curve, which, for a single-domain protein, signals the cooperative melting of the structure. In the case of multidomain proteins, individual melting peaks overlap and their deconvolution can provide information on the domain structure of the protein. The two basic parameters derived from the melting curve are the transition (or melting) temperature T_m and the enthalpy of melting (see Figure 3.2).

It intuitively follows that the absence of such a cooperative transition may signal the lack of globularity and, indirectly, disorder (Receveur-Brechot et al. 2005). DSC was used to demonstrate disorder in an acid-denatured globular protein, alpha-fetoprotein (AFP), which lacks a stable fold and behaves as a molten globule (MG) (Uversky et al. 1995). In the case of bona fide IDPs, there are only a few relevant studies, such as in the case of Df31 (Figure 3.2), a Drosophila protein of chromatin decondensation and remodeling activities (Szollosi et al. 2008); the nuclear co-activator binding domain (NCBD) of CBP (Demarest et al. 2004); the carboxy-terminal domain (CTD) of caldesmon (Permyakov et al. 2003); α-synuclein, β- and κ-casein, and tau protein (Syme et al. 2002); rGmD-19, a soybean group 1 LEA protein (Soulages et al. 2002); and the N-terminal prion domain

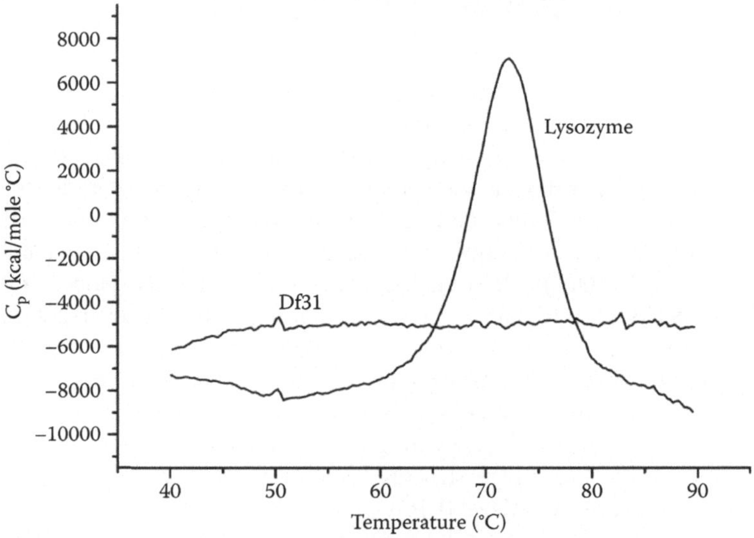

FIGURE 3.2 DSC of Df31 and lysozyme. The DSC curve of intrinsically disordered Df31 and globular lysozyme was recorded, and the change in heat capacity was calculated from the observed flow of heat. The curves show the distinct behavior of the two proteins: Lysozyme undergoes a cooperative structural transition (melting) with a T_m of 72°C, whereas Df31 lacks such a transition, which suggests its disordered structural state. Reproduced with permission from Szollosi et al. (2008), *J. Proteome Res.* 7, 2291–9. Copyright by the American Chemical Society.

of Ure2p (Baxa et al. 2004). DSC can also be used to study structural features of IDPs in more subtle ways, as demonstrated by the next two examples.

3.6.1 Transition to a More Ordered State

Rv3221c biotin-binding protein of *Mycobacterium tuberculosis* was characterized by a variety of techniques, such as CD, UV fluorescence, Fourier-transform infrared spectroscopy (FTIR), and DSC (Kumar et al. 2008). The protein is intrinsically disordered at physiological temperatures, whereas an increase in temperature induces its transition to a more structured state. By DSC, its endothermic folding has a transition temperature of 53°C, and it passes through several intermediate states toward a β-sheet structure, also shown by CD and FTIR. These data suggest that although Rv3221c is disordered at ambient temperatures, it has the potential to adopt more ordered structures at higher temperatures without oligomerization or aggregation, probably driven by increased hydrophobic interactions. The functional implication of these observations is that Rv3221c may participate in biochemical processes requiring biotin as a cofactor and adopt suitable conformations upon binding other folded targets.

3.6.2 Residual Structure in Calpastatin

DSC can also be used to characterize residual structure within IDPs (Hackel et al. 2000). This was demonstrated by studying pig (pCSD1) and human (hCSD1) calpastatin domain 1 by a combination of CD and DSC. By CD, Gnd-HCl-induced denaturation results in a positive shift in molar ellipticity around 222 nm, which suggests the loss of residual structure present under native conditions. By DSC, both pCSD1 and hCSD1 exhibit smaller heat capacities than those calculated by assuming the full disorder and hydration of their polypeptide chains. Theoretical heat capacities for the random coil state were derived from an increment system based on heat-capacity values of GXG tripeptides (Hackel, Hinz, and Hedwig 1999) and were compared with values measured for pCSD1 and hCSD1. It was found that the residues of calpastatin, on the average, are less hydrated than the same residues in short peptides. This suboptimal hydration probably results from intrachain contacts (i.e., residual structure within the chain). The lack of additivity of the CD spectra of the two halves of calpastatin also provided evidence that this is the case (Csizmok et al. 2005).

3.7 ISOTHERMAL TITRATION CALORIMETRY

Isothermal titration calorimetry (ITC) is used to determine the thermodynamic parameters of interactions by directly measuring the affinity (K_a), enthalpy (ΔH), and stoichiometry (n) of binding between molecules. From these, changes in Gibbs energy (ΔG)

and entropy (ΔS) can be calculated. The sample is titrated with aliquots of the ligand, causing heat to be absorbed or released, which is measured by maintaining the sample and a reference cell at the same temperature. Heat flow spikes are then integrated to yield the total heat effect per injection, from which the thermodynamic parameters can be calculated (see Figure 3.3). The actual parameters may suggest and characterize disorder, as demonstrated by two examples.

3.7.1 The Energetics of Binding of a PPII Helix to Its Cognate SH3 Domain

A quantitative account of the energetics of the interaction between the SosY peptide (Ac-VPPPVPPRRRY-NH2) and Sem-5 C-SH3 domain is given by Ferreon and Hilser (Ferreon and Hilser 2004). Binding of the peptide occurs in a PPII helix conformation, with a much larger apolar than polar surface area buried (ΔASA_{ap} = –257 Å^2, ΔASA_{pol} = –155 Å^2), from which enthalpy and entropy changes of binding (ΔH = 1.85 kcal mol^{-1}, $T\Delta S$ = 6.0 kcal mol^{-1}) can be estimated. Actual ITC measurements (Figure 3.3), however, yield very different values, with large negative changes in both enthalpy and entropy (ΔH = –8.0 kcal mol^{-1}, $T\Delta S$ = –2.0 kcal mol^{-1}) upon binding.

This discrepancy suggests that the thermodynamic determinants of binding do not rest exclusively in the chemical nature of the interaction surfaces, but also in folding of the peptide induced upon binding. Quantitative estimates of the energy associated with folding the SosY peptide into PPII conformation (Ferreon and Hilser 2004) underscore the disordered state of the peptide prior to binding and its redistribution into the binding-competent PPII conformation, which causes the observed differences in thermodynamic parameters. Unlike enthalpy and entropy, free energy of binding is much less affected by folding upon binding (i.e., free energy alone might not provide a suitable description of the determinants of binding of IDPs).

3.7.2 Binding of the KID Domain of p27^{Kip1} to Cyclin A-Cdk2

ITC also provided mechanistic details on the binding of the KID domain of p27^{Kip1} to the Cyclin A-Cdk2 complex (Lacy et al. 2004). The functional importance of this interaction is detailed in Chapter 10, Section 10.2.3.1 and Chapter 15, Section 15.1.3, with the bound structure depicted in Figure 10.3. KID has two domains connected by a linker helix (LH, 38–60) and two primary functional determinants. A conserved Leu-Phe-Gly motif (32–34, within domain 1) binds in a hydrophobic patch on Cyclin A, connected by LH to the segment (domain 2) that binds Cdk2. Domain 2 has a short antiparallel β-sheet (62–70) followed by a 3_{10} helix (74–80) that leads to a short α-helical segment (86–90) that inserts a Tyr (Tyr88) into the nucleotide binding pocket of the kinase (Russo et al. 1996; Sivakolundu, Bashford, and Kriwacki 2005). A key mechanistic question of the interaction of KID with the Cyclin A-Cdk2 complex is the order

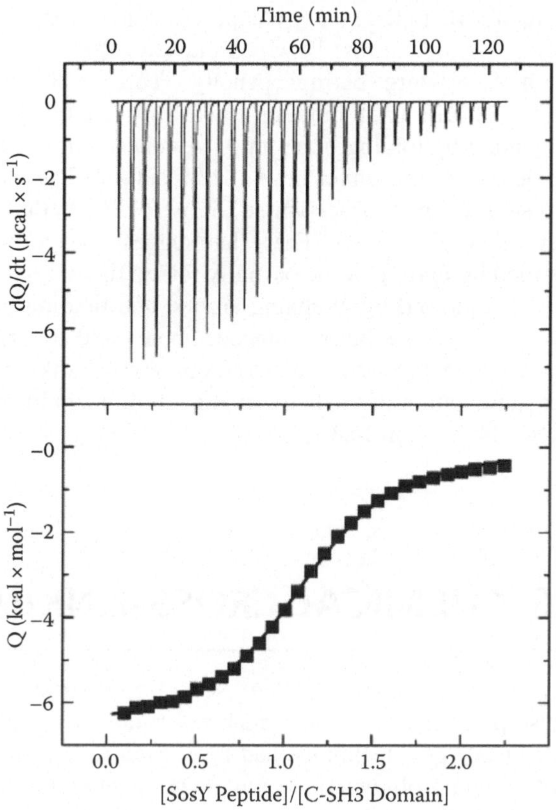

FIGURE 3.3 ITC titration of an SH3 domain with a Pro-rich peptide. The Sem-5 C-SH3 domain was titrated with the SosY peptide (Ac-VPPPVPPRRRY-NH2). The power (in μcal/sec) needed to maintain the reference and sample cells at identical temperatures is measured, from which molar heat (Q) is calculated. The thermodynamic parameters obtained by fitting the data suggest large negative enthalpy and entropy changes of binding, which contradict the sizeable hydrophobic surface buried upon the interaction and suggests that the peptide is disordered before binding (see Section 3.7.1 for details). Reproduced with permission from Ferreon and Hilser (2004), *Biochemistry* 43, 7787–97. Copyright by the American Chemical Society.

of binding of these two motifs and the role of the formation of LH. To this end, the interactions of KID with Cyclin A (mediated by domain 1), Cdk2 (mediated by domain 2), and the Cyclin A-Cdk2 binary complex were separately characterized by ITC (Lacy et al. 2004).

It was found that all three binding reactions are driven by enthalpy, which overcomes a large unfavorable decrease in entropy. Binding to Cyclin A ($\Delta G = -10.4$ kcal mol^{-1}) is slightly more favorable than binding to Cdk2 ($\Delta G = -9.8$ kcal mol^{-1}), with a very large entropic penalty for the latter, which reflects that the extent of folding upon binding is very different (estimated from the length of domains to be 29 residues vs. only 10). Binding of KID to the binary complex is much stronger than that of either of its fragments ($\Delta G = -11.6$ kcal mol^{-1}), which indicates that both domains favorably contribute to binding. A very large entropic penalty ($-T\Delta S = +28.6$ kcal mol^{-1}) suggests the extensive ordering of KID upon binding. It was estimated that binding of KID to Cyclin A is accompanied by folding of about 34 residues, which corresponds to both domain 1 (~12 residues) and the linker helix (~22 residues). The value for binding at Cdk2 is about 59 residues, which is accounted for by folding both domain 2 (~30 residues) and the linker helix (~22 residues). These data suggest that binding of KID to either partner is accompanied by folding of the partially folded linker region. KID binding is initiated by domain 1, followed by wrapping around and binding of domain 2 (Lacy et al. 2004). Ordering of the linker helix is induced upon—and not prior to—binding, in accordance with previous mutagenesis studies (Bienkiewicz, Adkins, and Lumb 2002), which showed that stabilization of the helix kinetically hinders formation of the complex (see also Chapter 14, Section 14.3.1).

3.8 CHEMICAL CROSS-LINKING

Chemical cross-linking is a technique often used to gain structural information on the spatial relationship of residues, from which restraints on tertiary and quaternary structure can be deduced. This technique had been used successfully in studying the topology of subunits in homo-oligomeric enzymes (Hajdu et al. 1979) or even the motility (disorder) of terminal tails of globular proteins (Gusev, Hajdu, and Friedrich 1979). Its potential application for IDPs is justified by the fact that data on the anomalously high apparent M_W of IDPs is often interpreted in terms of an oligomeric structure composed of several subunits. In these cases, cross-linking can be used to demonstrate that the protein is in fact monomeric.

This approach was directly used in the case of Df31, a Drosophila chromatin decondenstaion factor (Szollosi et al. 2008). The pattern of cross-linking by dimethyl-suberimidate was compared to that of a positive control, dimeric globular phosphoglycerate-mutase (PGM), and a negative control, monomeric disordered α-synuclein. Cross-linking products analyzed by SDS-PAGE showed that dimeric PGM was cross-linked, whereas Df31 and α-synuclein were not, which confirmed the monomeric nature of these two IDPs.

A special kind of cross-linking, oxidation of adjacent sulfhydryl residues following reduction, was used in the case of caldesmon (Lynch et al. 1987). After prolonged exposure of the 160-kDa protein, bands of smaller (140 kDa) and higher (with apparently no limit) M_W were generated. Concentration dependence of the formation of each form suggested that smaller-M_W species were generated by intramolecular cross-linking of the monomer, whereas higher-M_W species arose due to cross-linking of randomly colliding monomers, rather that due to cross-linking of subunits within a stable oligomer. This idea was also corroborated by the inability to cross-link the protein by disuccinimidyl-tartrate (Lynch et al. 1987).

3.9 H/D EXCHANGE

The exchange of amide protons with water hydrogens is impeded by structural elements in which backbone amides are H-bonded and/or buried; thus exchange rates carry information on local structural state. To extract this information, the protein is transferred into heavy water (D_2O), and the kinetics of the exchange of its amide protons to deuterium is followed. The rate of H/D exchange is orders of magnitude faster in unfolded and disordered proteins (or regions) than in folded proteins (Bracken et al. 2004). The kinetics of exchange can be followed by FTIR, whereas sequence-specific information on local structural state is usually obtained by analyzing the protein by NMR (see Chapter 6, Section 6.4.2) or MS (the latter is termed HXMS, which stands for hydrogen/deuterium exchange MS) (Ferraro, Lazo, and Robertson 2004). Due to the sensitivity of HXMS, small amounts of large proteins can be studied. MS can also be combined with protease digestion to determine the location and dynamics of flexible/disordered regions in proteins (Yamamoto, Izumi, and Gekko 2004).

Because H/D exchange is usually too fast to follow in the case of of IDPs, this technique has its use in ordered proteins mostly. For example, Yamamoto and colleagues used HXMS to study local flexibility in dihydrofolate reductase (DHFR) (Yamamoto et al. 2004). The backbone-fluctuation map was determined by matrix-assisted laser desorption/ionization mass spectrometry (MALDI-MS) coupled with H/D exchange of 18 digestion fragments generated by pepsin. H/D exchange was particularly fast within the fragment comprising residues 5–28, a loop region participating in substrate uptake suggesting local disorder of this region.

Another application of HXMS is the rapid high-throughput screening (HTS) of disordered protein regions to improve target selection for crystallographic structure solution (Pantazatos et al. 2004). Because disordered regions may hamper crystallization, considerable advantage can be gained by removing proteins that harbor them (outlined in Chapter 9, Section 9.9). To this end, HXMS analysis was carried out on 24 *T. maritima* proteins with varying crystallization and diffraction characteristics. In the case of targets of known structure, the HXMS method correctly localized regions of disorder, and truncations of proteins based on these results greatly improved crystallization of targets.

Hydrodynamic Techniques

4

A variety of techniques provide information on the size or hydrodynamic (diffusion) behavior of intrinsically disordered proteins (IDPs), collectively termed as hydrodynamic techniques. Because the dimensions of disordered proteins are much larger than those of globular proteins, these results are usually conclusive with respect to the gross structural status of IDPs. These techniques also enabled the development of low-resolution structural models, which is a critical step toward interpreting the function of IDPs in terms of structure. Pulsed field gradient nuclear magnetic resonance (NMR), which is used for determining hydrodynamic behavior, is also covered in this chapter. Small-angle X-ray scattering (SAXS), due to its capability of bridging low-resolution and high-resolution structural models, is given extended coverage.

4.1 GEL FILTRATION (SIZE-EXCLUSION) CHROMATOGRAPHY

Gel filtration (GF) or size-exclusion chromatography (SEC) is a technique in which the solution of a protein is passed through the stationary phase of a gel matrix, and its position of elution relative to other proteins is measured. The matrix is a cross-linked polymer, usually made of acrylamide, dextran, or agarose, with a pore size distribution commensurable with the proteins to be separated. Depending on their size, proteins may be fully or partially excluded from the matrix or they may freely penetrate its holes. The primary measurable feature is the elution volume (V_e, or retention time, R_t, at a given flow rate), which falls between the void volume (V_v), where large proteins that cannot enter the pores elute, and the total volume (V_t), where small proteins that can freely penetrate the pores appear.

V_e is in a linear relationship with the logarithm of molecular mass (M_W) (see Figure 4.1A); thus the column can be calibrated with globular proteins (Permyakov et al. 2003; Uversky 1993), and V_e can be used to determine the apparent M_W or other hydrodynamic parameters (R_s, V_H, R_H) of an unknown protein. Significantly higher values than that expected for a globular protein of the given M_W are indicative of the extended structural state (i.e., disorder) of the protein. From the actual value, the type of disorder, such as molten globule (MG)-type, pre-molten globule (PMG)-type, or random coil-type, can also be ascertained (Uversky 1993). As a rule of thumb, compact MG-type IDPs elute at an apparent M_W about 2 times their real value, whereas more

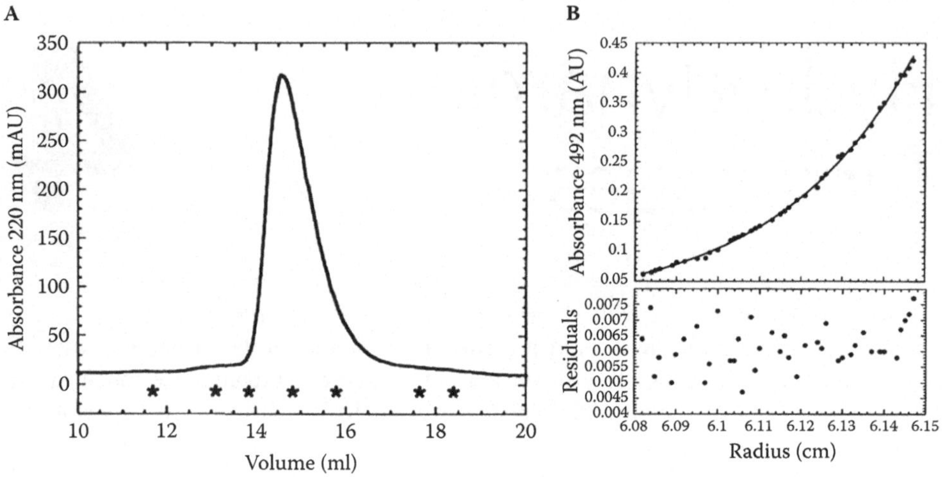

FIGURE 4.1 Hydrodynamic characterization of securin. (A) Size-exclusion chromatography elution profile monitored at 220 nm. Asterisks denote the positions of the molecular weight standards. From its position, securin has an apparent M_W of 105.5 kDa, which is five times larger than the theoretical value of 22.2 kDa, which indicates either its oligomeric structure or its disordered character. (B) Equilibrium sedimentation experiments of fluorescein-labeled securin demonstrate the monomeric status of the protein. Residuals correspond to fitting of the data to a single exponential model of a monomer of apparent M_W of 18.6 kDa. Reproduced with permission from Sanchez-Puig et al. (2005), *Protein Sci.* 14, 1410–8. Copyright by the Protein Society.

extended, random coil–type IDPs elute at an apparent M_W that is 4–6 times their real value (Csizmok et al. 2006).

It should be noted that an unexpectedly high apparent M_W of a protein can also be interpreted in terms of an oligomeric structure, as suggested in the case of Df31, for example (see Chapter 3, Section 3.8) (Crevel and Cotterill 1995). Disorder and oligomeric state can be distinguished by elution at high ionic strength and/or in the presence of denaturants, such as 8M urea, which can also provide evidence about the possible residual structure of an IDP. For example, the R_S of the carboxy-terminal domain (CTD) of caldesmon determined by GF (Permyakov et al. 2003) increases slightly in the presence of 6M Gnd-HCl (28.1 Å in buffer and 35.3 Å in Gnd-HCl), which suggests that the protein is in a PMG state (estimated values for the same M_W are: globular, 19.1 Å; MG, 21.7 Å; PMG, 27.4 Å; and random coil, 34.4 Å). The absolute M_W of the protein can also be determined by analytical ultracentrifugation (see Figure 4.1B and Section 4.3).

GF has provided many observations on the unusual hydrodynamic behavior of proteins and has basically contributed to the development of the concept of protein disorder. It has been applied in the case of DARPP-32 (Hemmings et al. 1984), α- and β-thymosin (Haritos et al. 1989), caldesmon (Lynch et al. 1987; Permyakov et al. 2003), PKI (Thomas et al. 1991), prothymosin alpha (ProTa) (Cordero et al. 1992), Df31 (Crevel and Cotterill 1995), α-synuclein (Weinreb et al. 1996), deoxyribonucleic acid (DNA) repair protein XPA (Iakoucheva et al. 2001b), Nup2p (Denning et al. 2002), AavLEA1 (Goyal et al. 2003), measles virus N_{TAIL} (Longhi et al. 2003), intracellular domain of gliotactin (Zeev-Ben-Mordehai et al. 2003), IF7 of glutamine synthetase (Muro-Pastor et al. 2003), T-cell receptor zeta cytD (Sigalov, Aivazian, and Stern2004), securin (Sanchez-Puig, Veprintsev, and Fersht 2005), glutamic acid-rich proteins (GARP) of rod photoreceptors (Batra-Safferling et al. 2006), and the C/EBP homolog CHOP (Singh et al. 2008), for example.

4.2 DYNAMIC LIGHT SCATTERING

Dynamic light scattering ((DLS), also known as photon correlation spectroscopy (PCS) or quasi-elastic light scattering (QELS)), provides information on the hydrodynamic behavior of proteins in solution (Bloomfield and Lim 1978; Langowski, Kremer, and Kapp 1992). The sample is illuminated by a laser light, and time-dependent fluctuations in the scattered intensity are detected. Because the molecules undergo Brownian motion in the time that elapses between absorption and emission of incident light, the fluctuations contain information on their diffusion coefficient *D*. There are two major measurement schemes of DLS, correlation spectroscopy, which directly measures fluctuations in light intensity, and spectrum analysis, in which scattered intensity is measured as a function of the frequency by which light intensity is modulated. The two schemes require different hardware but provide equivalent results, because the autocorrelation function (ACF) obtained in correlation spectroscopy is the inverse Fourier transform of the power spectral density function obtained in spectrum analysis.

Intensity fluctuations are usually converted to ACF (the treatment is analogous to that applied in fluorescence correlation spectroscopy [FCS]; see Chapter 5, Section 5.2.5), which is an ensemble average of the product of the signal with a delayed version of itself as a function of the delay time. Analysis of ACF in terms of *D* can be done analytically for a structurally homogeneous sample, whereas for an ensemble of conformations, the data need to be fitted with the assumed distribution of *D* values. From *D*, the R_S can be obtained via the Stokes–Einstein equation:

$$D = \frac{k_B T}{6\pi\eta R_S} \tag{4.1}$$

where k_B is Boltzmann's constant and η is the viscosity of the medium.

DLS has been used for demonstrating the structural disorder of ProTa (Gast et al. 1995), the NM region of yeast prion Sup35 (Scheibel and Lindquist 2001), measles virus nucleoprotein CTD (Longhi et al. 2003), rod photoreceptor GARPs (Batra-Safferling et al. 2006), plant stress protein acid stress ripening 1 (ASR1) (Goldgur et al. 2007), and the cytoplasmic tail of L1-CAM (Tyukhtenko et al. 2008).

4.3 ANALYTICAL ULTRACENTRIFUGATION

The sedimentation technique known as analytical ultracentrifugation (AU) is a versatile tool for studying the size, shape, and interactions of proteins through studying their hydrodynamic behavior (Lebowitz, Lewis, and Schuck 2002). In AU, the protein solution is spun at a high centrifugal field (above 100,000 rpm and 1,000,000 g), and the evolution of sample concentration profile versus the axis of rotation is monitored. The technique has two basically different and complementary implementations: sedimentation velocity (SV) and sedimentation equilibrium (SE) measurements, which provide slightly different information. Overall, the power of AU stems from the fact that the technique is firmly based on equilibrium and nonequilibrium thermodynamics, due to which it represents the gold standard for characterizing the hydrodynamic properties: M_W and binding constants of proteins.

In SV, the application of a centrifugal force causes the formation of a concentration boundary of the protein that moves toward the bottom of the centrifuge cell. The movement is characterized by the sedimentation coefficient, *s*, which is defined by the Svedberg equation:

$$s = \frac{u}{\omega^2 r} = \frac{M(1-\bar{\nu}\rho)}{N_A f} = \frac{MD(1-\bar{\nu}\rho)}{RT} \tag{4.2}$$

where u is the observed radial velocity of the macromolecule, ω is the angular velocity of the rotor, r is the radial position, $\omega^2 r$ is the centrifugal field, M is the molar mass, ν is the partial specific volume, ρ is solvent density, N_A is Avogadro's number, f is the frictional coefficient, and D is the diffusion coefficient. From D, R_s can be calculated by the Stokes–Einstein equation (Eq. 4.1). Values of s are commonly expressed in Svedberg (S) units, which correspond to 10^{-13} sec. A correction of experimental s values to a standard state of water (20°C) usually leads to the standard and easily comparable corrected $s_{20,w}$. An important parameter characterizing the size/shape of a protein is the ratio of the maximum s value (that of a sphere of the given M_W) to the actually observed value, $s_{\text{sphere}}/s_{20,w}$, which is equal to the ratio of the experimentally observed frictional coefficient to the minimum frictional coefficient expected for a sphere (f/f_0). This ratio characterizes the shape asymmetry of the molecule, which describes its possible deviation from globularity. SV is also very useful in the identification of the oligomeric state and the stoichiometry of heterogeneous interactions.

In SE, at centrifugal fields lower than those generally used in SV, sedimentation is eventually balanced by diffusion opposing the concentration gradient, resulting in a time-invariant concentration profile (Figure 4.1B). This experiment is insensitive to the shape of the protein and directly reports on its M_W and, for chemically reacting mixtures, on chemical equilibrium constants. Thus, analysis of SE data can also yield valuable thermodynamic and stoichiometric information on the interaction of molecules, and it is often used for studying self-association and heterogeneous interactions, such as protein–protein, protein–nucleic acid, and protein–small molecule binding.

AU has been used extensively for the initial characterization of IDPs, such as $p57^{Kip2}$ (Adkins and Lumb 2002), nucleoporin Nup2p (Denning et al. 2002), endocytic proteins AP180 and epsin 1 (Kalthoff et al. 2002), DARPP-32 (Hemmings et al. 1984), rod photoreceptor GARPs (Batra-Safferling et al. 2006), securin (Sanchez-Puig et al. 2005), and neuroligin 3 (Paz et al. 2008).

4.4 SMALL-ANGLE X-RAY SCATTERING

Small-angle X-ray scattering (SAXS) is formally analogous to small-angle neutron scattering (SANS), which are together termed small-angle scattering (SAS) techniques (Svergun and Koch 2002; Svergun and Koch 2003). Whereas SAXS has been extensively used for studying the size and shape distribution of IDPs, SANS has not yet been implemented for this purpose. In SAXS, a protein solution is exposed to X-rays (usually at a synchrotron producing radiation at a wavelength λ typically around 0.15 nm), and scattered intensity I is collected at small angles up to a few degrees. The random positions and orientations of protein molecules result in an isotropic intensity distribution, which is proportional to the scattering from a single particle averaged over all orientations. I represents the Fourier transform of the electron density distribution of atoms and is usually visualized in a reciprocal space as a function of the momentum transfer s (also denoted in the literature as scattering vector q, $s = 4\pi\,\lambda^{-1}\sin(\theta)$, where Z θ is the angle between the incident and scattered radiation):

$$I(s) = 4\pi \int_0^{D_{max}} r^2 \gamma(r) \frac{\sin(sr)}{sr} dr \tag{4.3}$$

where $\gamma(r)$ is the spherically averaged autocorrelation function of the excess scattering density, which is zero for distances exceeding the maximum particle diameter, D_{max}. From this relation, the histogram of interatomic distances within the molecule (i.e., the distance-distribution function $p(r)$) can be computed by the inverse Fourier transformation:

$$p(r) = \frac{r^2}{2\pi^2} \int_0^{\infty} s^2 I(s) \frac{\sin(sr)}{sr} dr \tag{4.4}$$

In principle, $p(r)$ contains the same information as the scattering intensity. This real-space representation is more intuitive, however, and information about the particle shape can often be deduced by simple visual inspection.

The basic problem in SAXS is to reproduce the three-dimensional structure from a one-dimensional scattering pattern, which, due to the reduced information content, usually does not lead to a unique solution. In general, I is sensitive to both the size and the conformational properties of the polypeptide chain, and a low-resolution image of the shape of the protein can be reconstituted from the measurements in a rather straightforward manner. SAXS patterns directly provide global hydrodynamic parameters such as R_G, V_H, M_W, and D_{max}. A historically very influential way of obtaining such global description is the linearization of $I(s)$, introduced by Guinier (see [Svergun and Koch 2002]), who showed that $\ln(I)$ vs. s^2 is linear at very small angles, with the slope corresponding to R_G. The Guinier plot represents a useful first stage of data analysis and the demonstration of structural disorder, which causes a discernible deviation from linearity (Figure 4.2A).

Much more advanced *ab initio* methods can be used to recover a rather detailed 3-D structure of structured proteins. Some of these approaches are also able to describe the structural ensemble of proteins (i.e., structural disorder). The tools developed by Svergun and colleagues (for reviews, see [Svergun and Koch 2002; Svergun and Koch 2003]) involve the application of an angular envelope function, an ensemble of dummy residues (DRs), and models built from high-resolution (NMR or X-ray) structures of individual domains or subunits. Novel implementations can model missing regions (e.g., loops) and multidomain proteins with linkers of potential structural heterogeneity, which are conceptually closely related to structural disorder.

An often-used simple approach for demonstrating the difference between disordered and ordered proteins is the Kratky plot (i.e., $s^2 \times I(s)$ vs. s (i.e., q)) (Figure 4.2B). For globular proteins, the plot is approximately bell-shaped and has a clear maximum, whereas in the case of IDPs, it increases monotonically and approaches linearity. A peak on this latter curve is indicative of residual structure; thus it can distinguish between MG, PMG, and random coil states of disorder (Receveur-Brechot et al. 2005). Similar information can be obtained by inspecting $p(r)$ (see Figure 4.3), which provides the D_{max} and low-resolution picture of the protein (Moncoq et al. 2004; Receveur et al. 2002).

SAXS also enables an explicit description of the structural ensemble of IDPs by the *ab initio* ensemble optimization method (EOM) (Bernado et al. 2007). In EOM, a pool of models covering the protein configuration space is generated first by the random selection of allowed amino acid conformations from a library of coil conformations of structured proteins. The SAXS scattering curve is then calculated for each model, and a subset of models is selected by an iterative genetic algorithm that satisfactorily fits the experimental data. The utility of EOM for characterizing IDPs was tested on denatured lysozyme and Bruton's protein tyrosine kinase, a multidomain protein with two IDR linkers. It was found that an ensemble size $N = 50$ is sufficient to describe the conformational ensemble of disordered state, representing a good compromise between accuracy and computational resources.

Due to its simplicity and increasing resolution, SAXS has provided important insight into the structure-function relationship of several IDPs. In most cases, a simple analysis of R_G suggested structural disorder, such as in the case of tau protein (Schweers et al. 1994), ProTa (Gast et al. 1995), caldesmon (Permyakov et al. 2003), ribonuclease E (Callaghan et al. 2004), Grb14 adaptor protein (Moncoq et al. 2004), and neuroligin 3

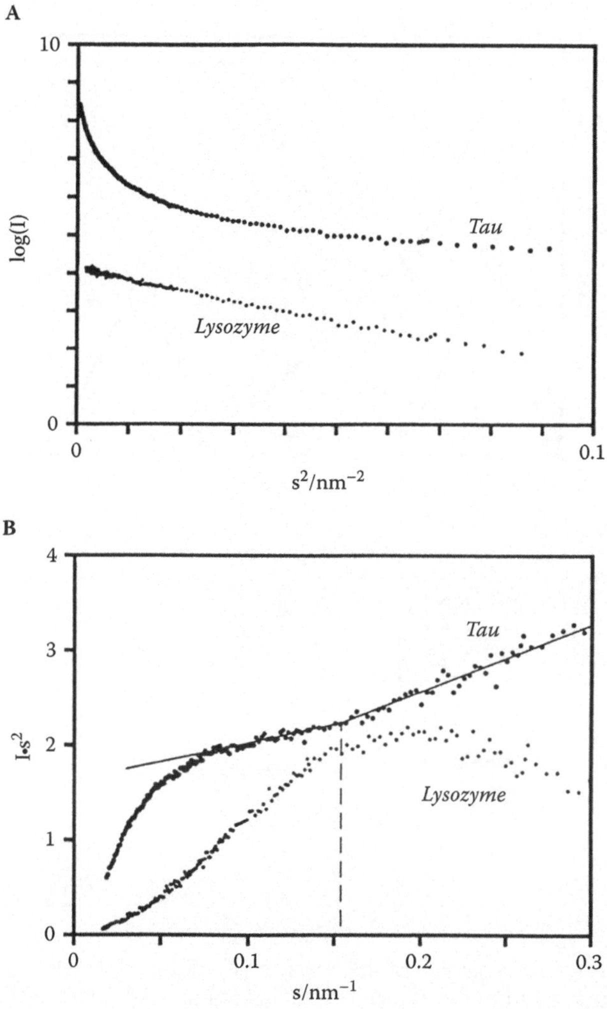

FIGURE 4.2 SAXS characterization of tau protein. Comparison of the SAXS data of tau protein and globular lysozyme. (A) The Guinier plot of tau is curved, indicating that no defined R_G can be assigned to this IDP. In contrast, the scattering curve of globular lysozyme follows a straight line. (B) The Kratky plot of tau increases monotonically, which typifies a fully disordered IDP. The hump on the lysozyme curve is characteristic of a globular structure. Reproduced with permission from Schweers et al. (1994), *J. Biol. Chem.* 269, 24290–7. Copyright by the American Society for Biochemistry and Molecular Biology.

(Paz et al. 2008). In some instances, a more advanced SAXS approach or the combination of SAXS with other analyses to limit the number of solutions has been used, as demonstrated by a few of the most influential cases, such as SAXS and electron paramagnetic resonance (EPR) (measles virus nucleoprotein); SAXS and molecular dynamics (MD), bacterial cellulase, and $p27^{Kip1}$ (see Chapter 14, Section 14.12); and SAXS, NMR, and MD (p53). These cases are discussed next.

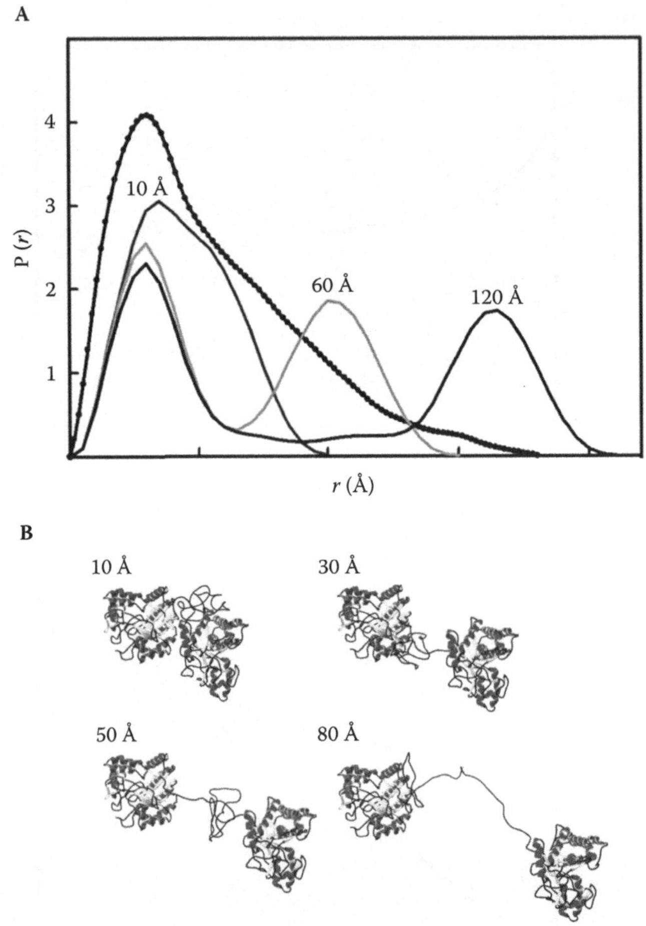

FIGURE 4.3 Structural ensemble of the linker region of bacterial cellulase. A chimeric cellulase was constructed from the catalytic domains of bacterial cellulases Cel6A and Cel6B, which is connected by a linker that is two times the length of the original. (A) The interatomic distance-distribution function (*P*(*r*)) of the structural ensemble of the chimera was determined by SAXS (trace with full-circle symbols). The *P*(*r*) profiles of models with intermodule separations of 10 Å, 60 Å, and 120 Å are also shown for comparison. MD simulations and modeling of the *P*(*r*) profile suggest a continuous distribution from a compact to a fully extended (linker stretched to 120 Å) state. (B) The α-carbon views of four typical molecular structures that were used for the weighted summation. Reproduced with permission from von Ossowski et al. (2005), *Biophys. J.* 88, 2823–32. Copyright by Elsevier Inc.

4.4.1 Measles Virus Nucleoprotein

SAXS was combined with data obtained from a range of other techniques in the study of measles virus nucleoprotein (Bourhis, Canard, and Longhi 2006; Bourhis et al. 2005; Longhi et al. 2003). Measles virus is an enveloped ribonucleic acid (RNA) virus with its negative sense, single-stranded genome packaged into a helical nucleocapsid by the viral nucleoprotein (N). Transcription and replication of the viral genome requires the action of RNA-dependent RNA polymerase (L) in association with another factor, phosphoprotein (P) (Curran and Kolakofsky 1999). In principle, N can self-assemble on cellular RNA in the absence of viral RNA, and the interaction of P with N prevents this illegitimate formation of nucleocapsid-like particles. The soluble N-P complex is the substrate of L polymerase, which initiates the encapsidation of genomic RNA.

The region responsible for self-assembly and RNA-binding is N_{CORE}, whereas interaction with P is mediated by the tail region N_{TAIL}. N_{TAIL} protrudes from the globular region and carries three short recognition elements (Box 1, 2, and 3) that are important for function. N_{TAIL} is disordered by many techniques (Longhi et al. 2003), also corroborated by SAXS, which suggests an R_G = 27.5 Å, compared to 15 Å expected for a globular protein. By this criterion, N_{TAIL} is largely, but not fully, disordered, because its R_G is smaller than that of a random coil-like chain (35–38 Å), its Kratky curve has a bump, and its *p(r)* shows a maximum dimension of 120–130 Å, which is smaller than that of a fully disordered random structure.

SAXS also provided structural details on the interaction of N_{TAIL} with its binding region within P, the XD domain (Bourhis et al. 2005). Box 2 of about 12 amino acids is the primary site of interaction undergoing induced folding to a local α-helix conformation. By SAXS, XD is globular with the expected R_G (12.1 Å) and maximum diameter (41 Å). The R_G of the XD-N_{TAIL} complex is much larger (32.7 Å), although its M_W is not much bigger than that of XD. Thus, the complex is not compact and is highly anisotropic, which is also underscored by a maximum at 20 Å, a shoulder at 30 Å, and a long tail up to 146 Å on the *p(r)* function. The overall envelope of the complex has a globular cluster and an elongated protuberance of varying shape, corresponding to the N-terminal segment of N_{TAIL}. Box 3, which also contributes to binding, tends to point out in different solvent-exposed conformations, without folding to a stable structure (as also demonstrated by EPR spectroscopy; see Chapter 5, Section 5.6). Overall, disorder of N_{TAIL} and the observed binding mode might play important roles in the processivity of virus replication, in which continuous rearrangement of complexes takes place.

4.4.2 Bacterial Cellulase

A thorough SAXS analysis of an IDP is exemplified by bacterial cellulase, also discussed with respect to the entropic chain linker function (see Chapter 12, Section 12.1.1) enabled by structural disorder. Because cellulose is insoluble and crystalline, its effectively degradation can be carried out by modular enzymes, composed of a catalytic domain connected to a much smaller cellulose binding domain (CBD) (Carrard et al. 2000). The catalytic domain and CBD are separated by a long flexible linker, the shortening or deletion of which results in a dramatic reduction of the activity of the enzyme (Srisodsuk et al. 1993). By SAXS, the linker exhibits an extended conformation, resulting in a maximum extension between the two domains that corresponds to about four cellobiose units on a cellulose chain (Receveur et al. 2002). Thus, flexibility of the linker is probably key in the appropriate positioning of the catalytic domain relative to CBD. Its detailed analysis was enabled by a fusion construct composed of the N-terminal half of cellulase, Cel6A, and the C-terminal half of cellulase, Cel6B (von Ossowski et al. 2005). The broad shoulder on the *p*(*r*) profile (Figure 4.3) is indicative of a distribution of conformations with a D_{max} of 178 Å and the linker adopting all the possible separations between the two globular domains. MD simulations of 1,000 linker conformations fitted to the experimental data show that the linker preferentially samples compact states, but it is able to undergo extension with a relatively low energy cost (i.e., it can position the catalytic module and the CBD at a distance comparable to one cellobiose unit, yet enabling processivity [see Chapter 14, Section 14.9] of cellulose degradation).

4.4.3 p53

SAXS results can be effectively combined with NMR residual dipolar coupling (RDC) data to obtain a self-consistent structural model of an IDP that combines both long- and short-range structural features (Bernado et al. 2005). In the analysis, intrinsic conformational sampling of an IDP, based on Φ,Ψ angles obtained from loop regions of folded proteins, are used to restrain SAXS calculations. This approach was applied to model both the bound and free forms of the tetrameric tumor suppressor p53 (Wells et al. 2008). As described in detail in Chapter 15, Section 15.1.2, p53 is a tumor-suppressor transcription factor of 393 amino acids, composed of four structural-functional domains: a trans-activator domain (TAD) subdivided into TAD1 (1–40), TAD2 (40–61), and a Pro-rich region (PRR, 64–92); a core DNA-binding domain (DBD, 93–293); a tetramerization domain (TD, 325–356); and a regulatory domain (RD, 367–393). Its modeling was achieved in three stages, which led to one of the major structural achievements of the IDP field (see Figure 15.2 and cover picture).

The structure of folded domains was solved by X-ray crystallography and NMR, and the quaternary structure of the protein was delineated by verifying the tertiary structures of individual domains in the intact protein, identifying domain–domain interactions by NMR, and determining the arrangement of domains in the full-length protein by SAXS. In the final stage of generating the structure, NMR, RDC, and SAXS data were combined with MD simulations (Bernado et al. 2005) to elucidate the ensemble of structures of the functionally important IDRs, TAD, and RD within the full-length tetramer. In complex with a specific DNA element, the folded domains are well-aligned and form a rather rigid single mass. The high degree of order of this region is efficiently propagated into PRR, due to the relatively long persistence length of the latter. In TADs, the local orientation sampling is less correlated due to their much shorter persistence length. The relative stiffness of PRRs projects the ensemble of TADs away from the main body of the protein, supporting a predominantly structural role of PRR. In the largely disordered TAD, a transient helix at the MDM2-interacing site (see Chapter 10, Section 10.2.3.6) is apparent. In the absence of DNA, the entire p53 structure is much more dynamic, with core domain dimers attached to tetramerization domains by flexible linkers and an apparent decoupling of the effective alignment of TAD and PRR, with this region experiencing more isotropic behavior.

4.5 PULSED-FIELD GRADIENT NMR

Pulsed-field gradient (PFG) NMR (see also Chapter 6, Section 6.2.3) is a convenient means for measuring translational diffusion, in which the attenuation of the echo signal from a Hahn spin-echo pulse sequence containing a magnetic field gradient pulse is used to measure the spatial displacement of the observed spins (Price 1998). The attenuation of a spin-echo signal results from the dephasing of the nuclear spins due to the combination of the translational motion of the spins and the imposition of spatially well-defined gradient pulses. In theory, both B_0 (i.e., magnetic) and B_1 (i.e., radiofrequency) gradient can be used. In practice, B_0 gradients are most commonly used to label the position of a spin in space. The most common approach is to use a simple modification of the Hahn spin-echo pulse sequence (pulse gradient spin echo, PGSE, or pulse gradient stimulated echo longitudinal encode-decode, PG-SLED, sequence [Jones et al. 1997; Price 1998; Wilkins et al. 1999]).

In practice, the sequence PG-SLED yields a series of 1-D spectra, each of which is recorded with a different gradient strength (Figure 4.4), due to the longitudinal replacement of the protein molecule (Jones et al. 1997; Wilkins et al. 1999). In most cases,

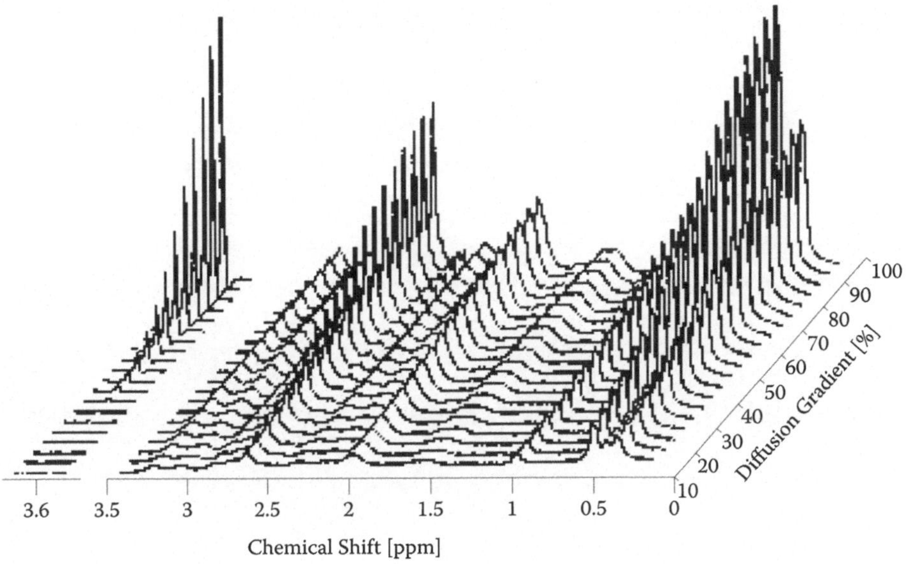

FIGURE 4.4 Pulsed-field gradient NMR spectra of fibronectin-binding protein. PG-SLED spectra of a peptide corresponding to residues 17–37 from D3 of FnBP. The diffusion gradient of magnetic field was varied between 5% and 100% of the maximal value (about 60 G cm^{-1}). The main aliphatic region (3.3–0.3 ppm) of the spectra is shown on the right, with the region of reference dioxin (3.6 ppm) shown on the left. The diffusion coefficient of the protein determined from the rate of decay suggests its disordered state. Reproduced with permission from Wilkins et al. (1999), *Biochemistry* 38, 16424–31. Copyright by the American Chemical Society.

fitting signal intensity as a function of gradient strength yields a decay rate, which is proportional to the diffusion coefficient D. From D, R_S and R_H can be calculated by the Stokes–Einstein equation (Eq. 4.1). Absolute values of D can only be obtained if the temperature and viscosity of the solution are precisely controlled, and an internal radius standard (e.g., dioxan) (Wilkins et al. 1999) is usually used instead, which provides the effective R_H of the protein by the following equation:

$$R_{H,\,protein} = D_{ref} / D_{protein} \times R_{H,\,ref} \tag{4.5}$$

The advantage of this technique is that the ensemble-averaged global hydrodynamic behavior of a protein can be studied under exactly the same conditions under which local structural characterization by NMR is accomplished. It has mostly been used in the field of protein folding (Casares et al. 2004; Pan, Barany, and Woodward 1997; Wilkins et al. 1999) and rather infrequently in the case of IDPs, such as the malaria surface protein MSP2 (Zhang et al. 2008), α-synuclein, and p53 TAD (Dawson et al. 2003). Its great inherent potential, which has not yet been exploited, is that it can also provide data *in vivo.*

Spectroscopic Techniques for Characterizing Disorder

5

The spectroscopic techniques described in this chapter provide both steady-state and dynamic residue-level structural information on intrinsically disordered proteins (IDPs). Because the conformational behavior of IDPs is significantly different from that of globular proteins, many techniques can be applied for their structural characterization. Due to its exceptional resolving power and importance in the IDP field, nuclear magnetic resonance (NMR) is covered in a separate chapter.

5.1 X-RAY CRYSTALLOGRAPHY

A description of the methodology of X-ray crystallography is beyond the scope of this book, but is covered in excellent reviews and monographs (Drenth 2006). X-ray crystallography can determine the arrangement of atoms in a protein by recording the intensity and pattern of the X-ray scattered by the electrons within the protein crystal. Diffraction appears as a pattern of regularly spaced spots known as reflections, from which the three-dimensional model of electron density can be recovered by using the Fourier transforms. The positions of the atomic nuclei are deduced from this electron density in a manner consistent with the covalent structure (sequence) of the protein.

The structure is characterized by its resolution (down to 1 Å in the best cases) and by B-factors of atoms (also termed temperature-factor, which describes the degree to which the electron density is spread out due to either static or dynamic mobility). The total number of structures solved by X-ray crystallography is about 44,000 today, which represents the majority of structures (more than 50,000) deposited in the Protein Data Bank (PDB). It is of special significance with respect to disorder that often the position of an atom cannot be precisely determined due to crystal defects, or actual multiplicity of positions, when it is missing from the electron-density map and is termed *disordered*. Whereas the underlying assumption is that their position can be ultimately determined, the term also has been applied for longer regions missing from the electron-density map, which may be caused by either ordered parts occupying multiple positions (wobbly domains) or "intrinsic" disorder, in a sense being used throughout this book (Dunker et al. 2001).

Initial identification of intrinsically disordered regions (IDRs) derived from such observations and the impact on the field is shown by 138 out of about 500 entries in the DisProt database having "X-ray crystallography" in their field of detection method. A few prominent examples illustrate the importance of such observations. The structure of a 10-subunit yeast *ribonucleic acid (RNA) polymerase II* (RNAP II), the enzyme responsible for transcription of protein-coding genes in eukaryotes, could be solved at 2.8 Å resolution (see Chapter 11, Figure 11.2) (Cramer, Bushnell, and Kornberg 2001). The 280 amino acid–long carboxy-terminal domain (CTD) of the largest subunit, Rpb1, is not seen in the structure. This region orchestrates a complex array of reactions in transcription and messenger RNA (mRNA) maturation, and is indispensable in the function of RNAP II (see Chapter 11, Section 11.2.3). *DNA topoisomerase I* (Topo I) is a nuclear enzyme involved in transcription, replication, recombination, chromosome condensation, chromatin remodeling, and DNA damage recognition. The enzyme catalyzes the ATP-independent breakage of single-stranded DNA, and it has an N-terminal region of about 170 amino acids missing from the X-ray structure (Redinbo et al. 1998). This IDR regulates Topo I activity through phosphorylation, and possible protein–protein interactions with other chromosomal proteins. *Calcineurin* is a Ca^{2+}-dependent calmodulin (CaM)-stimulated protein phosphatase, involved in many pathways, such as T-cell signal transduction, apoptosis, Wnt signaling, MAPK signaling, amyotrophic lateral sclerosis (ALS) pathway, and osteoclast regulation. The enzyme has a 95 amino acids-long region connecting its two subunits missing from the crystal structure (Kissinger et al. 1995). This region has an autoinhibitory element and a CaM-binding site, which cooperate in CaM-dependent regulation of the enzyme. *Core histones* form octamers, which make up nucleosomes (see Figure 5.1) in complex with DNA, which are the basic building elements of the chromatin (Luger et al. 1997). Their disordered N-terminal tails missing from the crystal structure provide multiple functions in epigenetic regulation, by mediating protein–protein interactions and posttranslational modifications (see Chapter 11, Section 11.4.2.1) (Bhaumik, Smith, and Shilatifard 2007; Hansen, Tse, and Wolffe 1998).

Structural disorder is rather widespread in the PDB, as shown by Dunker and colleagues by comparing data in PDB and the corresponding Swiss-Prot sequences (Le Gall et al. 2007). The complete Swiss-Prot sequence can be found in the PDB structure in only 7% of the cases, and more than 95% of the Swiss-Prot sequence in only 25% of the cases. Thus, a great majority of PDB proteins are shorter than their corresponding Swiss-Prot sequences, either because the construct has been truncated to enable crystallization or because the structure contains residues that do not have well-defined coordinates. Approximately 10% of the PDB proteins contain missing or ambiguous residues longer than 30 consecutive amino acids, and about 40% of the them have shorter regions (between 10 and 30 residues) missing. The failure of crystallization of a protein may also point to structural disorder. Of course, crystallization often fails even in the case of ordered proteins, and thus the lack of its success does not prove disorder. There have been some notable failures, though, which actually contributed to developing the concept of disorder (see Chapter 2, Section 2.2.5).

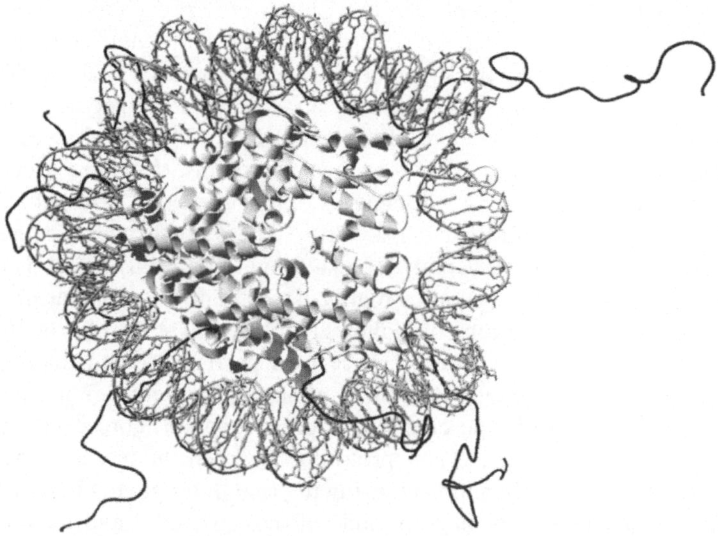

FIGURE 5.1 X-ray structure of the nucleosome. The structure of a core histone octamer with 146 base pairs of DNA winding around (i.e., the nucleosome [pdb 1kx5]), as solved by X-ray crystallography (Luger et al. 1997). The N-terminal tails of histones, as curved pieces, which are the primary sites of posttranslational regulation of chromatin, are missing from the actual structures.

5.2 FLUORESCENCE SPECTROSCOPY

Fluorescence spectroscopy has many applications in the field of the protein disorder. Fluorescence is easy to detect and is very sensitive, it can provide both structural and dynamic data, and it can be specifically detected even *in vivo* (Lakowicz 2006).

Fluorescence is the emission of a photon from an electronically excited state of the fluorophore, in which excitation occurs by the absorption of an incoming photon. The average period of the excited state (i.e., the lifetime [τ]) is typically on the order of 10 ns, whereas the quantum yield (q) (i.e., the probability that an absorbed photon is emitted) can be anywhere between 0.1 and 0.9. Fluorescence is usually recorded in the form of excitation (absorption) and emission spectra, which are intensities of light against wavelength; emission is shifted to higher wavelengths relative to absorption (Stokes' shift). If fluorescence is excited with polarized light, there is a selection of fluorophores in terms of their orientation, which results in an initial anisotropy of emitted light that gradually decays in time due to characteristic molecular motions. Thus, time-resolved anisotropy provides information on the dynamics of molecular motions. Various features of fluorescence can be exploited by applying continuous, time-averaged mode of measurement (steady-state fluorescence) or resolving fluorescence in time (time-resolved fluorescence).

For characterizing protein structures, one can apply intrinsic and extrinsic fluorophores. Intrinsic fluorescence usually emanates from aromatic amino acids, of which

Trp is the most highly fluorescent. The maximum of Trp excitation spectrum is at 280 nm, whereas its emission depends on its local molecular environment, which can be exploited in characterizing the disordered state of proteins. Extrinsic fluorophores have three basic types. The first is covalently attached small molecules, such as fluorescein and rhodamine isothiocyanate (FITC and RITC). These widely used extrinsic labels react primarily with Lys and Cys groups of proteins, they have excellent quantum yields, and, due to their long wavelengths of absorption and emission, biological samples do not interfere with their fluorescence. The second type of extrinsic probe is the non-covalent 1-anilino-8-naphthalene-sulfonic acid (ANS), which is practically non-fluorescent in water, but becomes highly fluorescent in an apolar environment. This makes ANS the classic probe of molten globule (MG) states of proteins, because its intensity is much higher in the presence of a partially folded than either fully folded or fully unfolded proteins (Goldberg et al. 1990; Greene, Wijesinha-Bettoni, and Redfield 2006). A third type of extrinsic probes, small fluorescent proteins, can be fused to the protein studied. Their prototype is green fluorescent protein (GFP), which has a highly visible, efficiently emitting internal fluorophore (Tsien 1998). Since its introduction into molecular biology and cell biology, GFP and its variants have become standard markers of gene expression and protein targeting in intact cells and organisms. Mutations that modulate its emission spectrum gave rise to variants of different spectra (e.g., GFP CyPet, and YPet), making GFP compatible with fluorescence resonance energy transfer (FRET) (see Section 5.2.4) applications (Nguyen and Daugherty 2005).

5.2.1 UV Fluorescence

The basic application of fluorescence in the IDP field derives from the spectral sensitivity of Trp residues to the local environment. Trp can be specifically excited at 295 nm (where Tyr practically does not absorb), and it emits fluorescence with a maximum around 350 nm if it is fully exposed to water. The fluorescence of a Trp shielded from the aqueous environment in the hydrophobic core of globular proteins is generally blue-shifted to around 320 nm (Schmid 1989). Trp residues of IDPs, depending on their exposure or transient burial in a local hydrophobic cluster, emit somewhere in between (see Figure 5.2), as shown in the case of α-casein. One should be aware, however, that this parameter is very sensitive to the exact position of Trp and local perturbations of its environment. Thus, emission at rather high wavelengths has been observed in the case of some globular proteins, such as nuclease (334 nm) and human serum albumin (342 nm) (Lakowicz 2006).

5.2.2 Fluorescence Quenching

Fluorescence quenching refers to any process that decreases the emitted light of a fluorophore, such as excited state reactions, energy transfer, complex formation, and collision with a quencher. A distinction may also be made between static and dynamic quenching, based on whether the fluorophore forms a stable complex, or only makes a transient

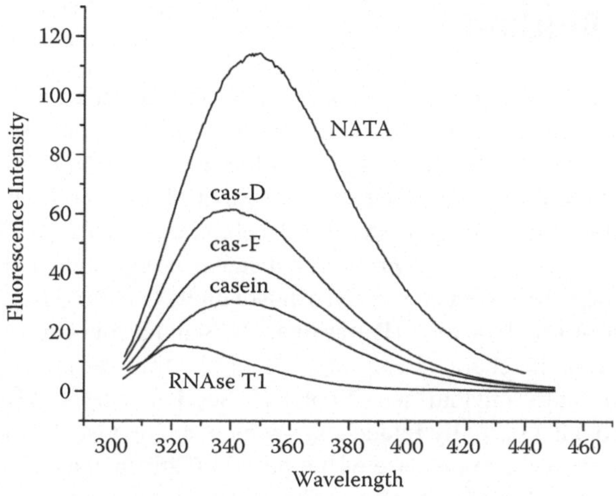

FIGURE 5.2 Fluorescence emission spectrum of α-casein. Fluorescence emission of α-casein was recorded at 295 nm excitation, in the absence and presence of crowding agents. The spectrum was recorded in buffer, in 400 g/l Ficoll 70 (cas-F), and 400 g/l Dextran (cas-D). For a comparison, the spectrum of two model compounds, RNase T1 and NATA, are also shown.

contact, with the quencher. Collisional quenching is especially useful for studying structural changes and dynamic processes of proteins. Its contribution to the deactivation rate of the excited fluorophore is described by the Stern–Volmer (SV) equation:

$$\frac{F_0}{F} = 1 + k_q \tau_0 [Q] = 1 + K_{SV}[Q] \tag{5.1}$$

where F_0 and F are the fluorescence intensities measured in the absence and presence of the quencher, applied at a molar concentration $[Q]$, k_q is the collisional rate constant, and K_{SV} is the SV constant, which is the product of a collisional quenching rate constant and the excited state lifetime of the fluorophore in the absence of the quencher ($K_{SV} = k_q$).

The collisional quenchers most often used are either charged (I^-) or neutral (oxygen, acrylamide) molecules, and quenching is visualized by plotting F_0/F as a function of $[Q]$, which gives a linear function, the slope of which is the SV constant. Values of K_{SV} reflect accessibility of Trp side chains, which are traditionally used to unveil structural rigidity of a protein, because transient "breathing" enables the quencher to reach internal Trp residues (Papp and Vanderkooi 1989). In the case of IDPs, Trp residues are much more exposed, and their actual accessibility can be used to address the presence of transient residual structure. This was addressed by comparing SV constants determined for IDPs with that of a fully folded protein (ribonuclease [RNase] T1) and the model compound of the fully exposed Trp (N-acetyl Trp amide, NATA), which indicated that accessibility of IDPs falls in between the two extremes (K_{SV} = 0.65, 3.99, and 24.37 for RNase T1, p21^{Cip1}, and NATA, respectively, Tompa unpublished results).

5.2.3 ANS Binding

Fluorescence of ANS dramatically increases upon its transfer from water to a hydrophobic environment, which makes it diagnostic for the MG-type disorder. ANS binding has been amply used in the field of protein folding and often in studying IDPs, for example, to show the partial structure of β-casein (Barzegar et al. 2008). It was also used to demonstrate the inside-out structural behavior of IDPs to a decrease in pH, which generally denatures globular proteins but causes compaction and partial structuring in IDPs (Uversky 2002a), such as prothymosin alpha (ProTa) (Uversky et al. 2000b) and the CTD of heat-shock factor 1 (Pattaramanon, Sangha, and Gafni 2007). A closely related application is the characterization of the structural transition of an IDP to the orderly aggregated state, amyloid (see Chapter 15, Section 15.5). The formation of amyloid proceeds through a partially folded state, which can be visualized by an increased ANS binding, as observed in the case of tau protein (Chirita et al. 2005), islet amyloid polypeptide (Kayed et al. 1999), κ-casein (Thorn et al. 2005), and α-synuclein (Uversky et al. 2001a).

5.2.4 Fluorescence Resonance Energy Transfer

Fluorescence resonance energy transfer (FRET, also called Forster resonance energy transfer), is the transfer of the excited state energy from one fluorophore (donor) to another nearby fluorophore (acceptor). The transfer occurs without the emission and reabsorption of a photon, and it results from the dipole–dipole interaction between the two fluorescent moieties. Efficiency of the transfer depends on the extent of overlap between the emission spectrum of the donor and the absorption spectrum of the acceptor, their relative spatial orientation, and distance. Within a given donor-acceptor pair, the rate of transfer of the energy is given by the equation

$$k_T = \frac{1}{\tau_d (\frac{R_0}{r})^6} \tag{5.2}$$

where τ_d is the lifetime of the donor in the absence of the acceptor, r is the distance between the two molecules, and R_0 is the Forster distance at which the efficiency of transfer is 50% (Forster 1948; Michalet, Weiss, and Jager 2006; Stryer 1978). Because transfer efficiency depends on the sixth power of distance, FRET can typically measure distances within the range 20–50 Å, which is well suited for studying the structure and structural changes (Corradi and Adamo 2007) or interactions (McIntyre et al. 2007) of proteins. By analyzing time-resolved decays of donor fluorescence, the relative rates of diffusion of donor-acceptor pairs can also be characterized, which provides information on the dynamics of structure.

Several applications have been presented in the IDP literature for the characterization of the distribution of distances and/or dynamics of disordered structures (see Chapter 10). For example, FRET was used to examine the spatial relationship of domains representing a preferred folding state of tau protein (Chapter 10, Figure 10.6) (Jeganathan et al. 2006), and to estimate the end-to-end distance distributions and persistence length of IDPs of various length, such as the charged-plus-PQ domain of ZipA, the tail domain of α-adducin, and the C-terminal tail domain of FtsZ (Ohashi et al. 2007). The structural and dynamic behavior of the NM domain of yeast prion Sup35p (Mukhopadhyay et al. 2007) was also studied by FRET (see also Section 5.2.5.2 and Chapter 10, Section 10.5.1).

5.2.5 Fluorescence Correlation Spectroscopy

Fluorescence correlation spectroscopy (FCS) measures the rate of diffusion of fluorescent molecules, and in this sense it is at the crossroad of spectroscopic and hydrodynamic techniques (Frieden, Chattopadhyay, and Elson 2002; Hess et al. 2002). The rate is deduced from fluctuations of fluorescence intensity that result from molecules diffusing in and out of the observation volume. For practical purposes, intrinsic Trp fluorescence and extrinsic fluorophores can also be used. Fluorescence is excited by a laser beam focused on a tiny volume, and emitted light is detected by a microscope. Due to the diffusion or conformational changes of individual molecules, the detected intensity fluctuates in time, which is analyzed by computing the ACF of fluorescence (see Chapter 4, Section 4.2 on DLS). The diffusion coefficient thus determined yields hydrodynamic measures, such as R_S by the Stokes–Einsten equation (see Chapter 4, Equation 4.1). FCS can also characterize several other processes that cause fluctuations in fluorescence intensity, such as chemical reactions, ligand binding, protein–protein interactions, and conformational changes. As demonstrated by the following examples, FCS has been used in the IDP field for several purposes.

5.2.5.1 Dimensions of an IDP and the effect of crowding

The basic use of FCS to quantify the hydrodynamic size of IDPs is demonstrated by studies on the conformational state of polyQ (Crick et al. 2006) involved in Gln-expansion diseases (see Chapter 15, Section 15.3.3). Its conformational behavior was approached by measuring the scaling of translational diffusion time (τ_D) with the length of the polymer (N), which resulted in $N^{0.32}$. Based on the R_S expectation of scaling for a random coil (an exponent 0.588; see Chapter 1, Section 1.7), water behaves as a poor solvent for polyQ, and its structural ensemble is made up of a heterogeneous collection of collapsed structures. An interesting pathophysiological implication of this finding is that the preference for collapsed structures arises in the absence of hydrophobic residues (i.e., its driving force might be similar to that of amyloid formation).

FCS was also used to study the conformational state of IDPs under the conditions of crowding. As discussed in Chapter 8, extreme macromolecular concentrations *in vivo* prefer compact states of proteins (Ellis 2001; Minton 2005) and may affect the folding state of IDPs. Via measuring their diffusion times, the compaction of denatured RNase T1 and model compound Fluorescein-PEG in 30% PEG 20000 and Ficoll 70 (Tokuriki et al. 2004), and that of three IDPs, β-casein, MAP2c, and p21[Cip1] in 40% Dextran (see Chapter 8, Figure 8.2), 40% Ficoll 70, and 3.6M TMAO (Tompa, unpublished results) was studied. In all cases, it was found that crowding causes compaction of IDPs, but without a cooperative transition to a folded state.

5.2.5.2 Internal protein dynamics

FCS can also be used to characterize conformational fluctuations in the unfolded ensemble of a denatured globular protein or an IDP. Intestinal fatty acid binding protein (IFABP) was unfolded under denaturing conditions, and its fluorescence self-quenching resulting from transient proximity of two fluorophores placed by site-directed mutagenesis was studied (Chattopadhyay, Elson, and Frieden 2005). Conformational dynamics of the protein appear as fluctuations in fluorescence, which have an apparent relaxation time, $\tau_R = 1.6\ \mu s$ in 3M Gnd-HCl, whereas in the MG state attained at pH 2, it is slower with $\tau_R = 2.5\ \mu s$. In the presence of 100 mM KCl, τ_R increaes to 8 μs, which suggests that ionic strength induces transient secondary structure in the protein that can prevent self-quenching. Thus, τ_R reflects the dynamics of formation and dissolution of ordered substructures.

Internal dynamics were also approached in the case of the disordered NM region of the yeast prion Sup35 (Mukhopadhyay et al. 2007). As also discussed in Chapter 10, Section 10.5.1.2, N is the amyloidogenic region of Sup35, whereas M is a charged region that keeps N in solution. Quenching by internal Tyr residues of singly labeled NM suggests fast conformational fluctuations on the 20–300 ns timescale. At least two well-separated components of the decay—a faster component in the range of 20–40 ns and a slower one in the range of 150–250 ns—can be distinguished. By observing the relative amplitudes of the two components, it could be ascertained that the faster decay component originates from short-range quenching due to relatively proximal Tyr residues, whereas the slower component originates from distant Tyr residues.

5.3 FOURIER-TRANSFORM INFRARED RESONANCE SPECTROSCOPY

Infrared (IR) spectroscopy deals with the infrared region of the electromagnetic spectrum, the most common form of which is absorption spectroscopy (Barth 2007). In IR spectroscopy, absorption arises from exciting rotational and/or vibrational modes of the molecule. Resonant frequencies are related to the strength (type) of the bond and the mass of atoms at either end of it. In relation to the visible spectrum, the infrared region is divided into three regions: the far-, mid-, and near-IR. Far-IR, approximately 400–10 cm^{-1}, is adjacent to the microwave region and deals with rotational transitions. The mid-IR is approximately 4,000–400 cm^{-1} and provides information on fundamental vibrations and associated rotational–vibrational structures of proteins. The highest energy near-IR, approximately 14,000–4,000 cm^{-1}, can excite overtone or harmonic vibrations. The most convenient implementation of IR is Fourier-transform infrared (FTIR) spectroscopy, in which an interferogram is recorded, from which the spectrum is recovered by Fourier transformation. Due to the short characteristic timescale on the order of 10^{-13} s, FTIR provides a snapshot of the ensemble of structures.

Many IR-active bonds occur in proteins, but the contribution of side chains is very complex and hard to interpret in terms of local structure (Barth 2007). Primary insight comes from spectral components related to the amide bond. Most informative on secondary structure is the amide I band near 1,650 cm^{-1}, which arises mainly from the stretching vibration of the C=O bond. The extent to which internal coordinates contribute to amide I normal mode depends on the backbone structure, which results in different typical values 1,656 cm^{-1} (α-helix); 1,633 and 1,684 cm^{-1} (β-sheet); 1,672 cm^{-1} (turns); and 1,647–1,654 cm^{-1} (disordered). Although secondary structure analysis of proteins is nearly exclusively done using the amide I band, amide II (around 1,550 cm^{-1}) and amide III (1,400–1,200 cm^{-1}) bands also make useful contributions. The deconvolution of the amide I band into components has been used several times for demonstrating the relative ratio (or absence) of repetitive secondary structural elements in IDPs.

For example, the most intense amide I band centered at 1,642 cm^{-1} in the spectrum of α-synuclein suggests a predominant random coil conformation, with minor contributions from antiparallel β-sheet and β-turn structures (Weinreb et al. 1996). Deconvolution of the amide I and amide II regions of the spectrum of α_{s1}-casein shows mostly extended (strand plus polyproline II helix (PPII)), turn, and coil conformations, with very little, if any, α-helix (Malin et al. 2001). Similar experiments suggest the dominance of structural disorder in the nuclear-pore protein Nup2p (Denning et al. 2002) and the *M. tuberculosis* Rv3221c biotin-binding protein (Kumar et al. 2008). In terms of the conformational changes that occur upon partner binding, it was demonstrated by FTIR that the AF1 domain of the androgen receptor has very little α-helix structure in isolation, with a significant increase in the presence of the partner of the protein, TFIIF (Kumar et al. 2004). The increase is due to the induction of a local helical element upon binding.

5.4 CIRCULAR DICHROISM

Circular dichroism (CD) spectroscopy is based on measuring the difference of absorption of left-handed and right-handed circularly polarized light, which results from the optical activity of molecules in the sample (Rodger and Nordén 1997). In circularly polarized light, the electric field vector rotates about the propagation direction and it interacts with chiral molecules, which makes the absorption of the two polarized lights differ. The signal recorded may be the difference in absorbance ($\Delta A = A_L - A_R = (\varepsilon_L - \varepsilon_R) \times c \times l$ in molar terms), but most often it is ellipticity of polarization (Θ), defined as $\tan \Theta = E_R - E_L / E_R + E_L$ (where E_R and E_L are the electric field vectors of the two circularly polarized light) converted to a molar ellipticity value. Generally, the difference in absorbance of the two polarized lights is extremely small, within the range 10^{-4}–10^{-6} times the actual absorbance of the sample. CD signals are observed in the same spectral region where absorption of the protein occurs. Typically, near-UV and far-UV regions are distinguished, which give different kinds of information about protein structure.

Near-UV CD in the 250–350 nm region (termed the aromatic region) provides information on the tertiary structure. Different aromatic residues tend to have distinct wavelength profiles. Phe mostly contributes at 250–270 nm, Tyr at 270–290 nm, and Trp at 280–300 nm, whereas disulfide bonds contribute broad, weak signals throughout the spectrum. The near-UV CD spectrum represents a detailed fingerprint of tertiary structure around these reporter residues, but it cannot be interpreted in terms of the actual structure. Proteins of stable 3-D structure are usually characterized by intense and detailed spectrum due to the asymmetric environment of their aromatic residues, whereas the spectrum of unfolded proteins or IDPs is of low-intensity and low-complexity, because their aromatic residues experience an isotropic environment. Further insight into residual structure may come from aromatic residues in certain IDPs experiencing local order, because they are part of an element of a residual structure/hydrophobic cluster.

The far-UV CD spectrum in the range 190–230 nm originates mainly from amide (peptide) bonds. It can be used to determine the relative amount of different secondary structural elements, because they have characteristic far-UV CD spectra, which is clearly distinguished in the case of α-helix, β-sheet, turn, PPII helix, and coil conformations (Figure 5.3A). An actual spectrum can be approximated as a linear combination of the contributions of different elements, but a major uncertainty of such deconvolution into components comes from the choice of basis spectra, which can be polyamino acids or actual proteins of well-known structure.

CD has been a dominant technique for identifying IDPs, and is also used to characterize their non-fully random structure (see Chapter 10, Section 10.2). The resemblance of the spectrum of a protein to that of the coil has been taken to indicate the random coil or disordered character of a protein (Figure 5.3B). In DisProt (Sickmeier et al. 2007), 156 out of about 500 proteins have the annotation CD in the field detection method, and some of the most prominent IDPs have been first shown to lack a well-defined structure by far-UV CD. To cite a few cases, CD was used in the case of MAP2 (Hernandez, Avila, and Andreu 1986), tau protein (Schweers et al. 1994), ProTa (Gast et al. 1995), α-synuclein (Weinreb et al. 1996), $p21^{Cip1}$ (Kriwacki et al. 1996), dehydrin Dsp16 (Lisse et al. 1996), and the high mobility group protein HMGA (Reeves and Beckerbauer 2001). Near-UV CD has been used less often (e.g., in the case of calpastatin [Konno et al. 1997]), but provided evidence for local residual structure in some cases. In the case of caldesmon, a rather intensive band around 275 nm suggests that the Trp residues of this protein are in an asymmetric environment, possibly in a hydrophobic cluster (Permyakov et al. 2003). In the case of the trans-activator domain (TAD) of transcription factor Vmw65 (Donaldson and Capone 1992) and the Potato virus A genome-linked protein VPg (Rantalainen et al. 2008), the contributions of Phe residues at 270–250 nm and Tyr residues at 270–290 nm are interpreted in terms of an asymmetric environment around the aromatic residues.

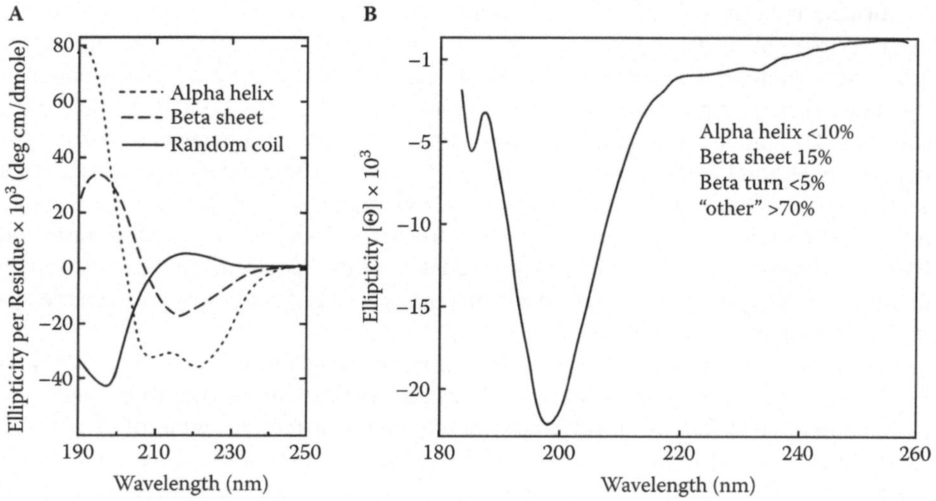

FIGURE 5.3 Typical circular dichroism spectra. (A) Typical CD spectra of α-helix, β-strand, and coil conformations. (B) CD spectrum of high-mobility groupA1a (HMGA1a, also termed HMG-I) protein, which shows that the protein largely lacks repetitive secondary structure. Calculations of the secondary structure composition of the protein (insert) suggest the presence of very little α-helix, β-sheet, or β-turn conformations, but the predominance of random coil or "other" structures. Reproduced with permission from Reeves (2001), *Gene.* 277, 63–81. Copyright by Elsevier Inc.

5.5 RAMAN OPTICAL ACTIVITY SPECTROSCOPY

Raman optical activity (ROA) spectroscopy is a vibrational spectroscopy that relies on measuring a small difference in inelastic Raman scattering of chiral molecules using polarized lights (Barron, Blanch, and Hecht 2002; Barron et al. 2000). In ROA, the complete vibrational spectrum from 100–4,000 cm^{-1} is available. The technique is sensitive to chiral elements in protein structure, and provides information on both structural and dynamic aspects of the molecule. The primary observables in ROA are the scattered intensities in right-handed and left-handed circularly polarized incident lights I^R and I^L, from which the dimensionless circular intensity difference (CID, Δ) is calculated:

$$\Delta = \frac{I^R - I^L}{I^R + I^L} \tag{5.3}$$

Although the relative intensities are sensitive to local secondary structure, the exact relation of spectral components and structural elements, such as α-helix and β-sheet, has not yet been established. The observed correlations of ROA band pattern with backbone conformation, however, can be used as sensitive indicators of secondary structural elements. Vibrations of the backbone in proteins are usually associated with three characteristic regions in the spectrum. Region I, spanning 870–1150 cm^{-1}, corresponds to backbone skeletal stretching that originates mainly from Cα-C′, Cα-Cβ, and Cα-N stretches. The amide III region, spanning 1,230–1,310 cm^{-1}, originates mainly from the in-plane N-H bond deformation with respect to Cα-N. The amide I region, spanning 1,630–1740 cm^{-1}, originates primarily from C=O stretches. Peak assignment is based on ordered proteins with well-characterized structures (Barron et al. 2000; Krimm and Bandekar 1986), and the large information content of the spectrum can be exploited by multivariate analysis, such as non-linear mapping (NLM), for the structural characterization of proteins (Zhu et al. 2007).

Because the timescale of the ROA scattering event (on the order of 10^{-14} s) is much faster than that of the fastest conformational fluctuations, the ROA spectrum is a snapshot of different chiral conformers present in the ensemble of structures. Thus, ROA can distinguish bona fide IDPs, which are in a state of dynamic disorder from proteins, in which all residues are in well-defined but non-repetitive local conformations (e.g., loopy proteins) (see Liu, Tan, and Rost 2002). This can be demonstrated by comparing the spectra of Bowman–Birk proteases inhibitor, lysozyme, and tau protein (Figure 5.4). A further attractive aspect of ROA is its ability to detect the presence of PPII conformation. The ROA spectrum of casein, α-synuclein, and tau protein (Syme et al. 2002), as well as the wheat gluten A-gliadin (Blanch et al. 2003) are very similar, dominated by a strong positive band centered at approximately 1,316–1,318 cm^{-1}. This band probably corresponds to the PPII-helix conformation, and such studies have led to the suggestion that rheomorphism (flowing shape, see Chapter 2, Section 2.2.4) (i.e., the ability of changing shape in a functional context suggested for caseins) (Holt and Sawyer 1993) might be a general feature of IDPs. A related study on lysozyme misfolding and amyloid formation (Blanch et al. 2000) suggested that PPII structure may be critically involved in pathological fibril formation (see Chapter 15, Section 15.3.3.2).

5.6 ELECTRON PARAMAGNETIC RESONANCE SPECTROSCOPY

Electron paramagnetic resonance (EPR, also termed electron spin resonance, ESR) spectroscopy studies chemical species, which have unpaired electrons (Morin et al. 2006). The basic physical concept of EPR is analogous to that of NMR (Chapter 6), but it consists of electron spins that are excited instead of spins of atomic nuclei. Because electrons have a spin quantum number $s = 1/2$ with magnetic moment $m_s = \pm 1/2$, their energy is split in an external magnetic field. An unpaired electron can be moved between the two energy levels by either absorbing or emitting electromagnetic radiation at a resonant frequency. In practice, the majority of EPR measurements are made with microwaves in the 9–10 GHz region, with fields corresponding to about 3,500 G (0.35 T). Because the source of

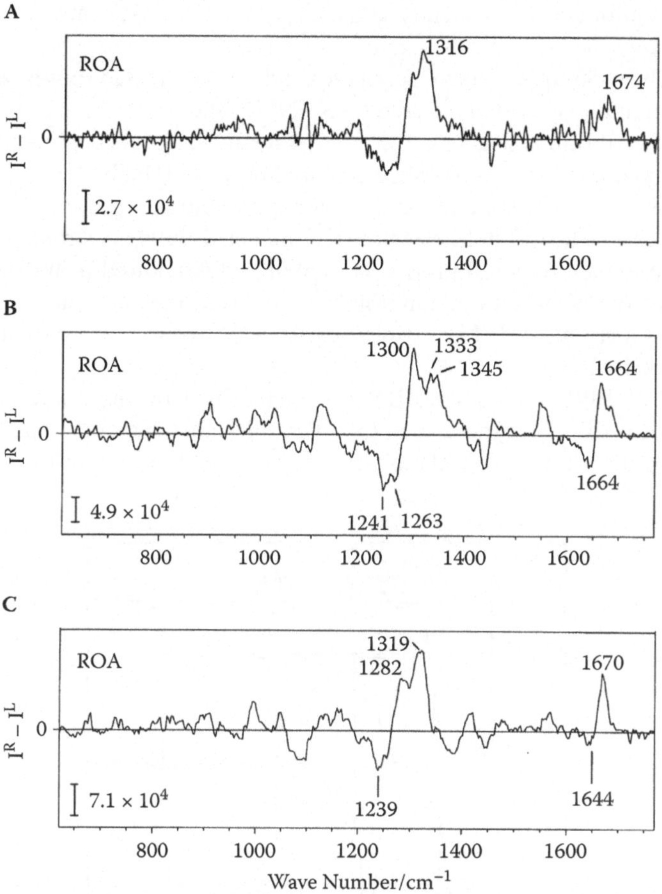

FIGURE 5.4 ROA spectra of ordered and disordered proteins. The spectra are shown to demonstrate the differences between the IDP tau (A), ordered lysozyme (B), and the Bowman–Birk protease inhibitor (BBI), which has an irregular fold (C). The spectrum of tau is dominated by the peak at 1,316 cm^{-1}, which is attributed to PPII-helical conformation, also seen in BBI, which has long loops that occur locally in this conformational state. Reproduced with permission from Syme et al. (2002), *Eur. J. Biochem.* 268, 148–156. Copyright by John Wiley & Sons, Inc.

absorption of energy is a change in the spin state of an unpaired electron, the EPR spectrum is expected to consist of a single line. Interactions with nearby nuclear spins results in splitting of allowed energy states and a multi-lined spectrum that contains information on local structure (see Figure 5.5).

The application of EPR for studying proteins requires the presence of either paramagnetic metal ions or organic free radicals. Thus, EPR is used either for studying metalloproteins (containing Mn^{2+}, Cu^{2+}, or Fe^{3+}), or proteins labeled by specific spin-labels (spin-probes). The label (e.g., HgR, 2,2,5,5-tetramethyl-4-(2-chloromercuriphenyl)-3-imidazoline-1-oxyl or MTSL [1-oxyl-2,2,5,5-tetramethyl-3-pyrroline-3-methyl] methaneethiosulfonate) is covalently attached to a Cys residue introduced at a given position. Because the label is rather small (7Å), it usually does not perturb the activity of the protein. The technique of introducing the label is site-directed spin-labeling (SDSL), thus their combination is often referred to as SDSL EPR.

The EPR spectrum is sensitive to conformational changes occurring upon substrate binding or enzymetic turnover within structured proteins (Hubbell et al. 2003), and its sensitivity can also be exploited to study folding and unfolding transitions, as demonstrated in the case of acetylcholinesterase (Kreimer et al. 1994). A strongly immobilized nitroxide label on a structured protein has small and structured peaks due to specific interactions of the label with its environment, whereas the spectrum in the unfolded state becomes a strong and sharp triplet, close to that of the freely rotating unbound radical (Figure 5.5).

The use of EPR for studying IDPs is exemplified by the study of tau protein (Jeganathan et al. 2006). In the case of this IDP, spin-label dynamics was quantitatively approached by determining correlation time τ_R values of the label introduced at several

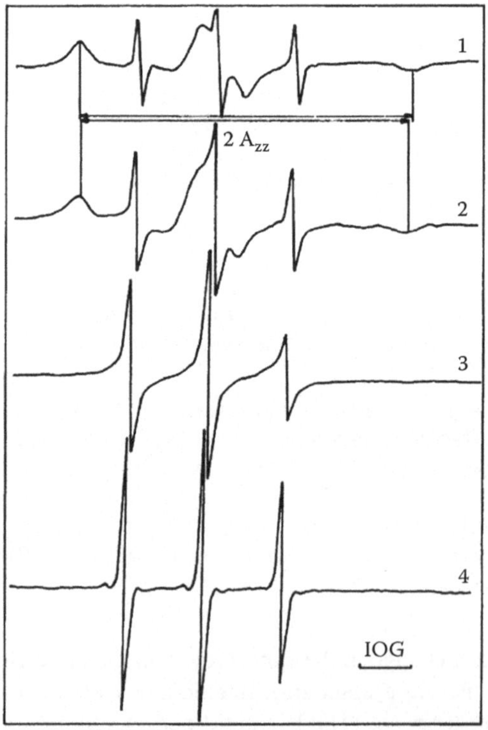

FIGURE 5.5 EPR spectra of acetylcholinesterase. The EPR spectrum of acetylcholinesterase (AChase) labeled with HgR recorded under various conditions (i.e., in native ((buffer, trace 1)), denatured ((at 1.5 M [trace 2])), and 5.0 M ((trace 3)) Gnd-HCl concentration states). The spectrum of the free label is shown by trace 4. Reproduced with permission from Kreimer et al. (1994), *Proc. Natl. Acad. Sci USA* 91, 12145–9. Copyright by the National Academy of Sciences.

positions. The shape of EPR spectra shows that the probe in the entire protein is in a large-mobility state, with characteristic τ_R values on the order of 0.2–0.6 ns. For comparison, the correlation time of the unbound probe is about 0.05 ns, whereas that of the label immobilized inside a well-folded protein is two orders of magnitude higher. These differences can also be exploited to study induced folding of IDPs upon partner binding, as demonstrated in the binding of the N_{TAIL} region of measles virus nucleoprotein to the XD domain of viral phosphoprotein (Morin et al. 2006). The mobility of the spin-probe at three different positions is significantly reduced in the presence of XD (for details, see Chapter 4, Section 4.4.1), whereas at a fourth position it is unaffected, which enables to delineate the regions involved in binding-induced ordering. A similar approach was used to show that a segment of myelin basic protein (MBP, 82–93), a membrane-bound protein in the central nervous system, forms an amphipathic α-helix, which lies on the surface of the membrane, partly embedded in it (Bates et al. 2004).

An influential and rather unique application of EPR concerns the structural changes that accompany amyloid formation. For example, free states and fibrils generated from 83 different spin-label derivatives of α-synuclein were studied by EPR (Chen et al. 2007). In the free state, all variants have sharp and narrowly spaced triplets, which is suggestive of a high degree of mobility that follows from the disorder of the protein. The situation is completely different in fibrils. Within about the N-terminal 30 and C-terminal 15 residues, the spectra are heterogeneous and slightly broader than in the free state, which is representative of high but somewhat restricted mobility. Spectra of the central core region (NAC, 35–95) become almost completely free of hyperfine lines, indicating spin-exchange narrowing. This fundamental change suggests spatial contacts between multiple spin labels, which can be best accounted for by a parallel in-register cross-β structure, which makes multiple molecules stack on top of each other in the amyloid (see Chapter 15, Section 15.5.3 and Figure 15.5).

5.7 ELECTRON MICROSCOPY

In electron microscopy (EM), electrons are used to illuminate a specimen and create an enlarged image (Watt 1997). Due to the much shorter wavelength of electrons than photons, the resolving power of EM is much better than light microscopy, with magnification up to 2 million times. There are several arrangements of EMs. In transmission EM (TEM), the high-energy electron beam is transmitted through the specimen, with the image created by electron intensities modified by the sample. Scanning EM (SEM) produces images by detecting low-energy secondary electrons emitted from the surface of the sample due to excitation by the primary electron beam. Reflection EM (REM) is operated as a TEM, but with elastically scattered electrons being detected. The most attractive feature of EM is that it can provide images of individual molecules, and thus it enables the direct visualization of structural disorder.

A critical step in EM is the processing of the sample to make it suitable for imaging, usually achieved by chemical fixation (chemical cross-linking), cryofixation (instantaneous freezing in liquid nitrogen in cryo-EM), freeze drying, embedding, sectioning (slicing), and staining in heavy metals (lead, uranium, or tungsten). A variant of this latter method, known as rotary shadowing, is very often used to increase the contrast of a protein sample, which proceeds by freezing the specimen very rapidly, vacuuming off loose and disorganized ice crystals, spraying the sample with metal vapor, and then applying acids to dissolve away the protein itself. The specimen that remains is a thin metal shell moulded in the shape of the protein. Probably due to the problems of potentially denaturing side effects of fixation, however, EM has only been used in a few cases in the IDP field.

The classic observation is on caldesmon, which is an 89–93 kDa protein of the contractile apparatus of muscle cells. Caldesmon is one of the first proteins to be recognized as intrinsically disordered by heat stability, CD, and GF (Lynch, Riseman, and Bretscher 1987). Long-angle rotary shadow images of the protein indicate an elongated flexible molecule with an average contour length of 146 ± 40 nm (Figure 5.6). Its large length variation and highly varied shape is best accounted for by structural disorder of the protein. A similar picture emerges in the case of the P/Q domain of the cell-division protein ZipA (Ohashi et al. 2002). Rotary shadowing EM of a construct of the protein containing two fused globular domains show a wide distribution of

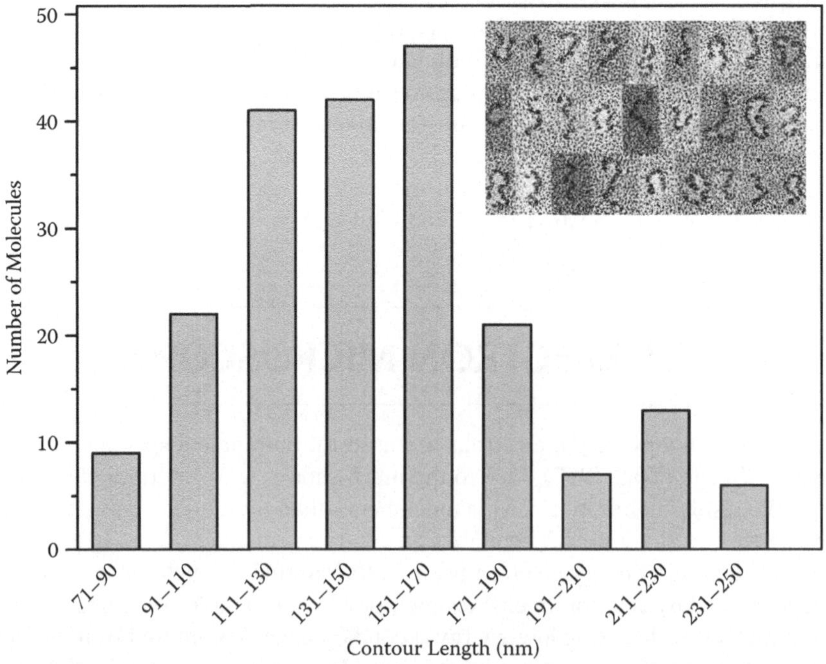

FIGURE 5.6 Electron micrographs of caldesmon. The histogram of the contour lengths of 208 molecules of caldesmon determined by EM has an average length of 146 ± 40 nm. The collection of low-angle rotary shadowed images demonstrates the extreme flexibility of the molecule (insert, magnification: 102,000 ×). Reproduced with permission from Lynch et al. (1987), *J. Biol. Chem.* 262, 7429–37. Copyright by the American Society for Biochemistry and Molecular Biology.

separations between the two domains, ranging from 7.8–19.5 nm, with an average of 12.4 nm. The EM of MAP2, which is 1,828 amino acids in length, shows a highly elongated shape of an apparent width of 3.5 nm and a rather varied length averaging 97 ± 17 nm (Wille et al. 1992a). It is highly flexible, as shown by the molecule folding back on several images to form antiparallel hairpin-like structures. Its juvenile form, MAP2c (467 residues), behaves in a similar fashion, forming rod-like structures 4 nm in width and 48 ± 7 nm in length, which are often curved/bent in shape (Wille et al. 1992b).

EM images can also provide functional insight. Cardiac titin is a giant muscle protein with multiple Ig domains and a highly repetitive disordered domain termed PEVK (Pro, Glu, Val, Lys-rich region) because of the preponderance of these four amino acids (see Chapter 12, Section 12.1.3 and Chapter 13, Section 13.3.1.3). A construct of Ig domains connected by a 186-amino acid long PEVK linker studied by rotary-shadowed EM (Li et al. 2001) has a wide length distribution ranging from 9–24 nm, with two peaks at 11 and 17 nm. The functional importance of this extended structural state (also addressed by atomic force microscopy (AFM) force-extension curves see Section 5.8) rests in its elastic behavior, which resists many stretch-relaxation cycles and is critical in its function as an elastic molecule in muscle. Rotary shadowing EM suggests a similar structural picture but somewhat different functional interpretation in the case of myosin VI (Rock et al. 2005). Myosin VI is a processive motor moving along actin filaments with larger than expected step size. Dimeric molecules have a distribution of the separation of head modules 27 ± 6 nm by EM, which suggests that the 80-residue long segment next to the tail is not rigid but highly flexible and allows a diffusive search for binding sites on F-actin (see Chapter 14, Section 14.9).

5.8 ATOMIC FORCE MICROSCOPY

Atomic force microscopy (AFM) is a scanning probe microscopy with resolution to the fractions of a nanometer. The AFM consists of a micro-scale cantilever with a sharp tip (tip radius of curvature in the nanometer range) at its end that is used to mechanically scan the surface of the sample. When the tip is brought into proximity with a sample surface, forces between the tip and the sample (e.g., mechanical contact forces, van der Waals forces, capillary forces, electrostatic forces) lead to the deflection of the cantilever, followed by interferometry of a laser spot reflected from its top. The two basic operation modes of AFM are imaging and force spectroscopy. In imaging (scanning) mode, the cantilever is externally oscillated, and the oscillation amplitude, phase, and resonance frequency monitored with respect to the position of the probe provide information about topology of the surface. In force spectroscopy, the AFM tip is extended toward and retracted from the surface, and the pico-Newton forces generated are monitored as a function of distance. The two different applications are exemplified by the direct visualization of the structural ensemble of matrix metalloproteinase 9 (MMP-9) and characterization of the folding/unfolding of α-synuclein.

5.8.1 Matrix Metalloproteinase 9

Matrix metalloproteinase 9 (MMP-9, also known as gelatinase B), is a processive enzyme of the extracellular matrix. MMP-9 has a Zn^{2+}-binding catalytic domain combined with three fibronectin type II exosite modules, connected to a C-terminal hemopexin C domain by a 54-amino acid Pro/Gly-rich linker. The enzyme can cleave a variety of substrates of distinct structures, and is known for high processivity, which enables it to move along an extended substrate at a rate of 4 μm/s (Overall and Butler 2007). Cross-linking to a silica layer enabled AFM visualization of a large number of individual molecules (Rosenblum et al. 2007). The images display two peaks corresponding to the globular domains that are separated by a variety of distances ranging from 55 Å to 85 Å, with two preferred states at 62 and 78 Å. This observation and small-angle X-ray scattering (SAXS) data suggest multiple enzyme conformations enabled by the flexible nature of the linker. The ensuing processivity of the enzyme is very similar to that of the bacterial cellulase (von Ossowski et al. 2005) and transport protein myosin VI (Rock et al. 2005) (see Chapter 14, Section 14.9).

5.8.2 α-Synuclein

AFM can also be used to measure the force required for the extension of an IDP (e.g., that of α-synuclein) (Sandal et al. 2008). To this end, force-extension curves of multiple unfolding events of a fusion construct of three N-terminal and three C-terminal titin immunoglobulin (I27) domains flanking a single α-synuclein molecule were recorded. About 30% of the molecules show unfolding typical of a fully disordered state, without any significant deviation from a worm-like chain behavior. About 60% of them display single or multiple small peaks superimposed on the purely entropic behavior, ascribed to additional mechanically weak interactions along the chain. Most interesting, about 7% of α-synuclein molecules display extension at a force very similar to that required to unfold the Ig-domains, which suggests a rather ordered structure dominated by β-type of interactions. The importance of this observation is underscored by the fact that the ratio of this structured component increases under conditions that promote the formation of α-synuclein aggregates, such as the presence of copper, the pathologic A30P mutation, and high ionic strength (see Chapter 15, Section 15.3.2.1).

Nuclear Magnetic Resonance

6

Nuclear magnetic resonance (NMR) spectroscopy has a special status among spectroscopic techniques because it can provide residue-level information on the structure and dynamics of disordered proteins. The NMR of intrinsically disordered proteins (IDPs) owes a lot to studies of protein denaturaion and folding, where many relevant experiments had been conducted. In the field of IDPs, NMR was initially used for demonstrating their disorder (i.e., for simply contrasting their behavior with that of folded proteins). Later, the emphasis shifted to characterizing residual structure and correlating it with (binding) function, which are the major assets of protein NMR. Combinations of NMR data with other methods or applying NMR to proteins in a living cell provide unprecedented insight into the structure and function of IDPs.

6.1 BASIC PRINCIPLES

The principles of NMR (Wutrich 1986) and their application to the unfolded/disordered state of proteins (Chatterjee et al. 2005; Dyson and Wright 2002a; 2004) are amply covered in excellent monographs and reviews; here, only basic aspects are surveyed briefly. NMR spectroscopy is based upon the existence of nuclear spins and the intrinsic magnetic moment of atomic nuclei such as that of ^{1}H, ^{13}C, and ^{15}N. These nuclei have two possible spin states, which are split in an external magnetic field (B_0). Transitions between the two states can be induced by the resonant absorption of electromagnetic radiation at a frequency that matches the energy difference between the states. Because the difference depends on the chemical environment, the NMR signal contains a wealth of information on the local covalent and spatial arrangement of atoms.

In current 1-D, 2-D, or multi-D NMR, often doubly (^{15}N, ^{13}C) or even triply (^{15}N, ^{13}C, ^{2}H) labeled proteins are used, and one or more radiofrequency excitation fields of different frequency are applied and are assembled into different time-domain pulse combinations. The combinations are designed to allow magnetization transfer between nuclei in coherent motion (spin systems), with the aim of detecting their interactions. Following excitation, how the out-of-equilibrium magnetization vector precesses about the external magnetic field and returns to the ground state is measured. These processes are characterized by two distinct relaxation events: Spin-lattice (longitudinal) relaxation (characterized by the relaxation rate R_1) describes the relaxation of the Z-component of the spin, parallel to B_0, toward equilibrium, which involves the exchange of energy with its environment. Spin–spin (transverse) relaxation (characterized by the relaxation rate R_2) corresponds to the relaxation of the XY-component of the spin, perpendicular to B_0, which occurs without exchanging energy with the environment and is usually much faster than R_1. The signal observed overall is the free induction decay (FID) that contains the sum of the NMR responses from all the excited spins.

Basically, there are two types of interactions: through-bond and through-space interactions between spin systems, the latter usually being a consequence of the nuclear Overhauser effect (NOE) (i.e., the transfer of spin polarization between spin populations via cross-relaxation). Fourier transform of FID provides the basic observable of NMR spectroscopy: the shift of the intrinsic NMR frequency due to the actual chemical environment, which is called the chemical shift (characterized by parts-per-million, i.e., ppm of the excitation frequency). Poor dispersion of chemical shift makes assignment (i.e., the process of identifying which resonance belongs to which residue of the protein) difficult. Whereas this step is critical for obtaining sequence-specific information on structure, several NMR approaches do not require assigned signals and provide a global description of the structural state of a protein.

6.2 GLOBAL CHARACTERIZATION BY NMR

6.2.1 1-D ^{1}H NMR

The simplest NMR experiment is to record a one-dimensional ^{1}H spectrum. The polypeptide chain of an IDP rapidly interconverts between multiple conformations, making the chemical shift dispersion of protons become rather poor, precluding direct assignment of resonances at this level (Figure 6.1). For example, the spectral region of amide protons in a globular protein typically spans the chemical shift range 6.5–10.0 ppm, whereas in the case of IDPs, these resonances are confined to a range of about 8.0–8.5 ppm. Thus, the spectrum shows characteristic differences between ordered and disordered proteins, which directly demonstrates structural disorder, such as in the case of fibronectin binding protein(A) (FnBPA) (Penkett et al. 1997), Dsp16 (Lisse et al. 1996), cyclic-AMP response element-binding protein kinase-inducible domain (CREB KID) (Hua et al. 1998), sialoprotein and osteopontin (Fisher et al. 2001), E-cadherin cytD (Huber et al. 2001), titin Pro, Glu, Val, Lys-rich (PEVK) domain (Ma, Kan, and Wang 2001), gliotactin (Zeev-Ben-Mordehai et al. 2003), and rod photoreceptor glutamic acid-rich protein (GARP) (Batra-Safferling et al. 2006). The analysis of ^{1}H spectra has also been incorporated into structural proteomics programs for the high-throughput screening (HTS) of proteins likely to crystallize (Peti et al. 2004).

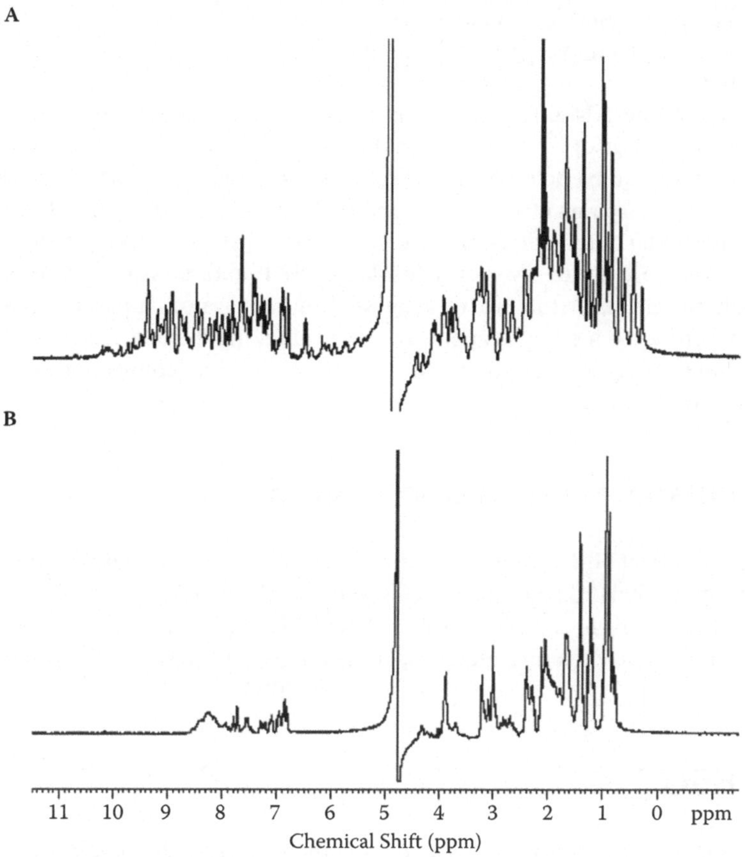

FIGURE 6.1 1-D ^{1}H NMR spectrum of the intrinsically disordered cytoplasmic domain of gliotactin. The 1-D ^{1}H NMR spectrum of the folded 9-kDa complex of α-bungarotoxin with a 13-mer peptide (A) compared to the spectrum of the cytD of gliotactin (B). The spectrum of Gli-cytD is typical of that of a protein in a random coil state. Reproduced with permission from Zeev-Ben-Mordehai et al. (2003), *Proteins* 53, 758–67. Copyright by Wiley-Liss, Inc.

6.2.2 Wide-Line NMR

The unfolded polypeptide chain of IDPs is largely exposed to the solvent, which is manifested in a high level of hydration. This can be directly visualized by measuring the FID of water protons, separating the signal coming from the hydrate layer from those of the protein and bulk water by freezing (Bokor et al. 2005). Water molecules in the hydrate layer remain motile below the temperature at which bulk water freezes out, and the phases of ice protons, protein protons, and unfrozen water protons become separated in the FID signal due to large differences in their spin–spin relaxation rates. Ice protons have a typical value of $R_2 > 200{,}000$ s^{-1}, which is completely buried in the dead time of the spectrometer. Protein protons also have a large $R_2 > 20{,}000$ s^{-1}, whereas water signals typically relax at a rate $R_2 < 2000$ s^{-1}. In practical terms, the temperature range can be divided into four regions, within which distinct observed behavior can be interpreted as weighted averages of the amplitude and dynamics of different unfrozen water fractions.

Thus, wide-line NMR relaxation can be used for demonstrating a large hydrate layer and structural disorder, as shown in the case of microtubule-associated protein 2 (MAP2) and calpastatin (Bokor et al. 2005), early responsive to dehydration (ERD10/14) (Bokor et al. 2005; Tompa et al. 2006a), and Df31 (Szollosi et al. 2008). Because transient intramolecular interactions effectively compete with hydration of the protein, a comparison of the hydrate layer of a full-length IDP and its segments also provides information on residual structure, as suggested in the case of calpastatin (Csizmok et al. 2005). Overall, the method is applicable for visualizing the interface region of IDPs, which is a surface representation of structure that is in direct connection with function (see Chapter 10, Section 10.6).

6.2.3 Pulsed-Field Gradient NMR

Pulsed-field gradient NMR directly measures the diffusion coefficient of proteins, from which their hydrodynamic parameters can be determined (Price 1998). The technique, detailed in Chapter 4, Section 4.5, provided evidence for the disorder of IDPs by their unusually large hydrodynamic radius, such as for α-synuclein and the trans-activator domain (TAD) of p53, for example (Dawson et al. 2003).

6.2.4 HSQC

Sequence-specific assignment of resonances is made possible by multidimensional triple resonance methods of ^{13}C- and ^{15}N-labeled proteins, because the chemical shifts of backbone ^{15}N and carbonyl ^{13}C resonances are well dispersed in the disordered state. Typically, the first experiment to be measured with an isotope-labeled protein is a 2-D heteronuclear single quantum coherence (HSQC) spectrum (Figure 6.2), which correlates the backbone amide nitrogen resonances with those of the directly attached protons. In a ^{1}H-^{15}N HSQC spectrum, the amide bond of each amino acid residue (with the exception of prolines) plus amide nitrogen-containing side-chains provide a signal. The peaks in the proton dimension of an IDP show the lack of dispersion apparent in the 1-D ^{1}H spectrum (Figure 6.1), spanning between 8.0 ppm and 8.5 ppm (Figure 6.2). In the nitrogen dimension, the spectrum is well spread out, spanning 105–130 ppm for backbone and side-chain amide groups. This difference in dispersion in the two dimensions is a reliable indicator of structural disorder, due to which HSQC is often recorded simply for characterizing the structural state of an IDP.

HSQC is also the starting point of resonance assignment, which is essential for a meaningful interpretation of more advanced NMR experiments. The problem of poor proton dispersion is overcome by using the dispersion of ^{15}N and ^{13}C nuclei in a variety of experiments that add further dimensions to the HSQC plane and help resolve ambiguous resonances. In the case of small unlabeled proteins, this is achieved by a set of two-dimensional homonuclear NMR experiments, such as homonuclear correlation

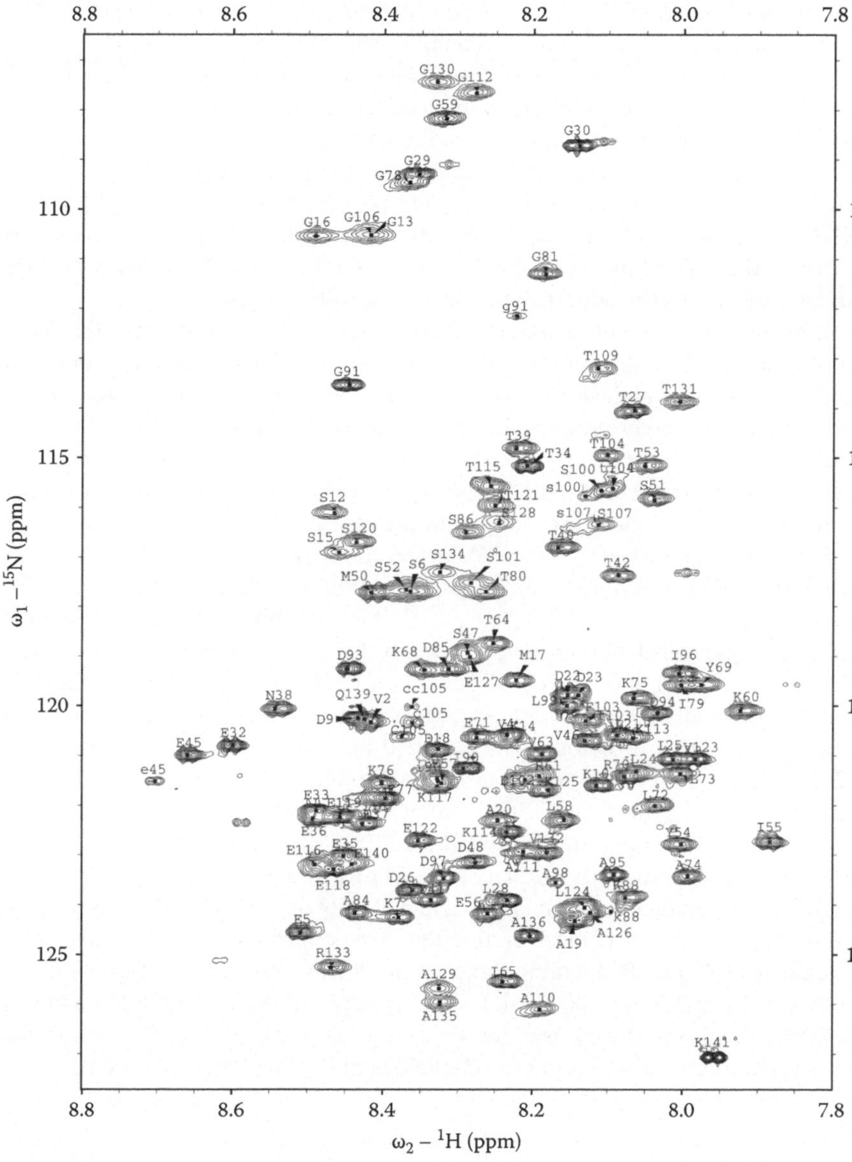

FIGURE 6.2 HSQC spectrum of calpastatin domain 1. The ^{1}H-^{15}N HSQC spectrum of uniformly ^{15}N-labeled calpastatin domain 1 of 141 amino acids (126 non-Pro), of which 121 expected backbone peaks could be assigned. See Kiss et al. 2008b for details.

spectroscopy (COSY), total correlation spectroscopy (TOCSY), and nuclear Overhauser effect spectroscopy (NOESY). With larger proteins, ^{15}N-edited three-dimensional experiments TOCSY-N-HSQC and NOESY-N-HSQC are performed. If the protein is both ^{13}C- and ^{15}N-labeled, it is possible to connect different spin systems through bonds, usually by using distinct combinations of HNCO, HNCACO, HNCA, HNCOCA,

HNCACB, and CBCACONH experiments, which add a carbon dimension to the HSQC plane. Assignment is achieved via a sequential walk through the backbone and along the side-chain on the basis of one-bond correlations. As to the state of the art, an IDP as large as 202 amino acids (human securin) could be fully assigned by a combination of proton-based and proton-less approaches (Csizmok et al. 2008), whereas by a lengthy procedure that combined the graphical analysis of the spectra of full-length wild-type protein, different isoforms and corresponding peptides, and subsequent (HA)CANNH and HNN experiments (Lippens et al. 2006; Lippens et al. 2004; Mukrasch et al. 2005; Mukrasch et al. 2007b; Smet et al. 2004), the sequence of hTau40, which is the longest human tau isoform (441 residues), could be almost fully assigned.

Following assignment of the peaks, HSQC is also the starting point of the determination of a variety of parameters for the sequence-specific characterization of transient structure and dynamics at the local level. The HSQC experiment is also useful for detecting interactions with other proteins, because a change in relaxation due to the interaction makes NMR parameters of residues directly involved shift or even disappear from the spectrum. For these reasons, the HSQC spectrum has been one of the most frequent experiments in the IDP literature, applied in the case of $p21^{Cip1}$ (Kriwacki et al. 1996), FlgM (Daughdrill et al. 1997), VP16 TAD (Uesugi et al. 1997), 4E-BP1 (Fletcher and Wagner 1998), D1–D4 of fibronectin binding protein(A) (FnBPA) (Penkett et al. 1998), CREB KID (Radhakrishnan et al. 1998), protein kinase inhibitor α (PKIα) (Hauer et al. 1999a), eukaryotic translation initiation factor 4G1 (eIF4G1) (Hershey et al. 1999), p53 TAD (Lee et al. 2000), α-synuclein (Eliezer et al. 2001), CP 12 (Graciet et al. 2003), Grb14 (Moncoq et al. 2003), Wiskott–Aldrich syndrome protein (WASP) (Panchal et al. 2003), Smad-anchor for receptor activation Smad-binding domain (SARA SBD) (Chong et al. 2004), thymosin β4 (Tβ4) (Domanski et al. 2004), IA3 (Green et al. 2004), myelin basic protein (MBP) (Harauz et al. 2004), T-cell receptor zeta cytoplasmic domain (cytD) (Sigalov et al. 2004), tau protein (Eliezer et al. 2005), BRCA1 (Mark et al. 2005), prion domain of Ure2p (Pierce et al. 2005), colicin E9 (Tozawa et al. 2005), UreG (Zambelli et al. 2005), rod photoreceptor glutamic acid-rich protein (GARP) (Batra-Safferling et al. 2006), phage lambdaN (Prasch et al. 2006), cystic fibrosis transmembrane conductance regulator (CFTR) R domain (Baker et al. 2007), β-synuclein (Bertoncini et al. 2007), Nogo (Li and Song 2007), Sic1 (Mittag et al. 2008), and MSP2 (Zhang et al. 2008). HSQC is also routinely used for screening the prospect of structure solution in various structural genomics programs (Oldfield et al. 2005c; Peti et al. 2004).

6.3 SEQUENCE-SPECIFIC STRUCTURAL INFORMATION

Once resonance assignment has been achieved, a variety of NMR parameters can be determined to characterize structural and dynamic behavior at the residue level. The values are usually compared to those expected on the assumption of the random coil state, and deviations are used for a detailed description of local structure (see Chapter 10, Section 10.2.3 and Table 10.1).

6.3.1 Chemical Shifts

The primary observable is the chemical shift of various nuclei. The deviation of chemical shift from random coil values (termed secondary chemical shift, SCS, or chemical shift index, CSI) for ^{1}Hα, ^{13}Cα, ^{13}Cβ, and ^{13}CO are all sensitive to local secondary structure, and their information content can be improved by appropriate corrections for local sequence (Schwarzinger et al. 2001). The approach has been initially used for studying protein folding, thus reference random coil states have been determined under different denaturing conditions, such as in 1M (Wishart et al. 1995) or 8M (Schwarzinger et al. 2001; Schwarzinger et al. 2000) urea. Because of the small values that are characteristic of the disordered state, slight differences between reference values may results in some ambiguities.

In an α-helix, ^{13}Cα and ^{13}CO values are shifted downfield, whereas ^{13}Cβ and ^{1}Hα values are shifted upfield (to smaller ppm values). In β-sheets, ^{13}Cα values are shifted upfield, whereas ^{13}Cβ and ^{1}Hα values are shifted downfield (see (Dyson and Wright 2002b; 2004)). In IDPs, the rapid conformational averaging cancels out the large deviations seen in ordered proteins and makes CSI values much smaller. The actual values carry information on the relative population of dihedral angles in the different local conformations, as exemplified by the solution structure of the KID domain of transcription factor CREB (see Chapter 10, Section 10.2.3.2). In short, CREB KID binds the KID-binding domain (KIX) of CBP in a phosphorylation-dependent manner by virtue of two helices, αA and αB (Radhakrishnan et al. 1997) (Figure 6.3A). CSI values in solution indicate that the regions also transiently populate helical conformations (Figure 6.3B), with a stronger preference within region αA (about 50%) than within αB (about 10%). Very little difference exists between the conformational preferences in the phosphorylated and nonphosphorylated forms (Radhakrishnan et al. 1998; Radhakrishnan et al. 1997), although these might be important (see Chapter 14, Section 14.3).

CSI values of different nuclei have been used for characterizing local structural propensities in the N-terminal domain (NTD) of prion protein (Donne et al. 1997), VP16 TAD (Uesugi et al. 1997), 4E-BP1 (Fletcher and Wagner 1998), CREB KID (Radhakrishnan et al. 1998; Zor et al. 2002), PKIα (Hauer et al. 1999a), the NTD of potassium channel (Wissmann et al. 1999), p53 TAD (Lee et al. 2000), α-synuclein (Eliezer et al. 2001), titin PEVK domain (Ma et al. 2001), Tβ4 (Domanski et al. 2004), tau protein (Eliezer et al. 2005), colicin E9 (Tozawa et al. 2005), IA3 (Ganesh et al. 2006), phage lambdaN (Prasch et al. 2006), CFTR R domain (Baker et al. 2007), β-synuclein (Bertoncini et al. 2007), and MSP2 (Zhang et al. 2008), for example.

6.3.2 Dynamic Information from Relaxation Data

Backbone and side-chain dynamics of IDPs can be approached by measuring either ^{15}N R_1 and R_2 relaxation rates or ^{1}H-^{15}N NOE in uniformly ^{15}N-labeled proteins. Although all three relaxation parameters are sensitive to motions on the picosecond to nanosecond timescale, NOE is the most sensitive to the high-frequency motions of the backbone

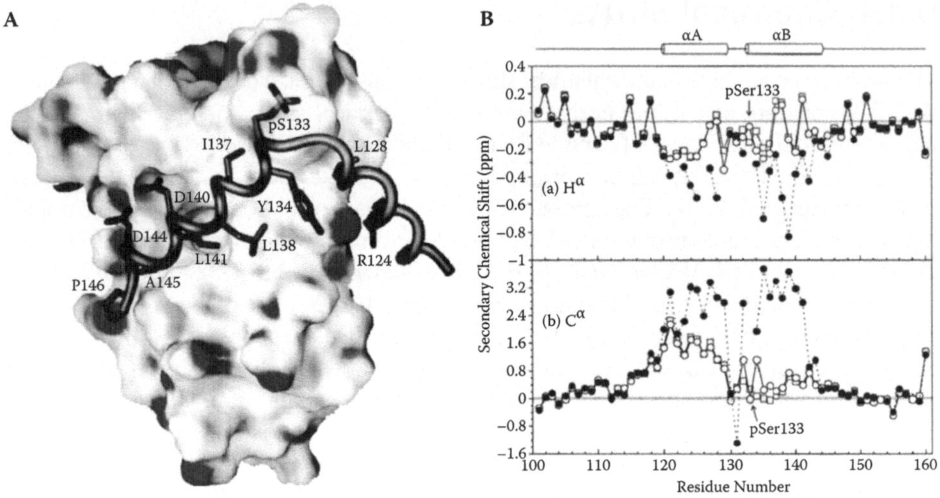

FIGURE 6.3 The structure of phosphorylated CREB-KID in CBP-bound and free states. (A) The structure of the phosphorylated KID domain of CREB in complex with the KIX domain of CBP (pdb 1kdx) that encompasses two helices connected by a wide turn harboring phosphorylated Ser^{133}, as solved by NMR (Radhakrishnan et al. 1997). (B) Hα and Cα CSI values of non-phosphorylated (□), Ser^{133}-phosphorylated (o), and KIX-bound (•) forms of KID. CSI values indicate helical preference within regions αA and αB, and very little difference between the non-phosphorylated and phosphorylated forms. Panel B reproduced with permission from Radhakrishnan et al. (1998), *FEBS Lett.* 430, 317–22. Copyright by Elsevier Inc.

whereas R_2 is also informative on microsecond to millisecond motions and conformational exchange processes. The data are usually measured by using HSQC-based methods.

For folded proteins, relaxation data is usually interpreted within the framework of a model-free formalism, in which three parameters, the overall rotational correlation time, internal correlation time, and order parameter describing amplitudes of internal motions are determined (Lipari and Szabo 1982). For IDPs and unfolded proteins, this approach is not valid because the whole molecule does not tumble with a single isotropic correlation time and the timescale of side-chain motions is commensurable with that of the backbone. In this case, a spectral density function assuming the distribution of correlation times is usually applied and is directly mapped on the relaxation data. In a reduced spectral density mapping approach, spectral density is mapped at three frequencies (i.e., $J(0)$, $J(\omega_N)$, and $J(0.87\omega_H)$), which provide information on internal molecular motions spanning a wide time range from milliseconds down to sub-nanoseconds (see Dyson and Wright 2002b).

Analysis of relaxation data is particularly informative on the local structural preferences, such as hydrophobic clusters, secondary structural elements, and transient long-range contacts. Due to its sensitivity to local conformational freedom, the method is also particularly well suited for studying interactions with macromolecular partners. The relaxation measurements have been applied in the case of FlgM (Daughdrill,

Hanely, and Dahlquist 1998), FnBPA (Penkett et al. 1998), sialoprotein and osteopontin (Fisher et al. 2001), histone messenger ribonucleic acid (mRNA) stem-loop binding protein (SLBP) (Thapar, Mueller, and Marzluff 2004), replication protein A (Olson et al. 2005), transcription factor ETS1 (Macauley et al. 2006), p53 (Veprintsev et al. 2006), β-synuclein (Bertoncini et al. 2007), CREB KID (Sugase, Dyson, and Wright 2007), and MBP (Libich and Harauz 2008).

6.3.3 Distance Information from NOE

In the case of folded proteins, 3-D structures are effectively elucidated by way of distance constraints determined by ^{1}H-^{15}N NOE, which occurs between atoms typically closer in space than about 5–6 Å. In the case of IDPs, local preferences of dihedral angles Φ,Ψ can be inferred from NOEs between sequential amino acids such as $d_{\alpha N}$ (i,i+1), $d_{\beta N}$ (i,i+1) and d_{NN} (i,i+1), whereas the presence of secondary structural elements can be ascertained by medium-range NOE such as $d_{\alpha N}$ (i,i+2), $d_{\alpha N}$ (i,i+3) and $d_{\alpha\beta}$ (i,i+3). Long-range NOEs characteristic of tertiary interactions are usually not observed in IDPs due to fluctuations and heterogeneity of the structure.

Thus, NOE can be primarily used to demonstrate the presence of transient local structure in denatured globular proteins, such as a hydrophobic cluster in urea-denatured 434-repressor (Neri et al. 1992), and the lack of appreciable interatomic contacts in the case of IDPs, such as the prion protein (Riek et al. 1997), 4E-BP1 (Fletcher and Wagner 1998), FnBPA (Penkett et al. 1998), the third cytoplasmic loop of cannabinoid 1 receptor (Ulfers et al. 2002), IA3 (Green et al. 2004), and the N- and C-terminal regions of human Nogo (Li and Song 2007). Medium-range NOEs ascertain local transient secondary structure, as in the case of FlgM (Daughdrill et al. 1998), SLBP (Thapar et al. 2004), and the KID domain of $p27^{Kip1}$ (Sivakolundu, Bashford, and Kriwacki 2005), and they may also suggest the presence of PPII elements, as in titin PEVK domain (Ma et al. 2001).

Long-range distance information can be obtained by the application of paramagnetic nitroxide spin labels, which cause broadening of nuclear spins (paramagnetic resonance enhancement, PRE) within a radius of about 15 Å (Gillespie and Shortle 1997). The method of inserting paramagnetic probes is the same as in the case of electron paramagnetic resonance (EPR) (Chapter 5, Section 5.6), and it requires the site-directed introduction of a single Cys residue to which the spin label, such as PROXYL, can be attached. HSQC spectra are then recorded with the spin-label in the paramagnetic (oxidized) and diamagnetic (reduced) states, and differences in one of several parameters, such as line width, relaxation rate, or intensity, are determined.

This approach has been used quite extensively to elucidate medium-range and long-range structural contacts in IDPs. In the case of α-synuclein (Figure 6.4), for example, deviations from random coil values suggest transient long-range contacts between its middle hydrophobic region and the C-terminal domain (CTD) (Dedmon et al. 2005), which need to be released for the molecule to undergo amyloid formation (Bertoncini et al. 2005). In general, PRE results may be analyzed either qualitatively by comparing the intensities of cross-peaks, or more quantitatively, determining actual distance ranges

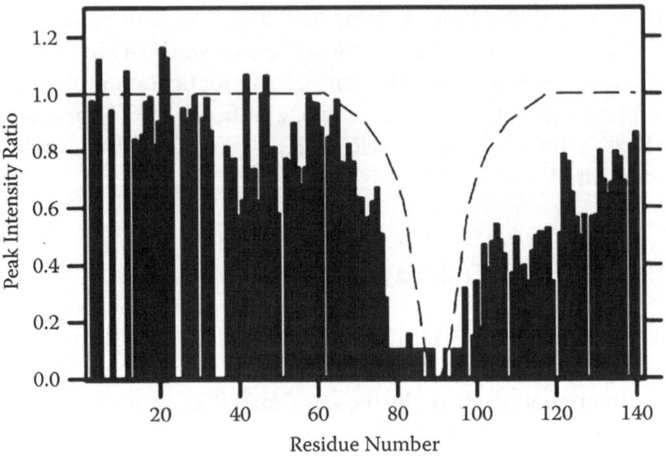

FIGURE 6.4 Paramagnetic resonance enhancement study of α-synuclein. A paramagnetic probe (MTSL) was attached at position 90 to α-synuclein. HSQC spectra in the paramagnetic (oxidized) and diamagnetic (reduced) states of the label were recorded, and intensity ratios were calculated (black bars). The dashed line indicates the paramagnetic effect expected for a random coil polypeptide. The observed difference suggests long-range tertiary-type structural communication between the middle segment and C-terminal regions of the molecule (see Figure 10.7). Reproduced with permission from Bertoncini et al. (2005), *Proc. Natl. Acad. Sci.* USA 102, 1430–5. Copyright by the National Academy of Sciences.

that can then be fed into molecular dynamics (MD) simulations to obtain a realistic picture of the structural ensemble of the IDP (Dedmon et al. 2005). This approach was used in the case of p53 (Vise et al. 2007) and α-synuclein (see Figure 10.7) (Dedmon et al. 2005), for example.

6.3.4 Coupling Constants

The coupling constant $^3J_{HN\alpha}$, which is sensitive to the dihedral angle Φ, can also provide insight into local backbone structural preferences. In α-helices, its value is on the order of 3–5 Hz, whereas β-sheets are characterized by much higher values, around 8–10 Hz. In the disordered state, conformational averaging results in intermediate values, which are not generally considered too informative on IDPs (Dyson and Wright 2002b). In partially aligned media, however, the slight orientation leads to incomplete cancellation of coupling, and the resulting residual dipolar coupling (RDC) contains sufficient information on the relative orientation of the vector between the two nuclei to be converted into a set of orientational restraints. Partial alignment may be accomplished in a number of ways, such as in dilute solutions of lipid bicelles or stressed polyacrylamide gels.

The majority of RDC applications have focused on the refinement of high-resolution 3-D structures and on the structural characterization of denatured globular proteins, such as staphylococcal nuclease (Shortle and Ackerman 2001) and eglin C (Ohnishi et al. 2004). For bona fide IDPs, RDC has been used much less frequently than CSI values or relaxation data. The few examples that should be noted are FnBPA (Penkett et al. 1998), histone mRNA binding protein SLBP (Thapar et al. 2004), Sendai virus phosphoprotein (Bernado et al. 2005), α- and β-synuclein (Bertoncini et al. 2005; Bertoncini et al. 2007), tau protein (Mukrasch et al. 2005; Mukrasch et al. 2007a), and p53 (Wells et al. 2008).

6.4 SPECIAL APPLICATIONS

These applications provide special insight into the structural ensemble of IDPs. The combination of NMR data with those of other techniques offers almost residue-level description of the ensemble. Measuring H/D exchange enables one to characterize the exposure of residues to solvent, which is of utmost importance for understanding their recognition functions. With special preparation of sample, NMR can even provide information on the structure and function of an IDP in the living cell.

6.4.1 Combinations with MD

Distance information obtained by NOE/PRE NMR can be used to constrain MD simulations to arrive at a reasonably confined distribution of structural states in the ensemble of IDPs. Such studies show transient elements of residual structure similar to the bound state in the case of $p27^{Kip1}$ (Sivakolundu et al. 2005), long-range interactions inhibitory to aggregation in the case of α-synuclein (Dedmon et al. 2005), and the proximity of murine-double minute 2 (MDM2)- and RPA70-binding regions in the TAD of p53 (Lowry et al. 2008b). MD simulations can also be simultaneously restrained by RDC and small-angle X-ray scattering (SAXS) data, which can provide a structural description that combines global and local structural clues of the ensemble of IDPs. This approach was used for the description of both the free and bound states of p53 (see Chapter 4, Section 4.4.3; Chapter 15, Figure 15.2; and cover picture).

6.4.2 Amide Proton Exchange Rate

As also discussed in Chapter 3, Section 3.9, the exposure/structural state of amide protons can be probed by H/D exchange, usually detected by Fourier-transform infrared spectroscopy (FTIR) or mass spectrometry (MS). The exchange can also be followed by NMR, which has two implementations. Slowly exchanging protons can be followed by the recording of successive HSQC spectra after placing the protein into D_2O. Because HSQC is slow to record, usually none of the amides in a typical IDP exchanges slowly enough to be measurable (Thapar et al. 2004). This inability to resolve rapid exchange, nevertheless, serves to indicate structural disorder.

The exchange rate of fast-exchanging amide protons can be measured by magnetization transfer measurements (Spera, Ikura, and Bax 1991). In such CLEANEX experiments, water protons are selectively excited, and their exchange with amide positions is observed by following transfer of magnetization to the amide site. The method is capable of detecting exchange processes on the milliseconds timescale, and is potentially applicable for the characterization of IDPs. The results are usually expressed as values of protection factor, which is the ratio of the intrinsic exchange rate of an unprotected amide in the same chemical environment (same sequence in local random coil conformation) and the observed exchange rate (i.e., k_{int}/k_{obs}) of the given amide hydrogen. The former can be directly measured or calculated from the sequence (Bai et al. 1993).

Typically, protection factors on the order of 10^3–10^6 are observed for folded regions, whereas transient structural elements in IDPs/intrinsically disordered regions (IDRs) can only provide a protection up to about 10-fold. This approach has been used for showing the lack of structure in the N- and C-terminal regions of Nogo (Li and Song 2007) and segments of the HET-s prion amyloid (Ritter et al. 2005), whereas it demonstrated transient helical segments in securin (Csizmok et al. 2008) and the NTD of histone mRNA binding protein SLBP (Thapar et al. 2004).

6.4.3 In-Cell NMR

NMR is a non-invasive technique that can also provide data on proteins in a living cell. The observables in this application (in-cell NMR) are the same as in test-tube experiments outlined in this chapter, but the basic difference lies in sample preparation. Details are given in Chapter 8, Section 8.3.2.

7 Proteomic Approaches for the Identification of IDPs

Previous chapters outline how individual proteins can be characterized in terms of structural disorder. This chapter focuses on how high-throughput studies are implemented for identifying disordered proteins *en masse*. In these, the basic approach is the enrichment of cellular extracts for intrinsically disordered proteins (IDPs) and their separation by two-dimensional (2-D) electrophoresis (2DE) followed by mass spectrometry (MS)-based identification.

7.1 EXPECTATIONS AND LIMITATIONS OF PROTEOMIC STUDIES

Experimental evidence points to the disorder of about 500 proteins (Sickmeier et al. 2007), which is far short of the thousands of proteins predicted to harbor significant disorder in the human proteome alone (see Chapter 13, Section 13.1.1). Thus, critical progress in the field is expected from large-scale and rapid identification of fully or mostly disordered proteins by dedicated high-throughput screening (HTS) approaches. As outlined in this chapter, several studies have met these goals, but their results are not without limitations. A serious complicating factor comes from our lack of a clear definition of disorder (see Chapter 2, Section 2.1), due to which there can be no universal solution to the large-scale identification of IDPs, and the results need confirmation by independent approaches. A further level of ambiguity comes from the definition of disorder at the residue level, not the level of whole proteins. To succeed in identifying the disorder of regions as opposed to whole proteins, ordered and disordered regions should be delineated and separately studied. HTS studies, instead, work at the level of entire proteins and provide binary information on whether the protein behaves as ordered, or rather as disordered, within the resolution of the technique.

7.2 2DE-MS IDENTIFICATION OF PROTEINS IN EXTRACTS ENRICHED FOR DISORDER

The initial step of proteomic identification of IDPs is usually the enrichment of extracts for mostly disordered proteins (Table 7.1). Because IDPs are usually resistant to denaturing conditions (see Chapter 3, Sections 3.1 and 3.2), such as low pH and heat (Kalthoff 2003; Tompa 2002; Uversky, Gillespie, and Fink 2002a), the application of trichloro-acetic acid (TCA), perchloro-acetic acid (PCA), or boiling temperatures can lead to the precipitation of globular proteins and relative enrichment for IDPs. The extract thus enriched for IDPs can be best separated on 2DE, which consists of two eletrophoretic steps: an isoelectric focusing in the first dimension and an sodium dodecyl sulfate poly-acrylamide gel electrophoresis (SDS-PAGE) in the second dimension. Different conditions significantly differ in their effectivity and selectivity of separating IDPs from globular proteins (Cortese et al. 2005). In the proteomes of bacteria and yeast, PCA is more potent than TCA, as treatment with 3.0% TCA results in a similar yield of proteins as with 1.0% PCA. Due to better selectivity, 3.0% PCA is probably optimal for separating IDPs from globular proteins and their subsequent identification: The ratio of mostly disordered proteins in the PCA-treated fraction is about 60% (90/158 spots). Boiling results in fewer spots and a different distribution on the 2DE.

Heat treatment at various temperatures (60 and 98°C), precipitation by an organic solvent (10% n-butanol), and low pH (2.0) were also compared (Galea et al. 2006) to address the proteome of mouse, which is much larger than that of bacteria and yeast. Conditions were tested on $p27^{Kip1}$ and its globular partner Cdk2; best separation could be achieved by heat treatment at the highest temperature: 98°C applied for 1 hour. When the same treatment was applied to the mouse proteome, it caused a reduction of the number of observable 2DE spots from 584 to 269, and enrichment in IDPs from

TABLE 7.1 Proteomic approaches for the identification of IDPs*

METHOD OF ENRICHMENT	*SPECIES*	*PROOF OF DISORDER*	*REFERENCE*
3.0% PCA	*E. coli, S. cerevisiae*	Bioinformatics	(Cortese et al. 2005)
98°C × 1 h	*M. musculus*	Bioinformatics, limited proteolysis	(Galea et al. 2006)
100°C × 10 min	*A. thaliana*	NA	(Irar et al. 2006)
100°C × 10 min	*S. cerevisiae*	CD, GF, bioinformatics	(Csizmok et al. 2006)

* The table lists the few HTS approaches used to address the disordered complement of proteins (unfoldome or disorderome) from various organisms. If applicable, additional evidence for the disorder of proteins identified is given.

11.8% to 41.9%, as confirmed by bioinformatic analysis. IDPs identified are enriched in regulatory, signaling, and structural functions, and are depleted in metabolic functions, in agreement with the functional preferences of structural disorder (Iakoucheva et al. 2002; Tompa, Dosztanyi, and Simon 2006b; Ward et al. 2004). Disorder of the identified proteins could be also confirmed by their susceptibility to proteolysis, a diagnostic feature of IDPs (Csizmok et al. 2005; Tompa 2002).

A proteomic study of the stress-related phosphoproteome of *A. thaliana* raises possible functional implications of an HTS of IDPs (Irar et al. 2006). As described in Chapter 11, Section 11.7, largely disordered late-embryogenesis abundant (LEA) proteins are expressed in seeds and under dehydration stress conditions in other tissues of plants (Garay-Arroyo et al. 2000; Mouillon Gustafsson, and Harryson 2006; Tunnacliffe and Wise 2007). Thus, the proteomic study of plant seeds (Irar et al. 2006), which are by definition dominated by such proteins, provides direct functional insight. Out of 710 proteins in the total seed extract, 406 were found to be resistant to a heat treatment of 100 C × 10 min, which supports the notion that the seed is particularly rich in stress-related potentially disordered proteins (Figure 7.1). About half of them are LEA and storage proteins, which can be further enriched for phospho-proteins by affinity chromatography. Proteins thus obtained belong, without exception, to the LEA/storage categories, among which several previously characterized IDPs, such as Em (Soulages et al. 2002) and early responsive to dehydration 10/14 (ERD10/14) (Bokor et al. 2005; Tompa et al. 2006a) can be found. Thus, a combination of enrichment, separation, and identification steps in a proteomic study has the potential of providing functional insight into IDPs.

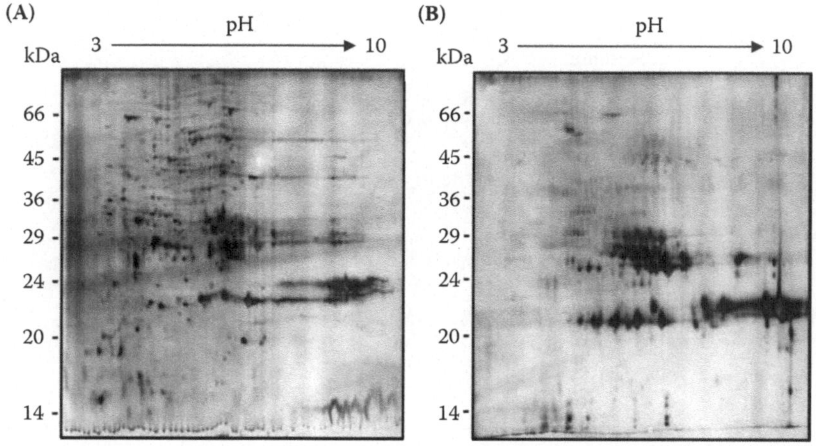

FIGURE 7.1 2-D electrophoresis of *A. thaliana* proteins enriched for disorder by heat treatment. *A. thaliana* seed extracts were boiled to select for heat-resistant dehydration stress proteins. The supernatant was run on 2DE, and spots were excised for MS identification. The enrichment is shown by comparing the 2DE gel before (A) and after (B) the heat treatment. Boiling reduced the number of spots from 710 to 406, and mostly preserved disordered LEA and storage proteins. Reproduced with permission from Irar et al. (2006), *Proteomics* 6, S1, S175–85. Copyright by Wiley-VCH.

7.3 NATIVE/UREA 2DE PROVIDES DIRECT INFORMATION ON DISORDER

The steps of enrichment and separation can be combined in a proteomic technique of identifying IDPs (Csizmok et al. 2006) in a way that directly provides evidence for the disordered character of the proteins identified. In this native/urea 2DE, an initial heat treatment of 100 C × 10 minutes enriches for IDPs and provides the first line of evidence of disorder. Heat-treated proteins are then separated on a 2DE, designed to provide the second part of the evidence. In this, a native electrophoresis in the first dimension (excluding SDS, which would denature proteins) and an 8M-urea electrophoresis in the second one, are combined. In both gels, proteins are separated according to their own charge/mass ratios, because urea only unfolds, but does not contribute charges to, proteins. In the second dimension, globular proteins unfold and slow down, whereas IDPs, which are structurally rather insensitive to denaturing conditions, cover about the same distance as in the first dimension. Thus, structural insight comes from IDPs running into the diagonal line of the gel, whereas rare heat-resistant globular proteins arrive above the diagonal line in the second dimension. This way, not only proteins in the two

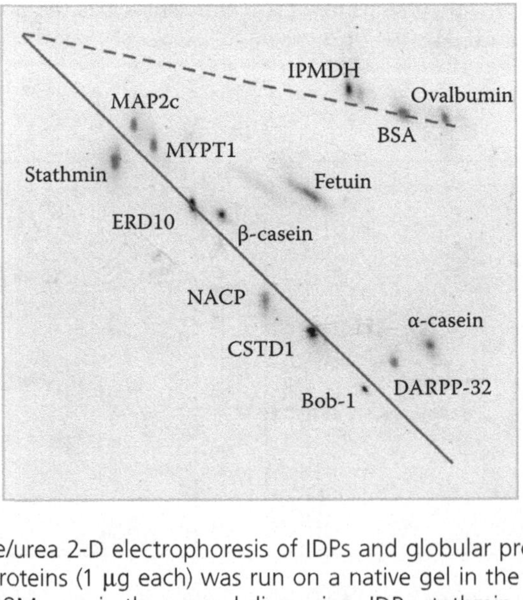

FIGURE 7.2 Native/urea 2-D electrophoresis of IDPs and globular proteins. A mixture of IDPs and globular proteins (1 μg each) was run on a native gel in the first dimension and on a gel containing 8M urea in the second dimension. IDPs stathmin, microtubule-associated protein 2c (MAP2c), MYPT1 304-511, ERD10, β-casein, α-synuclein (NACP), CSTD1, Bob-1, DARPP-32, and α-casein are aligned along the diagonal line of the gel (marked by a solid line). Globular proteins fetuin, IPMDH, BSA, and ovalbumin occupy off-diagonal positions (marked by a dashed line). Reproduced with permission from Csizmok et al. (2006), *Mol. Cell. Proteomics* 5, 265–73. Copyright by the American Society for Biochemistry and Molecular Biology.

TABLE 7.2 Comparison of various measures of disorder of proteins identified by native/urea 2DE*

PROTEIN	*2-D POSITION*	*PONDR®*	*CH PLOT*	*COIL (CD)*	$M_{W,APP}/M_W$
SRib	On	20.69	0.32	41.6	4.9
TFIIA	On	63.29	6.26	76.1	4.4
tropoM	On	49.25	12.46	12.6	2
Ubi6	On	51.7	18.13	30.9	2
CaM	Above	52.38	0.54	31.7	2

* Potential *S. cerevisiae* IDPs were identified by native/urea 2DE-MS. For each protein, its position relative to the diagonal line of the 2-D gel (on/above), percentage of disorder predicted by predictor of natural disordered regions (PONDR®), distance on the charge-hydropathy plot from the line separating disordered and ordered proteins (see Uversky et al. 2000a), percentage of residues in coil (disordered) conformation by CD, and the ratio of the apparent and real M_W determined by GF, are given. Adapted from Csizmok et al. 2006.

groups are separated from each other, but direct information on the structural status of IDPs is provided (Csizmok et al. 2006).

This separation principle can be illustrated by experimentally confirmed IDPs and globular control proteins (Figure 7.2). IDPs align along the diagonal line, whereas controls appear above. When applied for the proteomic-scale identification of IDPs from cellular extracts of *E. coli* and *S. cerevisiae*, bioinformatic, gel filtration (GF), and circular dichroism (CD) characterization showed that proteins tend to fall into two groups (Table 7.2). Some of them, such as transcription factor IIA (TFIIA), appear to be fully extended, devoid of appreciable secondary structure content, and might be classified as an IDP of random-coil- or pre-molten globule (PMG)-type. Other proteins, such as Ubi6, contain a significant amount of secondary structure, they are more compact by hydrodynamic criteria, and can probably be approximated as molten globules (MGs). The technique can also be used for studying proteins of limited purity and very low quantity (on the order of μg). Structural techniques traditionally used for characterizing IDPs, such as CD, nuclear magnetic resonance (NMR), or small-angle X-ray scattering (SAXS), require large amounts of purified proteins and cannot be applied to trace amounts of contaminated proteins. The position on the native/urea 2DE gel, however, is a dependable indicator of the disordered status of a protein under such conditions (Csizmok et al. 2006).

IDPs under Conditions Approaching *In Vivo*

8

A major challenge in the field of intrinsically disordered proteins (IDPs) is to assess to what extent *in vitro* observations on the structural state and function of these proteins can be extrapolated to the living cell. This chapter discusses *in vitro* experiments aimed at approximating *in vivo* conditions and also *in vivo* experiments that directly address the physiological state of IDPs. Most studies suggest that IDPs are probably more compact under such conditions, but they do not assume a unique folded state. Indirect considerations also support the notion that IDPs do not become folded by the crowded conditions encountered in the cell (i.e., disorder is their physiological state). The issue of their assumed fast intracellular degradation, which would apparently be contradictory to their existence and functioning, is also addressed.

8.1 MACROMOLECULAR CROWDING IN THE CELL

Our observations on IDPs are dominated by *in vitro* experiments, and it is implicitly assumed that the emerging picture is relevant with respect to their state and affairs in a living cell. The cell, however, has extremely high intracellular macromolecular concentrations that give rise to a crowding effect, which might bear direct relevance on the structural state of IDPs (Ellis 2001; Minton 2005). Typical concentrations of proteins and other macromolecules reach 300–400 mg/ml, which basically limits the available space for other macromolecules (Figure 8.1) and causes a severe excluded volume effect that increases chemical activity of the molecules. Theoretical and experimental estimates suggest that this effect can be of several orders of magnitude for a protein of average size, which may fundamentally affect structural transitions accompanied by changes in volume, such as protein–protein interactions and folding. In the case of unfolded/denatured globular proteins, crowding does promote their native-like compact states of at least partial activity (Baskakov and Bolen 1998; McPhie, Ni, and

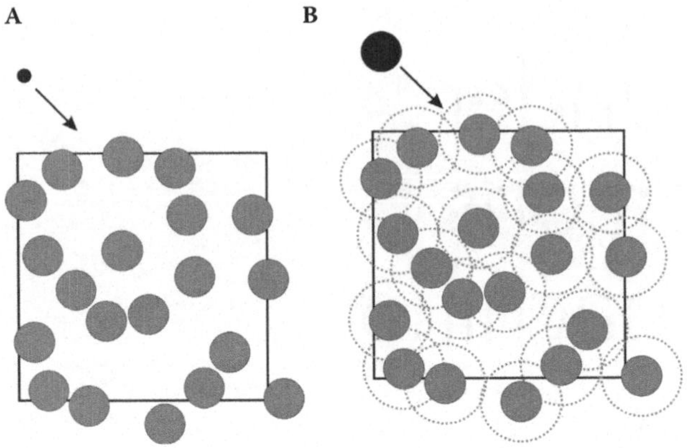

FIGURE 8.1 Volume exclusion by crowding. Macromolecules in this schematic cell occupy about 30% of the available space. (A) A small molecule has accessibility to virtually all of the remaining 70%. (B) A molecule of size similar to the "crowding" macromolecules is excluded from most of this volume, which gives rise to appreciable excluded volume effects. Reproduced with permission from Ellis (2001), *Trends Biochem. Sci.* 26, 597–604. Copyright by Elsevier Trends Journals.

Minton 2006; Qu and Bolen 2002). By analogy, crowding may also force IDPs to assume compact or even folded states, making this issue very critical with respect to their physiological structural state and function.

Basically, there are two direct approaches that can be used to address these issues. The first is to mimic crowding conditions in the test tube and study the structural state/function of selected IDPs. The second is to characterize the proteins in living cells by appropriate experimental techniques. Complementing these two is the collection of indirect observations, from which inferences can be made with respect to the likely physiological state of IDPs.

8.2 *IN VITRO* APPROACHES TO MIMICKING CROWDING CONDITIONS

In vitro, crowding is usually elicited by high concentrations of inert macromolecules or small hydrophilic molecules that affect the availability of water (i.e., hydration of proteins). The effect of applying high concentrations of different solutes may have at least three different components. They may cause an excluded volume effect, increase the viscosity of the solution, and indirectly affect the solvation/hydration of proteins. Intuitively, all three effects are encountered in the cell, and thus a solution that combines all three is justified. The problem is, however, that the cytoplasm is an extremely complex mixture of molecules of all sizes, not all of which are indifferent to the protein, and no single macromolecular compound can provide the right balance between these effects. Thus, the application of a variety of agents is needed to arrive at reasonable generalizations.

The most widely accepted approach is to apply the high molecular-weight polymers Dextran and Ficoll 70, which probably present a combination of all three effects. The application of another protein, such as bovine serum albumin (BSA), is also acceptable, but it may have unwanted aspecific interactions, for example. Small molecule osmolytes, such as sucrose, trimethylamine N-oxide (TMAO), and trifluoroethanol (TFE), primarily act upon viscosity and/or solvation of the protein backbone, which do occur in the cell, but do not directly pertain to crowding by definition. These small molecules are considered "chemical chaperones," and they do have the potential to induce and or stabilize the folding of proteins. Further, they do belong to nature's arsenal for fighting abiotic conditions, such as water deficit (Bray 1993) or unfolding of proteins (Baskakov and Bolen 1998), and thus they represent a fair alternative for approaching living conditions.

Studies carried out under a wide range of conditions mostly agree that crowding does elicit some compaction but not folding of IDPs. The conformational state of kinase inhibitory domain (KID) of p27^{Kip1} and trans-activator domain (TAD) of c-Fos was studied by circular dichroism (CD) spectroscopy or 1-anilino-8-naphthalene-sulfonic acid (ANS)-binding (Flaugh and Lumb 2001). Dextrans of various molecular mass (M_W) or Ficoll 70 (both up to 250 mg/ml) has no effect on either protein, whereas TFE at 30% concentration induces a significant amount of α-helix in both proteins. α-Synuclein and an unfolded globular protein, acid-denatured cytochrome *c* were studied by two methods—pulsed-field gradient (PFG) nuclear magnetic resonance (NMR) and CD (Morar et al. 2001)—in the presence of 1M glucose. Hydrodynamic measurements suggest only a slight compaction of α-synuclein (R_H decreases from 26.6 Å to 22.5 Å, compared to the value for a globular protein of 140 amino acids, 18–20 Å, and for a random coil, 33–37 Å), without any change in the secondary structure content. In the case of acid-denatured cytochrome *c*, 1M glucose does make it collapse to a state compatible with the compactness of the native conformation (R_H decreases from 30.6 Å to 17.7 Å). Fluorescence resonance energy transfer (FRET) in the case of the P/Q domain of ZipA (Ohashi et al. 2007) suggested that the average end-to-end distance in the ensemble slightly decreases in the presence of 20% Ficoll 70, which suggests a limited compaction under crowding conditions.

TMAO is often (and not fully justifiably) used to assess the structure of unfolded proteins under conditions approaching *in vivo*, with mixed effects. It has a significant effect on α-synuclein, as measured by various techniques (Uversky, Li, and Fink 2001d). It promotes the transition to a conformation dominated by α-helices, with a half-maximal TMAO concentration around 2.2M. ANS binding, fluorescence quenching, and small-angle X-ray scattering (SAXS) suggest a large compaction of the structure at 3.5M TMAO, compatible with the oligomerization/aggregation of the protein (Uversky et al. 2001d). A significant effect on secondary structure (i.e., development of local α-helices) and a slight compaction (i.e., blue-shifting of UV fluorescence) is seen in the case of the intermediate chain of cytoplasmic dynein at 2.4M TMAO (Nyarko et al. 2004), whereas the osmolyte has no effect on the secondary structure of myelin basic protein (MBP) at concentrations up to 2M (Hill et al. 2002). In the case of a mutant tau protein, TMAO promotes slight secondary structure formation and restores the ability of the protein to promote tubulin polymerization (Smith, Crowther, and Goedert 2000). The power of TMAO, on the other hand, is demonstrated by its effectivity in making unfolded globular proteins, such as carboxyamidated ribuonuclease (RNase) T1 (Qu and Bolen 2002), mutant staphylococcal nuclease (Baskakov and Bolen 1998), and apomyoglobin (McPhie et al. 2006) to refold to their native-like compact state.

Different crowding conditions were compared in the case of α-casein, microtubule-associated protein 2c (MAP2c), and $p21^{Cip1}$ (Tompa, unpublished results). The fluorescence spectra of the three proteins indicate that their structure is largely exposed but somewhat more shielded than N-acetyl tryptophane amide (NATA) (i.e., probably fall into locally restrained conformational environment). Crowding by 40% Ficoll, 40% Dextran, or 3.6M TMAO slightly changes these efficiencies (see Chapter 5, Figure 5.2), but the values remain far from those characteristic of a compact globular protein (RNase T1). Fluorescence correlation spectroscopy (FCS) studies also suggest that the proteins undergo slight compaction (Figure 8.2) by crowding but without a cooperative transition to a folded structure (i.e., they remain in a state of rapidly interconverting structural ensemble under crowding).

Probably the most important functional conclusion from such *in vitro* crowding studies is that the formation of local secondary structural elements of IDPs is promoted by crowding. Thus, structural concepts, such as preformed structural elements (PSEs,

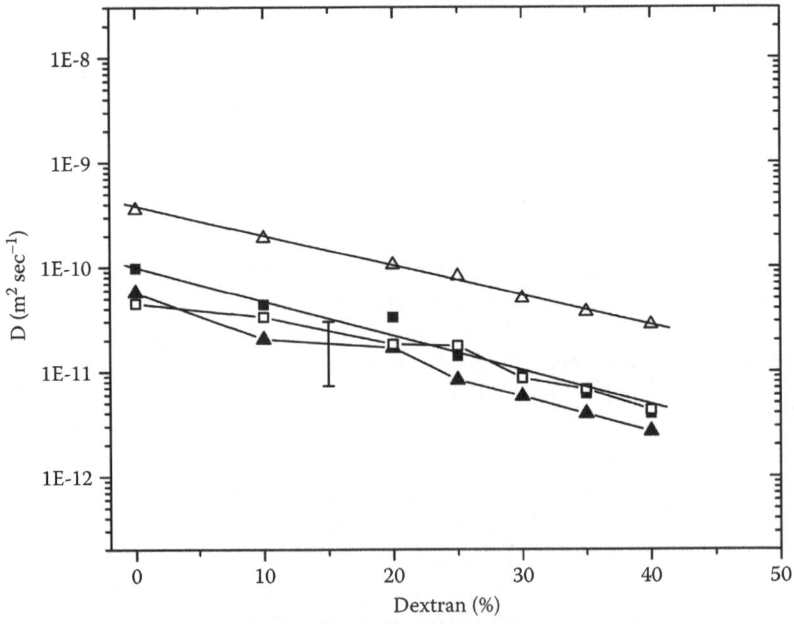

FIGURE 8.2 Fluorescence correlation spectroscopy measurements of the effect of crowding on IDPs. FCS measurements of IDPs MAP2c and $p21^{Cip1}$, and controls RNase T1 and Alexa 488 dye in buffer and under crowding conditions elicited by Dextran up to 40% (v/v) in concentration. The diffusion coefficient has been calculated from the autocorrelation function (ACF) of labeled proteins MAP2c (▲), $p21^{Cip1}$ (□), globular RNaseT1 (■), and the Alexa 488 dye (△). Data of RNaseT1 and Alexa 488 are fitted to a straight line, whereas those of MAP2c and $p21^{Cip1}$ are simply connected. No signs of cooperative transition are seen.

see Chapter 14, Section 14.2.1), may receive even more credit than suggested by *in vitro* structural studies. Local elements thus stabilized may be related to recognition functions, as suggested in the case of tau protein (Smith et al. 2000), cytoplasmic dynein (Nyarko et al. 2004), α-casein, p21^{Cip1}, MAP2c (Tompa, unpublished results), and perhaps α-synuclein (Uversky et al. 2001d).

8.3 THE STATE OF IDPS *IN VIVO*

Measurements *in vitro*, no matter how closely they come to mimicking conditions in the cell, can never fully ascertain how a protein behaves in the extremely complex intracellular milieu. Two approaches have the potential to provide information directly from within the cell: studying proteasomal degradation *in vivo* and applying in-cell NMR.

8.3.1 Proteasomal Degradation

An informative operational approach of the *in vivo* structural status of IDPs derives from the intimate link between structural disorder and degradation by the 20S proteasome. The proteasome is a large, barrel-like multi-protein complex that degrades ubiquitinated proteins in the cell, composed of a 20S (700 kDa) core (catalytic) unit and two 19S (700 kDa) regulatory units (Bochtler et al. 1999; Rape and Jentsch 2002; Voges, Zwicki, and Baumeister 1999). Active subunits of the proteasome have trypsin-like, chymotrypsin-like, and peptidyl-glutamyl peptide-hydrolyzing (PHGH) activities facing the inner chamber of the complex. In general, proteins have no access to this inner catalytic chamber because the structure is capped from both ends by the regulatory subunits, which serve to recognize ubiquitinated proteins, unfold, and feed them into the proteasome in an energy-dependent manner.

A significant fraction of the proteasomes in the cell occurs without the regulatory subunit, and this 20S proteasome is able to degrade unfolded proteins, such as oxidized, denatured globular proteins (e.g., hemoglobin and superoxide dismutase [Davies 2001]) and IDPs (e.g., α-casein [Davies 2001], α-synuclein [Tofaris et al. 2001], tau protein [David et al. 2002] and p21^{Cip1} [Sheaff et al. 2000]), without ubiquitination. In the case of p21^{Cip1}, degradation is promoted by the murine-double minute 2 (MDM2) E3 ubiquitin ligase (Jin et al. 2003), but it does not require ubiquitination, because mutation of all Lys residues comprising potential ubiquitinylation sites does not abolish its down-regulation in cells (Sheaff et al. 2000). This phenomenon of ubiquitin-independent proteasomal degradation is termed degradation "by default" (Asher, Reuven, and Shaul 2006), because it results from the inherent structural properties of proteins. Degradation by default has the power of providing evidence on the physiological structural status of IDPs, as shown by studies of p53, p73, and ornithine decarboxylase (ODC) (Asher et al. 2005). Because of the inherent susceptibility of disordered proteins to this mechanism, default degradation must be regulated (Figure 8.3),

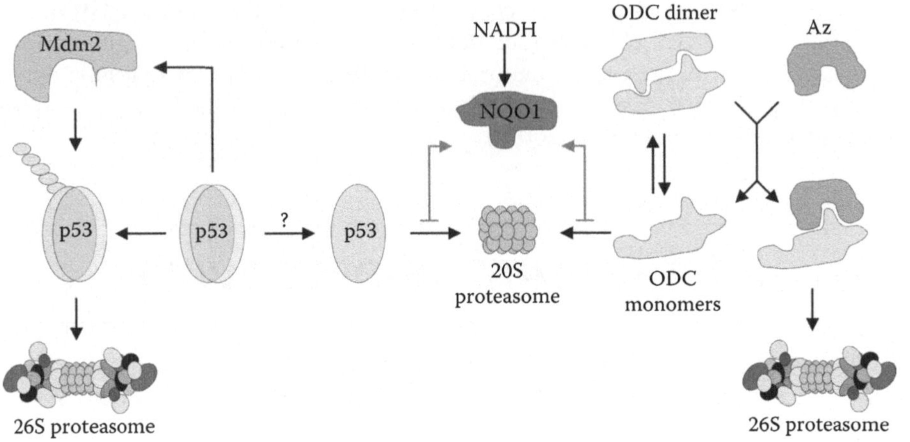

FIGURE 8.3 20S proteasome-mediated degradation of IDPs "by default." Schematic representation of the "default" degradation pathways of p53 and ornithine decarboxylase (ODC). These proteins are degraded by both the 26S and 20S proteasomes, but degradation by 20S proteasomes does not require prior poly-ubiquitination. It is suggested that disorder is important for this relation: p53 contains a significant level of disorder (Bell et al. 2002; Dawson et al. 2003), and ODC is a two-state dimer, preferentially disordered in the monomer state. Degradation of both proteins is regulated by interactions with NAD(P)H quinone oxidoreductase-1 (NQO1). (Az stands for Antizyme.) Reproduced with permission from Asher et al. (2006), *BioEssays* 28, 844–9. Copyright by Wiley-VCH.

which may occur by interaction of proteins with NAD(P)H quinone oxidoreductase 1 (NQO1) (Asher et al. 2006).

The common denominator in all proteins underlying such 20S-proteasome-mediated degradation is the presence of disorder, as shown in a systematic analysis of proteins containing extended disordered regions (VirE2, gliotactin, neuroligin, p21) (Tsvetkov et al. 2008). It should be added that even in the case of degradation initiated by ubiquitination, a disordered initiation site is required for the proteasome to start its action (Prakash et al. 2004), which is also corroborated by the endoproteolytic activity of the proteasome (Liu et al. 2003). Based on these and all prior observations, it is suggested that proteasomal sensitivity provides an "operational" definition of disorder, which can be exploited in studying the structural status of these proteins in the cell. *In vivo*, $p21^{Cip1}$ (Sheaff et al. 2000), p53, and p73 (Asher et al. 2005) are subject to this degradation mechanism. These proteins have been amply characterized for their disorder *in vitro* (Sickmeier et al. 2007), and their default degradation by the proteasome now strongly argues for their disordered state *in vivo*. The caveat to the general applicability of this approach is the inability of the proteasome to digest certain repetitive sequences, such as Gly-Ala repeats (Hoyt et al. 2006; Zhang and Coffino 2004) or polyQ regions (Venkatraman et al. 2004). Because an estimated 34% of all IDP sequences are made up of repeats (see Chapter 13, Section 13.3.1) (Tompa 2003b), it may seriously interfere with the proteasomal degradation of disordered proteins.

8.3.2 In-Cell NMR

The technique of in-cell NMR is the remedy in this situation, because it provides data on the structure, interactions, and function of proteins in the living cell (Selenko and Wagner 2007; Serber and Dotsch 2001). The technique requires deposition of labeled proteins within the cell, preferably at concentrations approaching their physiological values. Two approaches are used: overexpression in the presence of labeled precursors (e.g., $^{15}NH_4Cl$) in the cell of choice, *E. coli*, for example (Dedmon et al. 2002; McNulty, Young, and Pielak 2006), or microinjection into the cytoplasm of the cell, such as *Xenopus* oocytes (Selenko et al. 2006). Whereas microinjection enables better control over the intracellular concentration attained, a serious limitation of the technique stems from the relative insensitivity of NMR, in comparison to the typically very low concentrations of highly disordered regulatory/signaling proteins.

Overexpression was used to study FlgM, the inhibitor of σ^{28} transcription factor that regulates the synthesis of flagellar proteins in *Salmonella*. FlgM is one of the first proteins described as disordered (Daughdrill et al. 1997; Daughdrill et al. 1998), it is about 97 amino acids long, and its structure from *A. aeolicus* in complex with σ^{28} has been solved by X-ray crystallography (Sorenson, Ray, and Darst 2004). The protein binds its partner via four helix regions, H1′–H4′ (Figure 8.4A). Alignment of different sequences has revealed that the N-terminal 40 amino acids encompassing H1′ and the H1′–H2′ linker is extremely variable and probably does not take part in complex formation in *S. typhimurium* (Daughdrill et al. 1997). When the whole protein is overexpressed in *E. coli* under ^{15}N-labeling conditions (Dedmon et al. 2002) cross-peaks in the heteronuclear single quantum coherence (HSQC) spectrum corresponding to C-terminal region H2′–H4′ (about 57 amino acids) disappear, indicating that this half becomes ordered or binds a partner in the cell (Figure 8.4B). The same region that binds σ^{28} in this species tends to sample transient helix conformations in solution (Figure 8.4C) (Daughdrill et al. 1998). Thus, intracellular conditions prefer a structural ensemble reminiscent of the bound conformation of the protein, with the N-terminal region remaining disordered, whereas the C-terminal region becoming more ordered.

Microinjection of tau protein into *Xenopus* oocytes (Bodart et al. 2008) suggests that the microtubule (MT)-binding regions of the protein (tubulin-binding domain, TBD) become ordered, but extended regions that probably function as entropic chains remain largely disordered. Unique functional information on tau is also provided by this experiment because specific phosphorylation of its projection domain *in vivo* can be observed. α-Synuclein shows a different behavior. As shown in Section 8.2 *in vitro* crowding experiments suggest only slight (by 1M glucose (Morar et al. 2001)) or more extensive (by 3.5M TMAO (Uversky et al. 2001d)) compaction of the protein, and the formation of a significant amount of secondary structure in the latter case, which might be relevant with respect to its putative membrane-binding function (Eliezer et al. 2001; Ramakrishnan, Jensen, and Marsh 2003). When the protein is overexpressed in *E. coli*, in-cell NMR experiments show that it remains largely disordered in the crowded intracellular milieu (McNulty et al. 2006), which points to the limitations of the *in vitro* approaches.

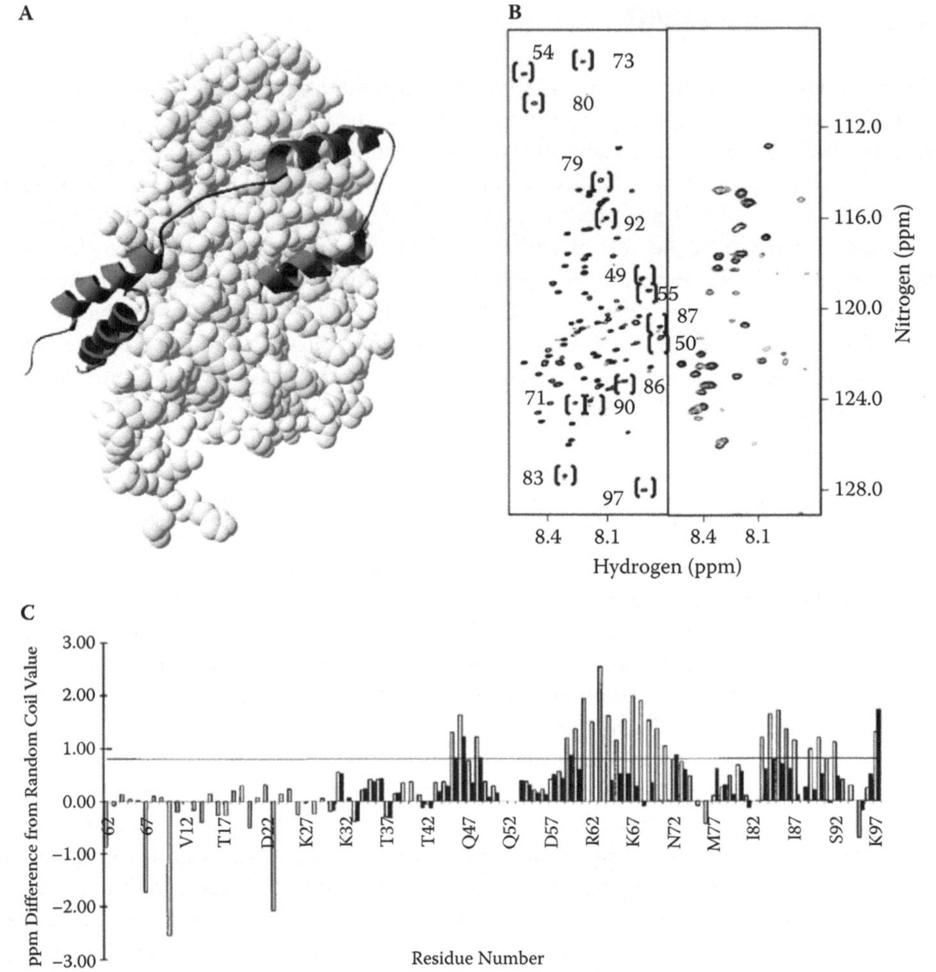

FIGURE 8.4 Bound structure, free state, and in-cell NMR of FlgM. FlgM is an inhibitor of the bacterial transcription factor σ^{28}. (A) σ^{28} is about 97 amino acids in length, and the structure of *A. aeolicus* protein bound to its partner (pdb 1rp3) has been solved by X-ray crystallography (Sorenson et al. 2004). (B) When FlgM is overexpressed in *E. coli*, its HSQC spectrum measured in solution (left panel) undergoes significant changes, which suggests that its C-terminal region assumes structure or is bound to a partner, in the cell (Dedmon et al. 2002). (C) In isolation, the protein from *S. typhimurium* tends to sample transient helix conformations in the C-terminal half, as shown by NMR chemical shift index (CSI) values (Daughdrill et al. 1998). Reproduced with permission from Dedmon et al. (2002), *Proc. Natl. Acad. Sci. USA*. 99, 12681–4, copyright by the National Academy of Sciences, and Daughdrill et al. (1998) *Biochemistry* 37, 1076–82, copyright by the American Chemical Society.

8.4 PHYSIOLOGICAL HALF-LIFE OF IDPS: NO SIGNS OF RAPID DEGRADATION

An issue pertinent to the physiological state of IDPs is their degradation *in vivo*, given their extreme proteolytic sensitivity *in vitro* (see Chapter 3, Section 3.4). Analysis of this feature is made possible by data on the *in vivo* half-lives of 3,750 yeast proteins determined in an HTS encompassing the entire yeast proteome (Belle et al. 2006). The distribution of half-lives is approximately log-normal, with a mean and median half-life of about 43 min and an unexpectedly large number of very unstable proteins (161 proteins with half-lives less than 4 min). These half-lives show weak correlation with different measures of predicted disorder (Figure 8.5), such as the number of disordered amino acids, average disorder score, and the number of proteins with long intrinsically disordered regions (IDRs) ≥30 residues (Tompa et al. 2008). Half-lives do correlate with the length of the polypeptide chain but not with classical degradation signals (destruction-box, KEN-box, low-complexity regions, the N-terminal residue, or ubiquitination site), with the exception of a slightly elevated level of PEST regions (regions enriched in Pro, Glu, Ser, and Thr [Rechsteiner and Rogers 1996]) in proteins with very short (<4 min) half-lives.

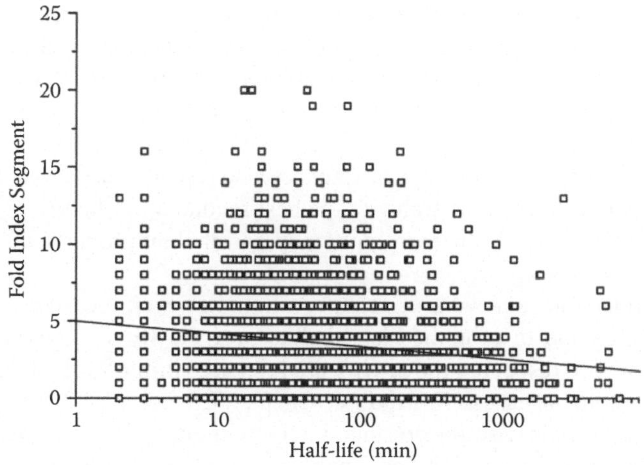

FIGURE 8.5 Low correlation of physiological half-lives and protein disorder. Data on the physiological half-life of 3,750 proteins were determined by epitope tagging and quantitative Western-blotting following arrest of translation in yeast (Belle et al. 2006). The half-lives thus determined indicate a very low level of correlation with the number of predicted IDRs ≥30 consecutive residues. Reproduced with permission from Tompa et al. (2008), *Proteins* 71, 903–9. Copyright by Wiley-Liss.

The most likely interpretation of these findings is that protein degradation is not determined by a single characteristic, and the proteolytic systems in the cell are highly regulated. This can be formally demonstrated by showing that proteases/proteolytic systems of the highest copy number in the cell are all tightly regulated by various means, such as ubiquitination, localization, special substrate requirement, or post-translational modification (Tompa et al. 2008).

8.5 INDIRECT CONSIDERATIONS UNDERSCORING DISORDER OF IDPS *IN VIVO*

A range of indirect considerations also suggests that disorder is the likely physiological state of IDPs. Although these often do not directly address the structural status of IDPs, their results are hard to reconcile with a compact, folded state of these proteins *in vivo.*

The physiological relevance of the disordered state of IDPs probably most directly follows from their effective functioning observed *in vitro.* IDPs in the test tube are definitely disordered, yet they can be extremely effective, as witnessed by pM-nM inhibitors (see Chapter 12, Section 12.4.1), for example. It is rather compelling to conclude that disorder is the native and functional—and thus physiological—state of these proteins.

The intrinsic difference of IDPs from globular proteins *in vivo* is also strongly argued by their distinct amino acid compositions (see Chapter 10, Section 10.1). The discontinuity of structural states between ordered proteins and IDPs is manifested in IDPs being enriched in disorder-promoting amino acids and depleted in order-promoting amino acids (Dunker et al. 2001), as well as the distinct sequence attributes of IDPs (Lise and Jones 2005). These differences also provide the basis of bioinformatic predictors, which perform comparably to the best secondary structure prediction algorithms (Chapter 9). Thus, disorder is apparently an inherent property of IDPs, not just an artifact.

This argument is somewhat circumferential, however, because predictors are trained on a collection of proteins identified as intrinsically disordered *in vitro* (i.e., their assessment with respect to the *in vivo* situation is of limited value). The situation is basically different in the case of IUPred (see Chapter 9, Section 9.4.2), which is based on estimating the total pair-wise interresidue interaction energy of sequences (Dosztanyi et al. 2005a; Dosztanyi et al. 2005b). The disorder score of this algorithm, which is based on low-resolution force fields derived from globular structures, suggests that the potential of IDP sequences to form favorable interactions is significantly smaller than that of globular proteins. Because IUPred was not trained on *in vitro* observed IDPs, its assessment of disorder underscores that the lack a folded, stable structure is the inherent property of certain proteins. The success of other predictors relying on contact potentials, such as FoldUnfold (Garbuzynskiy, Lobanov, and Galzitskaya 2004) and Ucont (Schlessinger, Punta, and Rost 2007b), also point into this direction.

The question of the possible compact fold of IDPs *in vivo* is irrelevant in the case of extracellular IDPs, which do not experience a crowded environment in their natural physiological habitat. There are many such IDPs in DisProt (Sickmeier et al. 2007). For example, casein(s) function as scavengers of calcium-phosphate seeds in milk (Holt, Walgren, and Drakenberg 1996), salivary proline-rich glycoproteins serve to neutralize plant polyphenolic compounds (i.e., tannins) in saliva (Lu and Bennick 1998), whereas bacterial fibronectin-binding proteins that protrude from the surface of the cell tether bacteria to the extracellular matrix (ECM) of the host (Penkett et al. 1997).

The issue of *in vivo* order or disorder is probably also irrelevant with respect to IDPs that perform their function directly by disorder, which is by definition incompatible with a single, stable, conformational state (entropic chains, see Chapter 12, Section 12.1). The function of entropic chains stems directly from the ability of their polypeptide chain to rapidly fluctuate between a large number of alternative conformational states. For example, the function of the Pro, Glu, Val, Lys-rich (PEVK) region of titin, an elastic protein that provides passive tension in muscle (Trombitas et al. 1998), the projection domain of MAP2, which ensures entropic spacing in the cytoskeleton (Mukhopadhyay and Hoh 2001), and the repeat regions of FG nucleoporins (Nups), which regulate gating of transport through the nuclear pore (Patel et al. 2007), cannot be rationalized in terms of a folded, well-defined structure.

As outlined in Chapter 12, many IDPs function by molecular recognition, when they either transiently or permanently bind to a structured partner. The structures of these complexes are solved in many cases and demonstrate that IDPs often bind in an extended, open configuration (see Chapter 6, Figure 6.3; Chapter 10, Figure 10.3; and Chapter 11, Figure 11.5). Because strength and specificity of binding (often in the range of nM/pM) argue that the structure determined *in vitro* is a faithful reflection of the mode of binding *in vivo*, it is logical to assume that the protein was unfolded, instead of having to unfold, prior to binding. In addition, several IDPs do not become fully ordered even in the partner-bound state but remain partially or even fully disordered, as captured by the concept of fuzziness (see Chapter 14, Section 14.8). Such complexes limit the idea that the IDP would be folded in the cell.

Directly linked to the binding argument is the fact that certain IDPs can bind several different partners in a process termed binding promiscuity (Kriwacki et al. 1996), or one-to-many signaling (Dunker et al. 2001), in which the IDP may adopt different structures. Such an adaptability (see Chapter 14, Section 14.6) has been reported in the case of the Cdk inhibitor p21^{Cip1}, which can bind different cyclin-Cdk complexes (Kriwacki et al. 1996), the C-terminal domain (CTD) of RNA polymerase II (RNAP II), which can recognize both RNA guanylyl transferase Cgt1 and peptidyl-proline isomerase Pin1 (Fabrega et al. 2003), the HIF-1α interaction domain bound to either the TAZ1 domain of CBP (Dames et al. 2002) or the asparagine hydroxylase FIH (Elkins et al. 2003), and Tβ4, which can bind G-actin (Irobi et al. 2004) (see Chapter 11, Figure 11.5) as well as integrin-linked kinase (ILK) and PINCH (Bock-Marquette et al. 2004). This level of adaptability can be best interpreted in terms of the disorder of the protein in the unbound state.

Prediction of Disorder

9

This chapter describes basic bioinformatic approaches of predicting protein disorder from sequence. Prediction is a classification problem, which can be approached from three distinct directions: (1) from simple propensities reflecting some basic physical or sequence features; (2) from machine-learning algorithms, which are trained to recognize sequences; and (3) from the tendency of amino acids to make or avoid contacts with each other. The resulting distinctions between predictors are not absolute, because several of them incorporate more than one of these features. The predictors can be combined into meta-predictors, and their performance can be compared in a statistically rigorous way. Their application addresses many structural/functional issues and improves target prioritization in structural genomics programs.

9.1 GENERAL POINTS

It should be made clear that although different predictors are based on different principles and apply different computational approaches, in one way or another they all rely on the biased sequence features of intrinsically disordered proteins (IDPs) (see Chapter 10, Section 10.1), with their basic feature being an enrichment in disorder-promoting amino acids and depletion in order-promoting amino acids (Dunker et al. 2001). There are about 20 predictors and also some meta-predictors (Table 9.1), which combine the assessment of several individual servers. An update of links of the predictors is available at the DisProt Web site (http://www.disprot.org), and the subject is covered in several reviews (Bracken et al. 2004; Dosztanyi et al. 2007; Ferron et al. 2006).

9.2 PROPENSITY-BASED PREDICTORS

A predictor is termed propensity-based if it relies on some simple statistics of the physical/chemical features of amino acids or on a preliminary concept on the physical background of disorder. Various predictors incorporate the low-complexity of sequence, biased amino acid composition, or the lack of secondary structural elements. These elements also often appear in other, more sophisticated algorithms.

TABLE 9.1 Disorder predictors*

PREDITCOR	*URL*	*PRINCIPLE*	*PSI-BLAST*	*REFERENCE*
Propensity-Based Predictors				
CH plot	N/A	Calculation of net charge and hydrophobicity of chain	No	(Uversky et al. 2000a)
FoldIndex	http://bip.weizmann.ac.il/fldbin/findex	Amino acid propensity	No	(Prilusky et al. 2005)
GlobPlot	http://globplot.embl.de	Amino acid propensity, preference for ordered secondary structure	No	(Linding et al. 2003b)
PreLink	http://genomics.eu.org/spip/PreLink	Amino acid propensity + hydrophobic cluster analysis	No	(Coeytaux and Poupon 2005)
NORSp	http://cubic.bioc.columbia.edu/services/NORSp	Secondary structure propensity	Yes	(Liu and Rost 2003)
Machine-Learning Algorithms				
PONDR® (VL-XT)	http://www.PONDR.com/	NN based on amino acid features	No	(Li et al. 1999)
RONN	http://www.strubi.ox.ac.uk/RONN	Bio-basis function NN	No	(Yang et al. 2005)
DISOPRED2	http://bioinf.cs.ucl.ac.uk/disopred	SVM, NN for smoothing	Yes	(Ward et al. 2004)
DisEMBL	http://dis.embl.de	Neural network	No	(Linding et al. 2003a)
DISpro	http://www.ics.uci.edu/~baldig/dispro.html	1-D-recursive NN with profiles	Yes	
Spritz	http://protein.cribi.unipd.it/spritz/	Two SVMs with non-linear kernel, for short and long disorder	Yes	(Vullo et al. 2006)
NORSnet		NN identifying long regions without secondary structure		(Schlessinger et al. 2007a)
PrDOS		PSI-BLAST-generated PSSM assessed by both SVM and similarity to PDB structures	Yes	(Ishida and Kinoshita 2007)

VSL2 (VSL1)	http://www.ist.temple.edu/disprot/predictorVSL2.php	SVM with linear kernel (logistic regression model)	Yes	(Obradovic et al. 2005; Peng et al. 2006)
Predictors Based on Inter-residue Contacts				
FoldUnfold	http://skuld.protres.ru/~mlobanov/ogu/ogu.cgi	Single amino acid propensity	No	(Garbuzynskiy et al. 2004)
IUPred	http://iupred.enzim.hu	Estimated pairwise interaction energy	No	(Dosztanyi et al. 2005a)
Ucon	http://www.predictprotein.org/submit_ucon.html	Amino acid contact potential	No	(Schlessinger et al. 2007b)
Metaservers				
MeDOR	http://www.vazymolo.org/MeDor/	Graphical output of 12 disorder predicors, plus that of secondary structure and HCA analysis	No	(Lieutaud et al. 2008)
metaPrDOS	http://prdos.hgc.jp/meta/	SVM integrating the output of 7 disorder predictors	No	(Ishida and Kinoshita 2008)

* The table lists the most often used disorder predictors, their URL addresses, the principles on which they are based, and whether they use sequence alignment in the prediction process.

9.2.1 Prediction of Low-Complexity Regions

The prediction of low-complexity regions is not equivalent with predicting disorder. The two are clearly related, however, and low-complexity regions are often associated with non-globular regions of proteins. As detailed in Chapter 10, Section 10.1.2, low complexity is defined by the informational entropy function of Shannon (Shannon 1948), adapted to protein sequences by Wootton (Wootton 1994a, b). Wootton observed that non-globular segments of proteins deviate significantly from the observed random composition of globular proteins/domains, because of the dominance of a few amino acids and/or the repetitive nature of their sequences (see Chapter 10, Figure 10.2). Based on this principle, the SEG program was developed to identify such sequentially biased fragments (Wootton 1994a, b). SEG first calculates local complexity for segments of a given size by the Shannon entropy, extends and merges them, and reduces them to a single optimal low-complexity region. Because low-complexity and disorder are related (Romero et al. 2001), this practice has definite value in delineating non-globular and possibly disordered regions of proteins.

9.2.2 Charge-Hydropathy Plot

In a study comparing characteristic features of ordered and disordered proteins (Uversky et al. 2000a), it was observed that a combination of mean net hydrophobicity and mean net charge distinguishes best between the two classes. This simple principle is harnessed to predict IDPs by plotting these measures in a form usually denoted as charge-hydropathy (CH) plot or Uversky plot (Figure 9.1). IDPs on this plot cluster in the high net charge–low net hydrophobicity half of the plane, in manifestation of the physical rationale that a low level of hydrophobicity precludes the formation of a stable compact core and high net charge favors extended structural states due to electrostatic repulsion. The formula of the linear function best separating ordered and disordered proteins:

$$\langle R \rangle = 2.743 \langle H \rangle - 1.109 \qquad (9.1)$$

provides a classification accuracy 83% overall—76% for fully disordered proteins and 91% for fully ordered ones (Oldfield et al. 2005a). The CH plot cannot provide sequence-specific assessment of disorder, but it enables a classification at the protein level, suggesting disorder or order for the entire polypeptide chain. Although rigorously not proven, the distance from the separating line may carry information on the extent and type of disorder for the whole chain (Oldfield et al. 2005a).

This idea was substantiated in the evaluation of proteins identified in a native/urea two-dimensional electrophoresis mass spectrometry (2DE-MS) analysis (Csizmok et al. 2006), in which the metric of distance from the border line was found to correlate with classification by other techniques, such as percent disorder predicted by predictor of natural disordered regions (PONDR®), apparent molecular mass (M_W) by gel filtration (GF), and secondary structure content by circular dichroism (CD) (see Chapter 7,

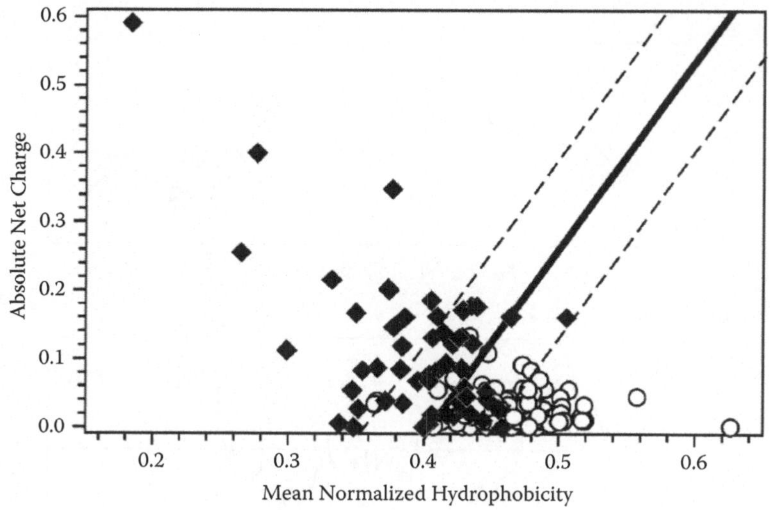

FIGURE 9.1 Charge-hydropathy plot of protein disorder. Net charge vs. mean hydrophobicity is plotted for intrinsically disordered (full diamonds) and ordered (empty circles) proteins. The two sets are separated by a straight line <charge> = 2.743 <hydropathy> – 1.109, with dashed lines delimiting the zone with a prediction accuracy of 95% for disordered proteins and 97% of ordered proteins, at the expense of discarding 50% of all proteins. Reproduced with permission from Oldfield et al. (2005), *Biochemistry* 44, 1989–2000. Copyright by the American Chemical Society.

Section 7.3 and Table 7.2). For example, identified *E. coli* and *S. cerevisiae* proteins segregate into three groups: disordered (distance on CH plot: 11; PONDR®%: 58.5%; apparent M_W relative to real M_W: 3.0), slightly disordered (3;40%;2.35), and ambivalent (–0.2;35.2%;2.9), by these distinct characteristics.

The CH-plot was developed into a sequence-specific predictor by calculating the CH values for a segment of the sequence within a sliding window (Prilusky et al. 2005). The predictor developed on this principle, FoldIndex (Figure 9.2A), is based on rearranging and optimizing the original formula (Uversky et al. 2000a) in the form

$$I_F = 2.785\langle H\rangle - \left|\langle R\rangle\right| - 1.151 \qquad (9.2)$$

to yield I_F, the "Fold Index," which is used to assign order to residues with positive values and disorder to residues with negative values.

9.2.3 Prediction of Globularity and Disorder

A simple propensity-based predictor is GlobPlot, developed with the primary intent of identifying globular domains (Linding et al. 2003b). As opposed to the CH plot, which is based on the summation of two parameters (charge and hydrophobicity), GlobPlot

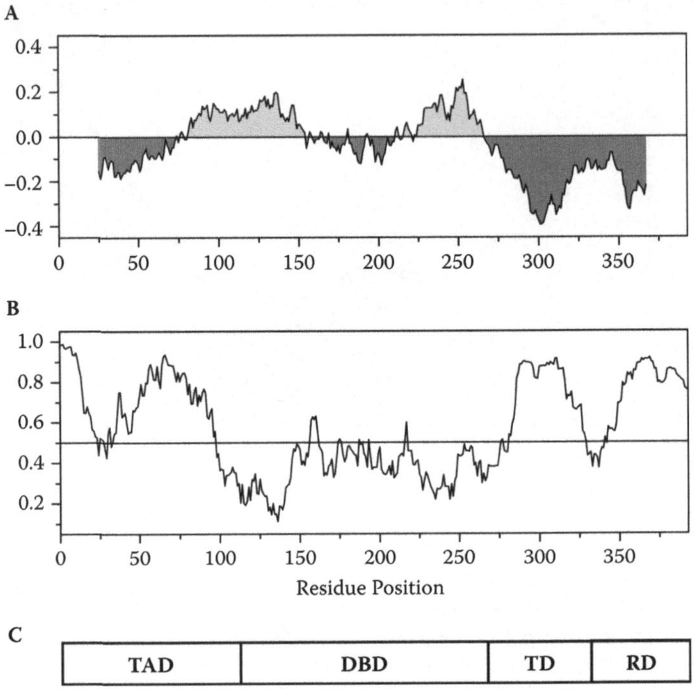

FIGURE 9.2 Predicted disorder in tumor suppressor p53. (A) FoldIndex score was calculated within a sliding window of 51 residues. Ordered (light gray) and disordered (dark gray) regions are color coded (Prilusky et al. 2005). (B) IUPred score was calculated by a window of 100 residues. A score above the threshold 0.5 is considered disordered (Dosztanyi et al. 2005a; Dosztanyi et al. 2005b). (C) The domain structure of p53 shows that its TAD, T(etramerization)D and R(egulatory)D domains are disordered, whereas its DNA-binding domain (DBD) is ordered (for further details, see text).

only takes into consideration a single parameter, which is optionally either the Russell/Linding scale (the difference of the propensity of an amino acid to be in "secondary structure" or "random coil" region) or a scale based on the preference of residues to be missing in the ATOM records, according to Remark465 in the Protein Data Bank (PDB). The values are integrated along the sequence following Savitzky–Golay smoothing, which also provides first derivative estimates. Putative globular and disordered segments are selected using a simple peak-finder algorithm, when the first derivative shows positive (disorder) or negative (order) values over a continuous stretch of amino acids with a minimum length.

9.2.4 Composition and Hydrophobic Cluster Analysis

A variant of propensity-based prediction is PreLink (Coeytaux and Poupon 2005), which is based on disordered linker regions connecting globular domains having a biased amino acid composition and usually containing no or only small hydrophobic clusters.

To quantify these two properties, amino acid distributions in disordered (residues missing from PDB) and ordered (ensemble of PDB structures) data sets were computed, and the distance to the nearest hydrophobic cluster is calculated by automated hydrophobic custer analysis (HCA). Disorder prediction is based on amino acid composition and maximal distance from the nearest hydrophobic cluster.

9.3 MACHINE-LEARNING ALGORITHMS

More sophisticated approaches to predicting disorder are algorithms trained to distinguish sequences that encode for disorder from those that encode for order, termed machine-learning algorithms (MLAs). MLAs have two main computational implementations: neural networks (NN) and support-vector machines (SVM). Whereas they are usually superior in performance to the propensity-based predictors, they do not provide clear insight into the physical principles underlying disorder. On the other hand, they may incorporate non-trivial amino acid features and hidden sequence properties in their assessment.

9.3.1 Neural Networks

An NN is a mathematical/computational mimic of a biological network of neurons. It consists of an interconnected group of artificial neurons, usually an input and an output layer connected by one or several hidden layers. An NN is an adaptive system that changes its weights based on information that flows through during the phase of learning (training), in which a given set of example pairs (e.g., sequences of ordered and disordered proteins) are presented, and parameters defining the interaction of neurons are optimized to provide an output that matches the input as closely as possible. A common implementation of training is an iterative back-propagation that uses the average error between the network's output and its target value to modify internal parameters to bring the difference toward the global minimum.

The classical MLA predictor is PONDR® and its variants, which employ feature selection algorithms using a set of features based on biological knowledge. The first version of PONDR® analyzed the frequency measures of eight amino acids (His, Glu, Lys, Ser, Asp, Cys, Trp, and Tyr) and two average attributes (hydropathy and flexibility), from which a feed-forward NN model with six hidden units was constructed (Romero et al. 1997). Later, NNs were also trained on disorder data derived from either X-ray crystallography or nuclear magnetic resonance (NMR) (Garner et al. 1998). By predicting disorder separately for the N-terminal, C-terminal, and internal regions of the sequence, the NN was also optimized against MLA models of logistic regression (LR) and discriminant analysis (DA) (Li et al. 1999). PONDR® was developed into several directions, by increasing the underlying amount and type of data and varying the algorithm, such as implementing a linear predictor using ordinary least-squares regression

(VL2) or an ensemble of feed-forward neural networks (VL3) incorporating Position-Specific Iterative Basic Local Alignment Search Tool (PSI-BLAST) homology models (VL3-H) (Obradovic et al. 2003). One version is specific for predicting short recognition elements (VL-XT (Iakoucheva et al. 2002); see Section 9.7 and Chapter 14, Section 14.2), whereas another combines two predictors optimized for the recognition of short- and long-disordered regions (VSL2 (Peng et al. 2006); see Section 9.5),

Another often-used NN algorithm is DisEMBL (Linding et al. 2003a), which uses the optional parameters Remark465 (missing coordinates from PDB) and "Coils/Loops" (propensity of being in local structure other than α-helix, 3_{10}-helix, and β-strand), as well as "Hot Loops," which is a refined subset of Coils/Loops, including only those that have a high degree of mobility characterized by Cα temperature factors (B-factors). DisEMBL is directly aimed at predicting disorder by an NN algorithm trained on datasets with high and low values of the above parameters.

Regional order neural network (RONN) is based on a different premise, being a special application of the bio-basis function neural network (BBFNN) pattern recognition algorithm (Yang et al. 2005). The underlying idea of BBFNN is that if two proteins have similar biological functions, or specifically, if they have similar tendency to be ordered or disordered, their sequences are also likely to show significant similarities. The level of similarity, and the likelihood of similar function, can be judged by sequence alignment. This principle is exploited to predict disorder by comparing the query sequence to a series of preselected prototype sequences of known folding state (ordered, disordered, or mixed) and classifying the sequence by a suitably trained NN based on the alignment scores.

NN predictions may also rely on statistics of the propensity of amino acids to belong to different secondary structural elements, such as α-helix, β-strand, and coil (see Chapter 1, Figure 1.1). It was suggested that regions devoid of elements of predicted secondary structure (i.e., regions of no regular secondary structure [NORS] longer than 70 consecutive amino acids with less than 12% of their residues in helix, strand, or coiled coil, and with at least one 10-residue-long region exposed to solvent) might be related to structural disorder (Liu and Rost 2003). The caveat of this approach is that there are well-ordered proteins, which are composed entirely of non-repetitive local structural elements (loopy proteins [Liu, Tan, and Rost 2002]), and IDPs, which are not fully devoid of regular secondary structures but have preformed and predictable structural elements (Fuxreiter et al. 2004). Therefore, NORSnet (Schlessinger, Liu, and Rost 2007a) was developed to distinguish well-structured and disordered (NORS) loops. The method can be useful for recognizing certain types of protein disorder that are not yet collected in databases.

9.3.2 Support-Vector Machines

SVMs are supervised learning methods that can also be used for classification. The input data is viewed as two sets of vectors in an N-dimensional space (e.g., 20-D for amino acid composition), in which an SVM constructs a hyperplane, which will maximize the margin of separation between the two data sets (in this case, the SVM is called a linear classifier, or one mounted with a linear kernel). In more sophisticated versions, separation is sought by higher-order functions, when we speak about non-linear (e.g., Gaussian) kernels. Intuitively, a good separation is achieved by the linear or non-linear function, which has the largest distance to the neighboring datapoints of both classes, because the larger the margin the better the generalization error of the classifier is.

The classical SVM to predict disorder is DISOPRED2 (Ward et al. 2004), which is mounted with a linear kernel and is trained on a database of amino acids missing from PDB structures. Unbalanced class frequencies (i.e., the actual data usually contain much more ordered than disordered residues) is handled by formulating the SVM to place asymmetric costs on points, with a greater cost to margin breaches by points from the minority (disordered) class than from the majority (ordered) class. Prediction is also supported by sequence profiles generated by three iterations of PSI-BLAST.

An SVM can also be combined with predictions based on propensities, as shown by the predictor PrDOS (Ishida and Kinoshita 2007). Its first element is a position-specific score matrix (PSSM) representation of the sequence generated by PSI-BLAST. Propensity of the local sequence for disorder is judged by an SVM that evaluates the PSSM based on the Remark465 record of PDB. The other approach is a PSI-BLAST that searches for homology with PDB templates (i.e., structured proteins) and ascribes disorder tendency from the lack of apparent similarity. The results of the two independent predictors are then combined by optimized weighing.

9.4 PREDICTION BASED ON INTERRESIDUE CONTACTS

Predictions may also be based on the idea that IDPs are disordered because they cannot make sufficient interresidue contacts to overcome the large decrease in configurational entropy during folding. The idea may be harnessed either by simple statistics of contact numbers or via more sophisticated approaches of estimating the total stabilization energy of a polypeptide chain.

9.4.1 Contact Numbers of Amino Acids

A simple statistical analysis of residue contact numbers is FoldUnfold (Garbuzynskiy, Lobanov, and Galzitskaya 2004), which judges the tendency of a protein to be folded or unfolded by the summation of the interresidue contact numbers of its amino acids. The values characteristic of different residues are calculated from a collection of globular protein structures, considering two residues in contact if any pair of their heavy atoms is within 8.0 Å to each other. The values range from 17.11 (Gly) to 28.48 (Trp), and to predict sequence-specific disorder, the contact density profile is calculated within a sliding window and a default threshold density of 20.4 is applied (Galzitskaya, Garbuzynskiy, and Lobanov 2006).

9.4.2 Estimating Pair-Wise Interresidue Interaction Energies

The number of contacts does not necessarily reflect the energetics of interresidue interactions, although this latter feature is directly linked with the lack of ability of a polypeptide chain to fold. This point has been addressed by IUPred, which estimates the total pair-wise interresidue interaction energy of sequences (Dosztanyi et al. 2005a; Dosztanyi et al. 2005b). The algorithm is based on low-resolution force fields (statistical potentials), which are energy-like quantities derived from the observed frequencies of interactions in globular proteins (Thomas and Dill 1996). The interaction energy of proteins is estimated directly from the sequence by a simple algebraic formula, involving a 20×20 energy predictor matrix derived from globular structures. The procedure relies on recognizing that amino acid interactions can be better estimated by a quadratic formula of composition than their linear combination, because it can take into account the dependence of energetics of amino acids on the sequence environment (Dosztanyi et al. 2005b).

The elements in this matrix assign the predicted energy to each position, depending on its amino acid type and the amino acid composition of the flanking region. The energy thus estimated from sequence approximates the energy of globular proteins calculated from their structure (Dosztanyi et al. 2005b). In the case of IDPs, estimated total energies are shifted toward less favorable values (Figure 9.3), which provides evidence that the potential of IDP sequences to form favorable interactions is smaller than that of globular proteins. This feature distinguishes them from globular proteins. By limiting the calculation of energies to a preset sequence window (typically 100 residues), the algorithm serves as a sequence-specific predictor of disorder, IUPred (Dosztanyi et al. 2005a) (Figure 9.2B).

9.4.3 Predictor of Contact Potentials

The local contact potential of sequences can also be estimated by combining several features, as demonstrated by Ucon (Schlessinger, Punta, and Rost 2007b). The algorithm uses PSI-BLAST to create position-specific profiles, which are then combined with other sequence-specific information, such as predicted secondary structure and solvent accessibility, to constitute the input to an NN contact-prediction method, PROFcon (Punta and Rost 2005). PROFcon predicts 2-D contact maps, which are then multiplied by energy-like statistical potential values to create a position-specific score of disorder.

9.5 PREDICTION OF SHORT AND LONG REGIONS OF DISORDER SEPARATELY

The structural disorder of a region also depends on its length (i.e., short and long regions of disorder are principally different). The distinction between the two categories is not

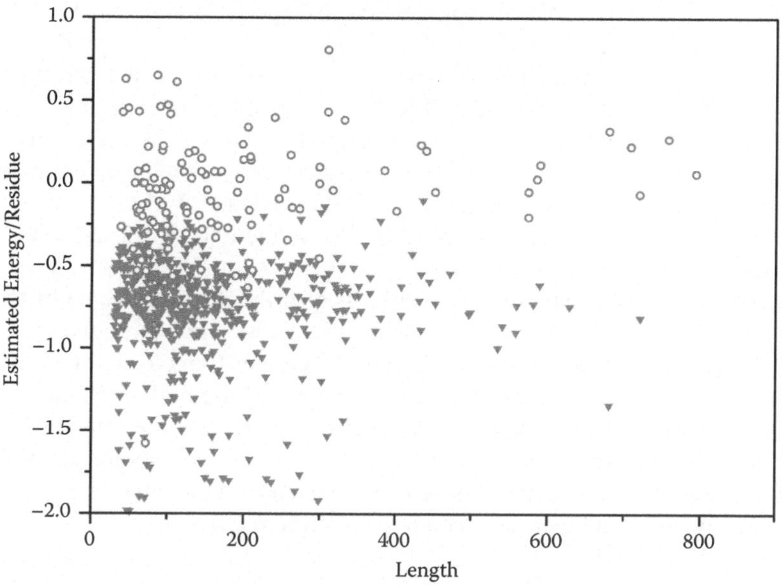

FIGURE 9.3 Pair-wise interresidue interaction energies of globular and disordered proteins. The total pair-wise interresidue interaction energy of globular proteins (inverse triangles) and disordered proteins (circles) is estimated from their amino acid composition and plotted as a function of the length of their polypeptide chains. Values that are more negative represent more stabilization due to amino acid interactions. The formula generating these energies forms the basis of the IUPred algorithm. Reproduced with permission from Dosztanyi et al. (2005), *J. Mol. Biol.* 347, 827–39. Copyright by Elsevier Inc.

easy to define. Either a length threshold (usually 30 amino acids) or a definition that short regions are the ones missing from PDB, whereas long ones are identified by other methods, is applied. Distinction between the two categories, nevertheless, is justified by significant differences in their amino-acid composition (Peng et al. 2006) and the existence of sequence clues of disorder located outside the region (see also Chapter 10, Section 10.3.3). This latter situation is underscored by a "twilight" zone between order and disorder (Szilagyi, Gyorffy, and Zavodszky 2008) and the difference in the performance of predictors on short versus long regions of disorder, with short regions usually being predicted less accurately (Bordoli, Kiefer, and Schwedel 2007; Jin and Dunbrack 2005; Melamud and Moult 2003). To handle this problem, predictors have been developed, which predict short and long regions separately and combine the results on disorder afterward.

The problem was addressed first by the development of VSL1 (Obradovic et al. 2005), which consists of three component predictors, each as an ensemble of logistic regression models, in a two-level architecture. At the first level, there are two specialized predictors for predicting long (VSL1-L, for >30 residues) and short (VSL1-S, for ≤30 residues) regions of disorder. VSL1-L was trained on DisProt sequences (112 amino acids in length on the average), whereas VSL1-S was trained on regions missing from

PDB structures (10 amino acids in length on the average). At the second level, VSL1-M (a meta-predictor) integrates outputs of the two predictors. In all three predictions, various attributes such as amino acid frequency, "spacer" frequency, K2-entropy, charge-hydrophobicity ratio, flexibility index, PSI-BLAST profiles, and predicted secondary structure are included. The algorithm was developed further as VSL2 (Peng et al. 2006). Its component predictors VSL2-S and VSL2-L are SVMs with a linear kernel, whereas for integrating data of the two, an optimized meta-predictor (VSL2-M2) is used. This latter is also a linear SVM that uses neighboring predictions of VSL2-S and VSL2-L as inputs.

Dataset-dependent prediction of disorder is also the defining theme of the Spritz algorithm (Vullo et al. 2006). Spritz includes two SVM predictors: one trained on a dataset of long disorder taken from DisProt and the other trained on a dataset of short disorder taken from the PDB. The two binary classifiers are both implemented with a non-linear Gaussian kernel, and unbalanced class frequencies are mitigated by using asymmetric costs (i.e., a larger penalty for disorder misclassification). The two classifiers use residue attributes such as amino-acid frequencies computed from PSI-BLAST multiple alignments combined by secondary structure prediction.

9.6 COMBINATION OF PREDICTORS: META-SERVERS

Different predictors rely on different principles and/or algorithms, each having strengths and weaknesses. As discussed in Section 9.8, there is no universal solution to comparing them and establishing the "best" predictor. Thus, it is recommended to compare predictions by different algorithms based on different physical and/or computational principles and seek a consensus of their scores. This practical point of view called meta-servers into existence, which either simply help carry out numerous parallel predictions or, in a more sophisticated way, integrate several outputs to produce a consensus by some predefined criterion.

A simple tool to speed up the process of disorder prediction is MeDor, which submits the query sequence to several servers simultaneously and provides a graphical output of disorder scores (Lieutaud, Canard, and Longhi 2008). Besides 12 different disorder predictions (e.g., different versions of IUPred, RONN, FoldUnfold, DisEMBL, FoldIndex, GlobPlot, DISPROT, and Phobius), it also adds the prediction of secondary structure and hydrophobic clusters.

A consensus is sought by metaPRDOS (Ishida and Kinoshita 2008), which sends the query sequence to seven individual predictors (PrDOS, DISOPRED2, DisEMBL, DISPROT, DISPro, IUPred, and POODLE-S), and compares their predictions. Because sensitivity and scaling of the component predictors differ, metaPRDOS does not use simple averaging, but takes the individual scores as a seven-element input vector to a SVM trained on a dataset of disordered proteins.

9.7 PREDICTION OF FUNCTIONAL MOTIFS IN IDPS

Prediction of disorder can also be used to generate functional insight. In particular, local anomalies in a disorder pattern can be interpreted in terms of short recognition motifs within IDPs/intrinsically disordered regions (IDRs), which can be directly associated with function within a region of disorder (discussed in detail in Chapter 14, Section 14.2).

Prediction of the intrinsic conformational preferences of IDPs can be optimized to pinpoint likely sites of disorder-to-order transition (Oldfield et al. 2005b) (i.e., sites of interaction with the partner molecule). It was noticed in the case of the original PONDR® predictor that interaction sites tend to be short regions of apparent order surrounded by regions of predicted disorder (Garner et al. 1999). For PONDR® VL-XT, the correlation between distinctive downward spikes in disorder scores and binding regions within IDRs was verified (Oldfield et al. 2005b). This local anomaly is caused by a deviation from the general amino acid composition of IDPs (Dunker et al. 2001), which is mostly a local enrichment in hydrophobic amino acids that is a hallmark of binding interfaces of IDPs (Meszaros et al. 2007).

This phenomenon can be illustrated by the complex of the eukaryotic translation initiation factor 4E (eIF4E) and the fully disordered 4E binding protein 1 (4E-BP1) (Figure 9.4). 4E-BP1 binds its partner by virtue of a short binding helix (see the functional details in Chapter 11, Section 11.5)—the location of which in the sequence is marked by a downward spike on the PONDR® VL-XT score that crosses the order/disorder boundary. This behavior was observed in the case of many other IDPs, such as the autoinhibitory region of calcineurin and regions of p53 binding to MDM2 and S100B (Oldfield et al. 2005b); the binding sites of RNA, helicase RhlB/enolase, and polynucleotide phosphorylase (PNPase) within RNase E (Callaghan et al. 2004); the binding region of Cdc42 in WASP and that of DIAP1 within Grim (Mohan et al. 2006); the recognition element of PCNA within $p21^{Cip1}$ and the region of Nup2p binding to karyopherin Kap60 (Vacic et al. 2007); and the BOX2 region of measles virus nucleoprotein, which is the binding region of phosphoprotein (Karlin et al. 2003; Oldfield et al. 2005b).

These observations have led to the concept of molecular recognition elements/features (MoREs/MoRFs, described in detail in Chapter 14, Section 14.2.3) and also to the development of algorithms for the identification of such regions (Oldfield et al. 2005b). A highly sensitive NN predictor (α-MoRF-PredII) to filter α-helix-forming propensity (α-MoRFs) within IDPs with predicted dips was developed based on actual α-MoRF examples and their cross-species homologs, as well as attributes from disorder predictions, secondary structure predictions, and amino acid indices (Cheng et al. 2007). Predicted α-MoRFs are much more abundant in eukaryotes than prokaryotes, and they are typically enriched in proteins of signaling and regulatory functions (Oldfield et al. 2005b). MoRFs can also adopt β-strand and irregular (coil) structures in the bound state (β-MoRFs and ι-MoRFs). These short recognition

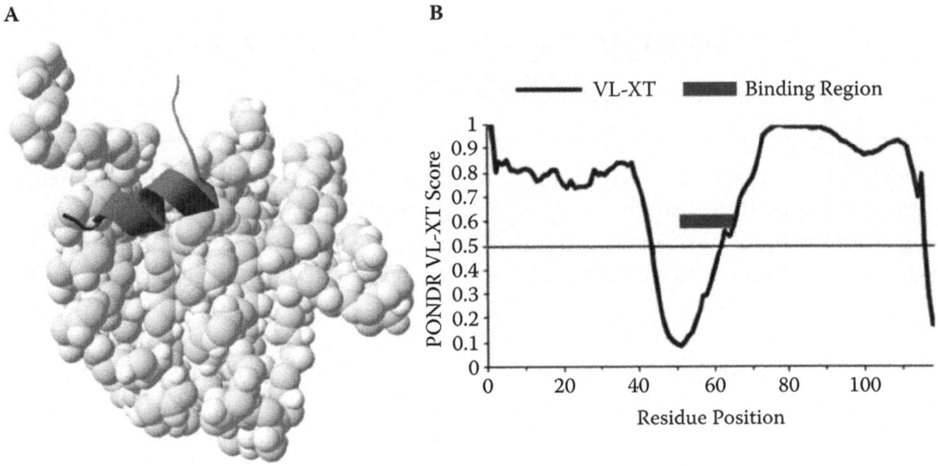

FIGURE 9.4 Prediction of the binding region of an IDP by PONDR® VL-XT. A binding region within an IDP can be recognized as a downward spike on the disorder score generated by PONDR® VL-XT. (A) The structure of the complex (pdb 1ej4) of the eukaryotic translation initiation factor 4E (eIF4E, light gray), and 4E binding protein (4E-BP1, dark gray). (B) The binding region of completely disordered 4E-BP1 is outlined by PONDR® VL-XT, which has a downward spike at the location of the binding site (marked by a horizontal bar). Reproduced with permission from Oldfield et al. (2005), *Biochemistry* 44, 12454–70. Copyright by the American Chemical Society.

elements in IDPs/IDRs are also captured by other related concepts, such as eukaryotic linear motifs (ELMs), preformed structural elements (PSEs), short linear motifs (SLiMs), and primary contact sites (PCSs)—the recognition of which is achieved by different bioinformatic approaches, such as SLiMDisc and DILIMOT (see Chapter 14, Section 14.2.2).

9.8 COMPARISON OF THE ACCURACY OF PREDICTORS: THE CASP EXPERIMENT

The performance of disorder prediction algorithms was assessed in the critical assessment of methods of protein structure prediction (CASP) experiments CASP5 (Melamud and Moult 2003), CASP6 (Jin and Dunbrack 2005), and CASP7 (Bordoli et al. 2007). Because the performance of disorder predictors depends critically on both the type of disorder and evaluation criteria, their comparisons are rather biased. The dataset for evaluation in CASP experiments is mainly restricted to regions of proteins missing from X-ray structures, which qualify mostly as short disorder, and the amount of the underlying data of order and disorder vastly differs.

To handle the limitations immanent due to such unbalanced class frequencies, the performance of predictors is usually compared by several approaches, which usually rely on the measures true positive (residues predicted and observed disordered, N_{TP}), false positive (residues predicted disordered but observed ordered, N_{FP}), true negative (residues predicted and observed ordered, N_{TN}), and false negative (residues predicted ordered but in fact disordered, N_{FN}). Considering these parameters, a standard approach is to generate a receiver operating curve (ROC), which answers the dilemma of overprediction and consequent loss of discrimination between ordered and disordered residues. The ROC curve is a plot of N_{TP} versus N_{FP} at a given prediction threshold (Figure 9.5), and the accuracy of the predictor is approximated by the area under the curve (maximum of true positives at a minimum of false positives). The ROC approach has been very often used to compare predictors (Garbuzynskiy et al. 2004; Ishida and Kinoshita 2008; Jin and Dunbrack 2005; Jones and Ward 2003; Li et al. 2000; Peng et al. 2006; Vullo et al. 2006; Ward et al. 2004).

For binary predictors, which associate either order or disorder with a residue, the assessment of prediction accuracy is usually based on combining sensitivity (the

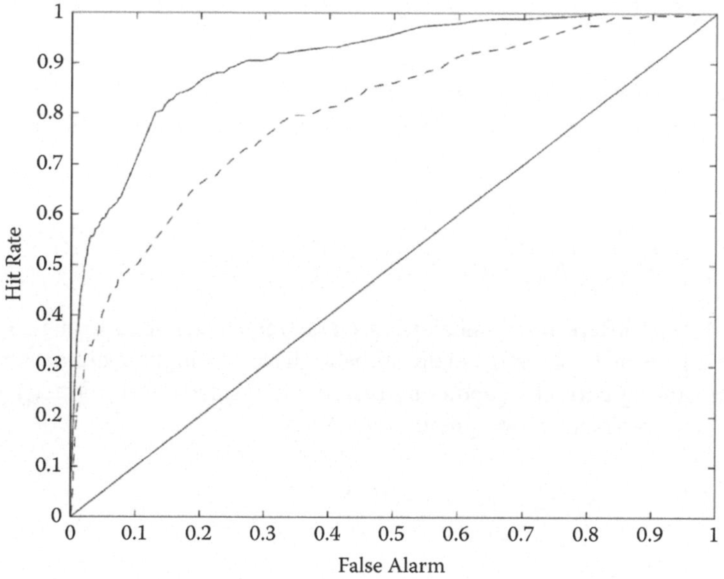

FIGURE 9.5 ROC curve of disorder prediction. True positive (hit rate) of disorder prediction is plotted against false positive (false alarm) rate to give ROC curves for predictions by the DISOPRED algorithm applied in two different modes, including (solid curve) or not including (dashed curve) information on the structure of homologs of known 3-D structure. The result expected for a completely random predictor is also shown as a solid diagonal line. Reproduced with permission from Jones and Ward (2003), *Proteins* 53, S6, 573–8. Copyright by Wiley-Liss.

efficiency of predicting disorder) and specificity (the efficiency to discriminate it from order), which is defined as follows:

$$S_{sens} = \frac{N_{TP}}{(N_{TP} + N_{FN})} = \frac{N_{TP}}{N_{disorder}} \tag{9.3}$$

$$S_{spec} = \frac{N_{TN}}{(N_{TN} + N_{FP})} = \frac{N_{TN}}{N_{order}} \tag{9.4}$$

where $N_{disorder}$ and N_{order} are the total number of residues observed as disordered and ordered, respectively. A predictor performs better if it has both higher sensitivity and specificity (Jin and Dunbrack 2005; Melamud and Moult 2003), which is sometimes assessed by calculating their product or combining them into an overall accuracy (ACC):

$$S_{product} = S_{sens} S_{spec} = \frac{N_{TP} N_{TN}}{N_d N_o} \tag{9.5}$$

$$ACC = \frac{S_{sens} + S_{spec}}{2} \tag{9.6}$$

Another measure of performance is the overall percentage of accuracy ($Q2$), which is also often used in evaluating secondary structure predictors (Jin and Dunbrack 2005):

$$Q2 = \frac{N_{TP} + N_{TN}}{N_{TP} + N_{FP} + N_{TN} + N_{FN}} \tag{9.7}$$

which, however, suffers from unbalanced class frequencies, due to which a method predicting all residues ordered would probably have the highest $Q2$ accuracy. Thus, schemes rewarding correctly predicting disordered residues over correctly predicting ordered residues, such as the weighted score (S_W):

$$S_W = \frac{W_d N_{TP} - W_o N_{FP} + W_o N_{TN} - W_d N_{FN}}{W_d N_d + W_o N_o} \tag{9.8}$$

where W_d and W_o are weights assigned to experimentally defined disordered and ordered residues, and the Matthews correlation coefficient (S_{MCC}):

$$S_{MCC} = \frac{N_{TP} N_{TN} - N_{FP} N_{FN}}{\sqrt{(N_{TP} + N_{FP})(N_{TP} + N_{FN})(N_{TN} + N_{FP})(N_{TN} + N_{FN})}} \tag{9.9}$$

offer a more reasonable and fair comparison of methods. Depending on the underlying datasets and criteria of comparison, different predictors have different performance, with the top ones being PONDR®, PrDOS, DISPro, IUPred and DISOPRED2 (CASP6) and PONDR®-VSL2, CBRC-DR, DISOPRED2, DisPRO, and GeneSilicoMetaServer (CASP7). As a final note, it should be stressed that it is very difficult and maybe impractical to search for the "best" predictor of disorder, because any predictor can be sensitive to certain types of disorder but less sensitive to others. Thus, it is recommended that several algorithms be tried, and their results combined with caution.

9.9 A BETTER TARGET PRIORITIZATION IN STRUCTURAL GENOMICS

Disorder predictions have the benefit of addressing diverse structural and functional issues exemplified by many studies covered throughout the book, with a special application of improving the efficiency of target selection in structural proteomics/genomics programs. These HTS programs aim to solve the structures of possibly all proteins of biological/biomedical interest. Their critical element is target selection (Brenner 2000), which ensures that limited resources are focused on feasible and important targets. To this end, the preference is often set to proteins with little sequence similarity to proteins of known structures, which limits the risk of duplicating already available knowledge. At the same time, however, it makes it more likely that the structure of the novel target cannot be solved because it contains a significant level of disorder. This concern was substantiated by the retrospective application of bioinformatic filtering to the Center for Eukaryotic Structural Genomics (CESG) targets (Oldfield et al. 2005c), which showed that the presence of disorder is detrimental to solving structures. Of 71 proteins that reached the HSQC screening stage in this program, 44 were found to be predominantly disordered and 27 predominantly ordered. Retrospective removal of targets predicted to be disordered resulted in an increase in the yield of proteins folded by NMR HSQC from 0.36 to 0.41. A 1-D NMR study of 141 potential *T. maritima* targets showed that only 25% are properly folded and are likely to be suitable for structure determination (Peti et al. 2004). In the 79 targets of the Joint Center for Structural Genomics (JCSG), 63% were found likely to provide good quality crystals, whereas 37% either did not produce crystals or did not diffract suitably for structure solution (Page et al. 2005).

The potential impact of disorder prediction on target selection can also be illustrated by analyzing the overall progress of targets of worldwide structural genomics projects collected in the TargetDB database (Dosztanyi et al. 2007). Their pipeline of structure determination involves several steps from cloning through expression, purification, and crystallization to the final stage of structure determination, each of which represents bottlenecks for a large number of targets. The amount of predicted disorder steadily decreases upon going through successive stages. For example, 16%

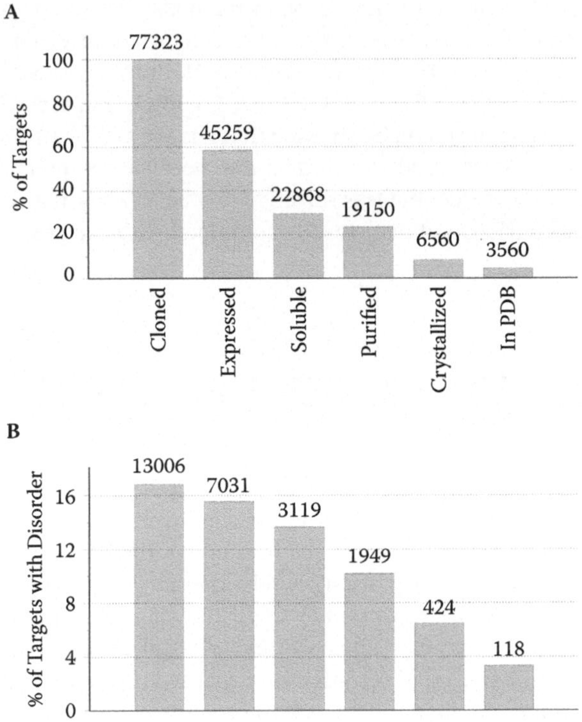

FIGURE 9.6 The number of targets and the percentage of proteins with long disordered regions. (A) The number and percentage of targets in the TargetDB database at various stages of the structure solution pipeline. (B) The number and percentage of proteins with an IDR ≥30 residues in length, as determined by the IUPred algorithm. For each bar, the number of proteins in the given stage was used as the 100%. Reproduced with permission from Dosztanyi et al. (2007), *Curr. Protein Pept. Sci.* 8, 161–71. Copyright Bentham Science Publishers Ltd.

of the targets contain at least one IDR ≥30 residues in the initial stage of cloning, whereas in the final stage of solved structures this value is only 3.5% (Figure 9.6). Thus, disorder correlates negatively with the success of solving novel structures—most significantly at the stages of crystallization and actual solution of structure.

Structure of IDPs 10

This chapter is central to the theme of the book, surveying ideas about the structure of intrinsically disordered proteins (IDPs). Because an IDP by definition has an ensemble of structures that differ for each individual protein, the "structure" of IDPs cannot be described in general. Rather, the general principles can be outlined, and it can be shown how these appear in some of the best characterized examples. As we will see, the description entails three different levels: the primary (sequence), secondary (local), and tertiary (global) structure. Often, the basic structural concepts are the same as in the case of globular proteins (see Chapter 1), and particularly close analogies apply with their denatured states.

10.1 PRIMARY STRUCTURE OF DISORDERED PROTEINS

The first level of structural characterization of IDPs is their primary structure (i.e., amino acid sequence). Because evidence is available for the disorder of only about 500 proteins and 1,100 disordered regions (Sickmeier et al. 2007), the data are insufficient to establish statistically meaningful extended features of sequence. Thus, descriptions of primary structure are dominated by results on amino acid composition and short sequence features.

10.1.1 Amino Acid Composition

The frequency of amino acids in disordered proteins significantly differs from that of ordered proteins (Dunker et al. 2001; Uversky, Gillespie, and Fink 2000a; Tompa 2002). Amino acid frequencies plotted as a function of the flexibility index of residues (Vihinen, Torkkila, and Riikonen 1994) show a distinctive pattern (Figure 10.1): IDPs are depleted in amino acids of low flexibility indexes and are enriched in amino acids of high flexibility indexes (Dunker et al. 2008). Amino acids in the former group (Trp, Cys, Phe, Ile, Tyr, Val, and Leu) are termed order-promoting, whereas those in the latter (Ala, Arg, Gly, Gln, Ser, Pro, Glu, and Lys) are disorder-promoting. The physical rationale of this trend comes from the ensuing high net charge and low net hydrophobicity, which ensure the extended state of IDPs (see Chapter 9, Section 9.2.2).

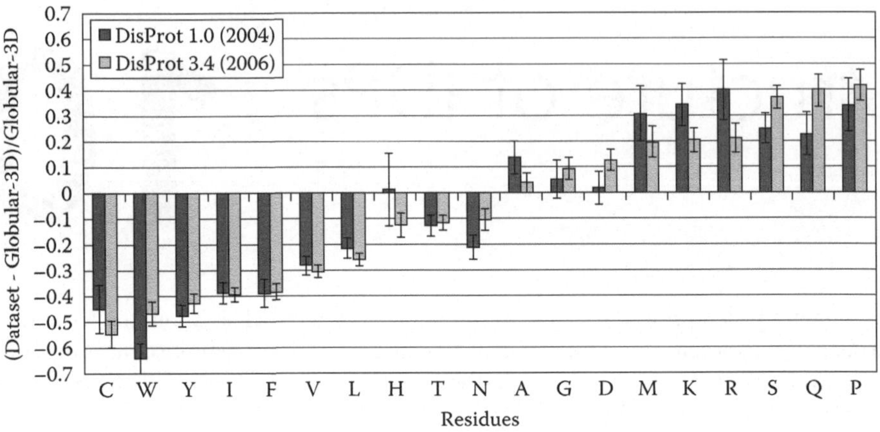

FIGURE 10.1 Amino acid composition of disordered proteins. The differences between the amino acid compositions of disordered datasets (DisProt 1.0 and DisProt 3.4) and that of an ordered dataset were plotted as a function of the B-factor estimates of flexibility of residues. There is a tendency for IDPs to be depleted in rigid (order-promoting) amino acids and enriched in the more flexible (disorder promoting) amino acids. Reproduced with permission from Dunker et al. (2008), *BMC Genomics* 9, S2, S1. Copyright by the BioMed Central Ltd.

Another underlying cause of this biased composition may come from the need to avoid amyloid formation (Tompa 2002). The ease of transition of the open structure of IDPs to a β-strand poses the danger of amyloidosis (Conway, Harper, and Lansbury 2000; von Bergen et al. 2000), as outlined in detail in Chapter 15. A likely selective force in the evolution of IDPs is to minimize this threat (Monsellier and Chiti 2007), as also manifested in the strong general correlation of β-strand-forming potential with order–disorder discrimination (Williams et al. 2001) (i.e., a negative correlation of β-aggregation tendency and disorder) (Linding et al. 2004).

The correlation of particular amino acid features with disorder supports these conclusions (Williams et al. 2001). Out of 265 features analyzed, in the 10 top-ranking ones discriminating between order and disorder: there are 2 contact scales, 4 hydrophobicity scales, 3 β-sheet propensity scales, and 1 polarity scale. Certain features are enriched in ordered (contact scales, hydrophobicty, and β-sheet propensity), whereas others in the disordered (polarity) dataset, which is in agreement with the functional requirement for an open structure that is capable of avoiding aggregation.

In accord with the foregoing points, ordered and disordered proteins can be discriminated by a reduced amino acid alphabet (Weathers et al. 2004), because the accuracy of IDP prediction by a support vector machine (SVM) is largely preserved when amino acids are clustered by chemical similarity (87% for 20 amino acids vs. 84% for 4 groups of amino acids). The weights associated with these four vectors (i.e., Phe-Trp-Tyr, Cys-Ile-Leu-Met-Val, Ala-Gly-Pro-Ser-Thr, and Asp-Glu-His-Lys-Asn-Gln-Arg) underscore the notion that simple general physicochemical properties and the avoidance of aggregate formation are critical in defining protein disorder.

10.1.2 Sequence Features Characterizing Disorder

Whereas the primary determinant of disorder is amino acid composition, higher-order sequence attributes contain additional information on the disordered state. As a statistically affordable step toward understanding these factors, Lise and Jones extracted simple but statistically significant local patterns of amino acids or amino acid properties in IDPs, such as hydrophobicity, polarity, size, aliphatic/aromatic nature, proline, and charge (Lise and Jones 2005). The two most highly significant regularities observed are simple Pro-rich patterns and charged patterns dominated by either positive or negative residues. For example, Pos(itive)-Pos-X-Pos and Neg(ative)-Neg-Neg, Glu-Glu-Glu, Lys-X-X-Lys-X-Lys, and Pro-X-Pro-X-Pro were found to occur most often. These sequence marks can be interpreted in terms of the preference for a locally extended nature of the polypeptide chain imposed by high local Pro content or electrostatic repulsion, and also an effective evolutionary mechanism of repeat expansion in the generation of repetitive motifs in IDPs (see Chapter 13, Section 13.3) (Tompa 2003b).

These features are also closely related to the low sequence complexity of IDPs, as mentioned in Chapter 9, Section 9.2.1. Wootton (Wootton 1994b) observed that globular proteins tend to be in a state of high-entropy (defined by Shannon (Shannon 1948), also termed high complexity), apparently random state. In contrast, about 25% of all amino acids in Swiss-Prot are in low-complexity regions, and 34% of all Swiss-Prot proteins have at least one such segment (Figure 10.2), which is repetitive in nature and/or use a reduced set of amino acids (Wootton 1994a). The exact relationship of low complexity and disorder was addressed in two studies. Romero and colleagues (Romero, Obradovic, and Dunker 1999) asked whether a minimal alphabet size (number of amino acids) and

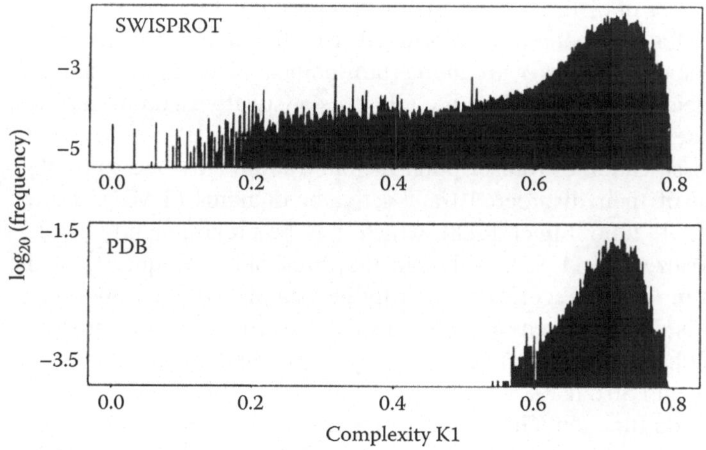

FIGURE 10.2 Distribution of Shannon's entropy in protein sequences. Wootton applied Shannon's entropy (Shannon 1948) to estimate the information content of sequences deposited in SwissProt and PDB. Sequences in PDB are all in a high-entropy (high-complexity) state, whereas a significant part of the sequences in SwissProt are in a low-complexity state. Reproduced with permission from Wootton (1994), *Computers Chem. 18,* 269–285. Copyright by Pergamon.

complexity are required for defining a domain. They found that proteins in SwissProt cover the entire possible range of alphabet size (1–20) and informational entropy range (K = 0.0 – 4.5), whereas globular domains only occupy a restricted region of values (alphabet = 10 – 20, K = 3.0 – 4.2). Regions of lower values (down to alphabet size = 3 and K = 1.5) correspond to fibrous structured proteins, such as coiled coils, collagens and fibroins. It was concluded that a minimal alphabet size of 10 and entropy around 2.9 are necessary and sufficient to define a sequence that can fold into a globular structure. Although complexity distributions of IDPs/intrinsically disordered regions (IDRs) are shifted to values lower than those of ordered proteins, there is a significant overlap (Romero et al. 2001), because disordered proteins cover the range K = 2 – 4.2, with occasionally exhibiting values as low as K = 1.0. Thus, disorder and low complexity are related but distinct phenomena.

Although amino acid composition and local sequence features are the primary determinants of disorder, several points suggest the importance of higher-level sequence attributes. For example, machine-learning predictors trained on sequences outperform simple propensity-based predictors (Chapter 9), and there is a twilight zone between order and disorder (Section 10.3.3). For example, in the case of short segments, disorder is encoded not only by local composition, but also sequence and environment (i.e., context (Szilagyi et al. 2008)). A critical increase in the amount of data on IDP sequences is required to enable the exploration of such higher-order features.

10.1.3 Flavors of Disorder?

The amino acid composition of IDPs represents averages, which may be realized by different subclasses of distinct characteristic composition. This would mean that the composition of IDPs does not follow a normal distribution in the 20-D hyperspace of amino acid frequencies, but cluster around certain points. Whereas the amount of sequence data (Sickmeier et al. 2007) does not allow a statistically rigorous analysis of this feature, the issue has been raised in different contexts.

The classification of transcription factors traditionally occurs by the amino acid composition of their disordered trans-activator domains (TADs) (Liu et al. 2006a; Minezaki et al. 2006; Sigler 1988), which may be predominantly acidic, Pro-rich or Gln-rich (Triezenberg 1995). Although the threshold of frequency of the "founder" amino acid in these categories can hardly be established, the importance of this categorization is demonstrated by the insensitivity of function of transcription factors to amino acid changes in their TADs, as long as its noted character is preserved (Hope, Mahadevan, and Struhl 1988). Mutations that change this (acidic) character do impair trans-activation function (Gill and Ptashne 1987) (i.e., in terms of composition, disordered TADs of transcription factors come in different flavors). A similar correlation of function with amino acid composition was reported in the case of linker histones (Hansen et al. 2006) and linker regions of multi-domain proteins (George and Heringa 2002). These latter display a high degree of flexibility and lack regular secondary structure, and they evolve rapidly (Daughdrill et al. 2007). Their rudimentary analysis suggests distinct categories, such as helical and non-helical linkers, with the former being mostly enriched in Leu, Arg, and Glu, whereas the latter in Pro. A previous study also suggested the presence of Gln-linkers in bacterial regulatory proteins (Wootton and Drummond 1989).

Clustering of IDPs in the composition space is also underscored by the observation that disorder predictors trained on one type of IDP often perform poorly on a different type (Vucetic et al. 2003). By a competition among increasing numbers of predictors, three "flavors" of disorder, designated V, C, and S, were identified. Flavor C is slightly enriched in His, Met, and Ala; flavor S is depleted in His; whereas flavor V has more of the least flexible amino acids Cys, Phe, Ile, and Tyr. The flavors have weak but discernible functional associations. For example, 9 out of 10 *E. coli* ribosomal proteins fall into flavor V, whereas IDPs that bind to the genomic ribonucleic acid (RNA) of viruses are excluded from this flavor. Protein-binding functions segregate with flavors V and S, while deoxyribonucleic acid (DNA) binding proteins are primarily associated with C and S.

10.2 SECONDARY STRUCTURE OF DISORDERED PROTEINS

The distinction between unbound (free) and bound states is most meaningful at the secondary structural level of IDPs. In the unbound state, most spectroscopic techniques provide evidence that IDPs are often not fully random but they have transient secondary structural elements. The exact place and propensity of these can be unveiled by nuclear magnetic resonance (NMR). In the bound state, the structure can be solved by X-ray crystallography or NMR, and much functional insight can be gained by comparing it to the unbound structures.

10.2.1 Secondary Structure in Solution State: Signs of Transient Order

The realization that IDPs are basically different from ordered proteins and the lack of detailed structural models have led to initial claims that they are featureless random coil-like proteins (see also Chapter 2). To simply contrast them with ordered proteins, this is an appropriate approximation. Reports of random coil behavior in the case of myelin basic protein (MBP) (Sedzik and Kirschner 1992), ProTa (Gast et al. 1995), tau protein (Schweers et al. 1994), and casein (Williams 1989) basically contributed to challenging the classical structure-function paradigm. Later, it became increasingly apparent that many IDPs have transient local organization, the description of which is critical toward reassessing the structure-function paradigm. Their deviation from a random coil-like state was also suggested by observations and arguments that a true random coil does not exist, not even in the highly denatured states of globular proteins (Shortle 1996). Thus, the term *residual structure* was adopted from the literature of protein folding and is used somewhat misleadingly for the structural phenomenon of transient local structure in IDPs.

A significant level of secondary structure in IDPs is often reported by circular dichroism (CD), although the amount is rather uncertain due to deconvolution being carried out with standard spectra derived from ordered proteins. A small amount of α- or β-structure on the order of 10–20% has been ascertained in the case of many IDPs, such as $p21^{Cip1}$ (Kriwacki et al. 1996), λN (Van Gilst et al. 1997), Dsp16 (Lisse et al. 1996), CST (Hackel, Konno, and Hinz 2000), caseins (Holt and Sawyer 1993), α-synuclein (Kim et al. 2000b), tau protein (Schweers et al. 1994), stathmin (Wallon et al. 2000), 4E-BP1 (Marcotrigiano et al. 1999), and CREB TAD (Richards et al. 1996), for example. Fourier-transform infrared spectroscopy (FTIR) has also suggested similar local organization in the case of α-synuclein (Weinreb et al. 1996) and tau protein (Schweers et al. 1994). Some limited structural order can also be inferred from the shift of CD spectra toward the random coil state upon heating of the protein or the addition of a denaturant, as in the case of caseins (Holt and Sawyer 1993), calpastatin (Csizmok et al. 2005; Hackel et al. 2000), $p21^{Cip1}$ (Kriwacki et al. 1996), and fibronectin binding protein(A) (FnBPA) (House-Pompeo et al. 1996).

A local deviation from a fully random structural state can also be demonstrated by indirect techniques. For example, specific phosphorylation of tau protein makes it become extended and stiff, as shown by changes in paracrystal structure (Hagestedt et al. 1989). An epitope specific to Alzheimer's disease in tau protein only forms when kinases are applied in a certain order, which suggests that certain local conformational states follow in succession upon modification (Zheng-Fischhofer et al. 1998). Discernible local structure is also inferred from the nonrandom cleavage by proteases, which is indicative of the nonuniform accessibility of different but sequentially equivalent sites. The most notable examples for this behavior are tau protein (Steiner et al. 1990) and microtubule-associated protein 2 (MAP2) (Wille et al. 1992b), which are both cleaved preferentially within a narrow region separating their tubulin-binding domain (TBD) and projection domains (see also Chapter 3, Section 3.5).

10.2.2 A Lot of PPII Helix Conformation

Polyproline II (PPII) helix conformation is recognized as a distinct secondary structural element (Chapter 1, Section 1.5.1). By analyzing the main-chain conformations of ordered proteins for local structures that do not fall into the known classes of α-helix, 3_{10}-helix, and β-strand, it was found that PPII helices of three consecutive residues occur 120 times in 80 proteins, which is significantly more frequent than expected by chance (Adzhubei and Sternberg 1993). Pro is frequent in these PPII regions, but many times it is entirely missing. PPII conformation can be mostly found on the surface of globular proteins, with only a few main-chain hydrogen bonds established with the main body of the protein. They correspond to mobile segments of the molecule often involved in protein–protein interactions (Williamson 1994).

The presence and prevalence of PPII conformation is also apparent in many IDPs, as shown by CD analysis in the case of tau protein (Uversky et al. 1998), casein (Andrews et al. 1979), stathmin (Wallon et al. 2000), Bob1 (Chang et al. 1999), the Pro, Glu, Val, Lys-rich (PEVK) region of titin (Ma, Kan, and Wang 2001; Ma and Wang 2003; Makowska et al. 2006), MAP2 (Csizmok et al. 2005) and the C-terminal domain

(CTD) of RNA polymerase II (RNAP II) (Bienkiewicz, Woody, and Woody 2000). The characteristic signature of PPII in the CD spectrum, a positive peak at 217 nm, however, is obscured by large negative contributions from the α-helix and β-strand, which limits the insight provided by CD. Conclusive and quantifiable data on PPII conformation in IDPs is only provided by Raman optical activity (ROA) (see Chapter 5, Section 5.5), which shows the prevalence of PPII in several IDPs, such as casein, α-synuclein, and tau protein (Syme et al. 2002); some wheat gluten proteins (Blanch et al. 2003); and in Ala-repeat oligopeptides (McColl et al. 2004).

The importance of PPII helix conformation also derives from its probable involvement in conformational diseases (Blanch et al. 2000). It is suggested that disorder of the PPII type may enable the formation of regular fibrils, whereas more dynamic or random coil-type disorder may instead lead to amorphous aggregates. In accord, PPII conformation appears to dominate in the partially unfolded states of amyloidogenic globular proteins (Blanch et al. 2000), the neurodegenerative IDPs α-synuclein and tau protein (Syme et al. 2002), and repetitive amyloidogenic peptides/proteins, such as polyQ stretches (Chellgren, Miller, and Creamer 2006), and oligoAla peptides (Chen, Liu, and Kallenbach 2004; Shi et al. 2002).

10.2.3 Secondary Structure in Solution State: Sequence-Specific Information

Information on the location and dynamics of transient secondary structural elements noted in Section 10.2.2 is provided by multi-dimensional NMR, sometimes in combination with molecular dynamics (MD) analysis. There are more than a dozen such cases in the literature (Table 10.1). Because of the importance of this issue to the extension of the structure-function paradigm, the most instructive examples are discussed in some detail.

10.2.3.1 p27^{Kip1}

The Cdk-inhibitor p27^{Kip1} is one of the best characterized IDPs and is detailed for its function and involvement in disease in Chapter 15, Section 15.1.3. Its nonrandom solution state characterized by NMR provides one of the most insightful examples of the transient organization of an IDP, also reflecting its partner-bound state. Its backbone Hα, Cα, and amide-N chemical shift index (CSI) values suggest extensive deviations from the random-coil behavior (Lacy et al. 2004). Most notably, positive values in the D37–K59 region are consistent with transient local helical conformations within a segment termed linker helix (LH; for domain definitions, see Chapter 3, Section 3.7.2). These results were corroborated and extended by a combination of NMR studies and MD simulations (Sivakolundu, Bashford, and Kriwacki 2005), which provided a detailed and almost high-resolution structural description of the structural ensemble of the kinase-inhibitory domain (KID) of p27^{Kip1} in the unbound state (Figure 10.3).

In the KID domain, no nuclear Overhauser effect (NOE) can be observed in the specificity-determining domain 1, whereas proximity of sequential residues result in

TABLE 10.1 Secondary structural elements in IDPs*

PROTEIN	*DISPROT*	*LENGTH*	*DISORDERED REGION*	*SECONDARY STRUCTURE OBSERVED (RESIDUES)*	*REFERENCE*
p27^{Kip1}	DP00018	198	1–198	α-helix (38–60) α-helix (37–59), β-hairpin (65–75), single turn of helix (87–90)	(Lacy et al. 2004) (Sivakolundu et al. 2005)
α-synuclein	DP00070	140	1–140	α-helix (18–34), partial α-helix (1–100), possible β-turn (C-terminal region)	(Bussell and Eliezer 2001)
Potassium channel shaker	DP00267	401	1–62	α-helix (2–10, 44–52, 56–61)	(Wissmann et al. 1999)
Tau protein F	DP00126	441	1–441	β-strand (274–284, 305–315, 336–345)	(Mukrasch et al. 2007a)
Stem-loop binding protein, SLBP	DP00144	276	1–175	α-helix (28–45, 50–57, 66–75, 91–96)	(Thapar et al. 2004)
CREB	DP00080	341	1–265	α-helix (119–130) α-helix (120–129, 134–144)	(Hua et al. 1998) (Radhakrishnan et al. 1998)
FlgM	DP00027	97	1–97	α-helix (60–73, 83–90)	(Daughdrill et al. 1998)
p53	DP00086	393	1–73	α-helix (18–24), mixture of α-helix, β-strand and random coil (39–59)	(Vise et al. 2005)
Neh2 domain of Nrf2		98	1–98	α-helix (39–71) β-strand (74–76, 82–85)	(Tong et al. 2006)

FnBP	DP00025	1018	745–874	β-strand (757–762, 774–779, 795–800, 817–822, 845–850, 851–856)	(Penkett et al. 1998)
Calpastatin	DP00196	141 (domain1)	1–141	α-helix (12–30, 87–105) probable β-strand (50–70)	(Kiss et al. 2008b)
Securin	DP00256	202	1–202	α-helix (150–159)	(Csizmok et al. 2008)
MBP	DP00236	176	1–176	α-helix (33–46, 83–92, 142–154)	(Libich and Harauz 2008)
Tβ4	DP00357	43	1–43	α-helix (5–17)	(Domanski et al. 2004)
IA3	DP00179	68	1–68	α-helix (44–60)	(Green et al. 2004)
Human replication protein A70 (hRPA70)	DP00061	616	105–182	α-helix (105–123)	(Olson et al. 2005)
Cl channel ClC-0		810	603–699	α-helix (637–646)	(Alioth et al. 2007)

* IDPs are enlisted for which deviations of NMR observables from random coil values allowed the characterization of local structural preferences in the solution state. DisProt number, total length of the protein, the region known to be disordered, and the region of transient secondary structure are shown.

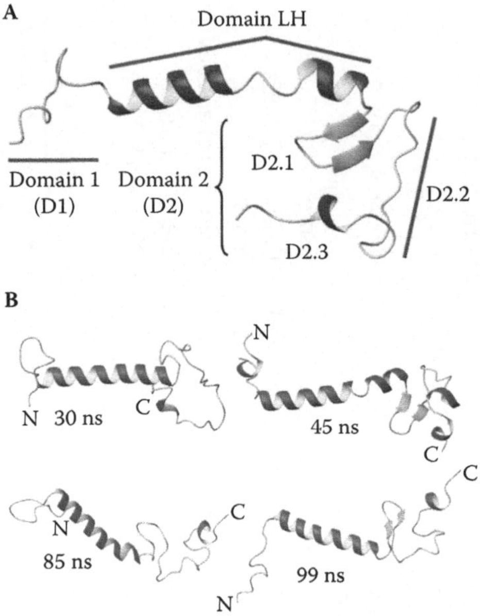

FIGURE 10.3 Structure of the KID domain of p27^{Kip1} in the bound and free states. (A) The structure of p27-KID bound to the Cyclin A-Cdk2 complex (Russo et al. 1996). (B) The structure of unbound p27-KID was characterized by MD simulations restrained by NMR NOE distance values. The MD trajectory from 14 ns to 100 ns was analyzed by the dictionary of protein secondary structure (DSSP) algorithm for secondary structure, and respective samples are shown. Reproduced with permission from Sivakolundu et al. (2005), *J. Mol. Biol.* 353, 1118–28. Copyright by Elsevier Inc.

observable correlations for residues 38–60 (domain LH), residues 62–70 within domain 2 (domain 2.1), 74–80 within domain 2 (domain 2.2), and 86–90 within domain 2 (domain 2.3, see Figure 10.3). To extract details, NOE-constrained MD simulations were carried out with starting coordinates taken from the crystal structure (Russo et al. 1996). The trajectory confirms the previously characterized helical conformation in LH, and also significantly populated locally folded conformations in several other regions, such as a short antiparallel β-sheet in region 62–70, and a transient α-helix at 86–90. These regions, termed intrinsically folded structural units (IFSUs), closely match the structural features of bound p27-KID.

10.2.3.2 CREB KID

CREB is an important transcription factor involved in many key cellular processes (Sands and Palmer 2008). CREB acts in concert with the co-activator CREB-binding protein (CBP or p300), a multidomain protein described in detail in Chapter 11, Section 11.2.2 (Dyson and Wright 2005). CBP/p300 serves as a scaffold for the assembly of the transcriptional machinery and has an ordered domain, KID-binding domain (KIX), for interaction with the disordered kinase-inducible domain (KID) located within the TAD of CREB (Goodman and Smolik 2000). CREB-KID undergoes specific phosphorylation at Ser^{133}, which promotes its interaction with the CBP-KIX in a helix-turn-helix configuration (Radhakrishnan et al. 1997) (for details, see Chapter 6, Section 6.3.1). This interaction is essential for CREB function (Zor et al. 2002).

Heat resistance and CD analysis of the entire TAD of CREB (N-terminal 265 amino acids, also termed Act256) suggested that Act265 is largely disordered (Richards et al. 1996). Protein kinase A (PKA) phosphorylation has only a negligible effect on the CD spectrum, whereas NMR Hα and Cα CSI values and NOE connectivities suggest transient helical structures in the regions αA (120–129) and αB (134–144), which form stable helices in the bound structure (Radhakrishnan et al. 1998) (Figure 6.3).

10.2.3.3 Tau protein

Interest in the microtubule-associated protein tau derives primarily from its involvement in Alzheimer's disease, where it forms aggregates termed paired helical filaments (PHFs) deposited as neurofibrillary tangle in the neurons affected by the disease (see Chapter 15.3.1). Tau contains a C-terminal repeat domain (i.e., TBD), which binds microtubules (MT) and promotes MT assembly. TBD has 31-amino acid long microtubule-binding repeats (MTBRs, R1 through R4) (Buee et al. 2000; Mandelkow et al. 1996). A reasonably full assignment of tau of 441 amino acids could be achieved by a combination of several NMR procedures (see Chapter 6, Section 6.2.4).

Based on this assignment, Cα CSI values suggest a distinct pattern of small but significant deviations from random-coil values in MTBR. Several continuous stretches (containing 7–11 residues) with negative values can be observed, in particular Lys^{274}–Leu^{284} (R1/R2), Ser^{305}–Asp^{315} (R2/R3), and Gln^{336}–Asp^{345} (R3/R4). The values indicate a propensity for β-conformations populated 22%, 25%, and 19% of the time for the three regions. This local structural element is located at the beginning of repeat units R2, R3, and R4, but it lacks from region R1, most likely because of the presence of Pro residues there. This structural interpretation can also be confirmed by additional residual dipolar coupling (RDC), $^{3}J_{HN\alpha}$, and NOE measurements (Mukrasch et al. 2007a), which also suggest stable highly populated β-turn conformational elements immediately following the β-strand regions, and probably also at the end of the repeats as well.

10.2.3.4 Fibronectin-binding protein A

FnBPA is a member of the MSCRAMM (microbial surface components recognizing adhesive matrix molecules) family of proteins (Patti et al. 1994), which mediate adherence to host tissues by the specific recognition of host molecules. MSCRAMMs bind fibronectin, fibrinogen, collagen, and heparin-related extracellular matrix (ECM) proteins, and play important roles in the colonization and invasion of host tissues. The long extracellular segment of *S. aureus* FnBPA contains a 130-amino acid repeat region, Dl–D4, which is highly disordered by NMR (Penkett et al. 1997), but contains transient structural elements, which correlate with regions of binding (Penkett et al. 1998). Hα and amide-N CSI values, $^{3}J_{HN\alpha}$ coupling constant, and NOE data show that the region involved in binding preferentially samples extended conformations. This local structural state allows a number of charged and hydrophobic groups to be exposed and presented to fibronectin for highly specific binding, as shown by the structure of *S. dysgalactiae* FnBPA peptide bound to two tandem Fn1 modules of fibronectin (Schwarz-Linek et al. 2003). This binding mode represents the first example of a tandem β-zipper with one strand of the sheet donated by FnBP, ensuring high affinity and specificity in binding (Schwarz-Linek et al. 2004).

10.2.3.5 α-synuclein

α-synuclein (NACP) is involved in Parkinson's disease and other synucleinopathies (Chapter 15, Section 15.3.2). It is also not fully random, but contains elements of transient local order, most notably a short helical segment in the N-terminal region (residues 18–34) (Bussell and Eliezer 2001; Eliezer et al. 2001). Because this region is involved in membrane-association of the protein, this transient structure may be important for its physiological function and possibly also for its pathological aggregation. It should be noted that long-range structural order was also identified in α-synuclein (see Chapter 10, Section 10.4.2).

10.2.3.6 p53

p53 is a tumor suppressor gene product involved in the regulation of DNA repair and apoptosis (see Chapter 15, Section 15.1.2) (Joerger and Fersht 2008; Levine 1997). Its N-terminal TAD is disordered by CD (Bell et al. 2002), ultraviolet (UV) spectroscopy, gel filtration (GF), and NMR (Dawson et al. 2003), whereas its full structural characterization by a combination of small-angle X-ray scattering (SAXS), NMR, and MD is one of the triumphs of IDP research (see Chapter 4, Section 4.4.3, Chapter 15, Figure 15.2, and cover picture). Thorough NMR analyses based on CSI values, RDC constants, amide-N relaxation rates and NOEs suggested that TAD has a transient α-helix in the region 18–24, which is the site of murine-double minute 2 (MDM2) binding (Kussie et al. 1996), which is the regulatory protein (see Chapter 12, Section 12.6.2.2) that plays a key role in the regulation of p53 (Lee et al. 2000; Vise et al. 2005; Wells et al. 2008).

10.2.3.7 Calpastatin

Calpastatin is the inhibitor of calpain, the calcium-activated intracellular cysteine protease involved in various physiological and pathological processes (Wendt et al. 2004). Full assignment of 121 out of its 126 non-Pro residues (see Chapter 6, Section 6.2.4 and Figure 6.2) enabled its detailed structural characterization in the solution state (Kiss et al. 2008b). CSI values, amide-N relaxation rates, and heteronuclear NOE values indicate that the conserved subdomains A (Ser^{12}–Gly^{30}) and C (Ser^{87}–Cys^{105}) sample partially helical backbone conformations, whereas the primary determinant of inhibition, subdomain B (Met^{50}–Arg^{70}), also has non-fully random local conformational preferences (see Chapter 13, Figure 13.4). As shown by the X-ray structure of the calpain–calpastatin complex (Moldoveanu, Gehring, and Green 2008), the helical regions bind calpain in a calcium-dependent manner, whereas the turn region directly inhibits the enzyme.

10.2.4 Secondary Structure in the Bound State

The structure of IDPs in complex with their partners is known in many cases, which enables their detailed description in the bound state. The distribution of their elements of secondary structure and backbone torsion angles, and also predictability of their local secondary structure have been addressed in several studies, by considering the bound structures of ribosomal proteins, two-state complexes, and experimentally verified IDPs (Gunasekaran, Tsai, and Nussinov 2004; Fuxreiter et al. 2004; Meszaros et al. 2007). In terms of secondary structure, significant differences between IDPs and globular proteins can be found (Figure 10.4). Helices are almost equally populated in the two datasets, but extended structures in IDPs are about 50% less frequent than in globular proteins. In globular proteins, β-strands are mostly located in the hydrophobic core of the protein stabilized as part of sheets, whereas such extended structures are not preferred in IDPs due to entropic reasons. Likewise, as turns are mostly confined to β-hairpins in globular proteins, their population is reduced in IDPs. The most significant difference between the two groups is the increased level of coil conformation in IDPs, which even in their bound states show less regularity than globular proteins.

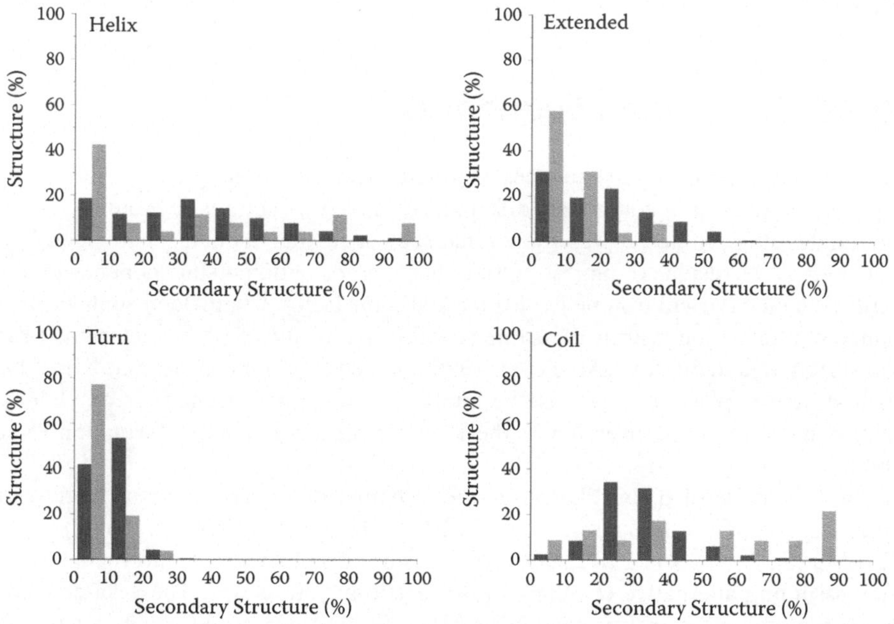

FIGURE 10.4 Secondary structure distribution of IDPs in the bound state. Distribution of the residues of IDPs in complex with their partner (light gray) and those of reference globular proteins (dark gray) in helix, extended, turn, and coil conformations. Reproduced with permission from Fuxreiter et al. (2004), *J. Mol. Biol.* 338, 1015–26. Copyright by Elsevier Inc.

The analysis of the distribution of secondary structures in MoREs/MoRFs (see Chapter 14, Section 14.2.3) in the Protein Data Bank (PDB) (Cheng et al. 2007; Mohan et al. 2006; Oldfield et al. 2005b; Vacic et al. 2007) is in agreement with these findings. Such short binding elements have 27% of their residues in α-helical conformation, 12% in β-strands, and approximately 48% in irregular conformations (13% missing from the atomic coordinates). Major insight comes from the excellent correlation between transient structural elements (Section 10.2.3) and the bound structures, and also from the predictability of the bound structure from sequence (preformed structural elements (PSEs); see Chapter 14, Section 14.2.1). In all, IDPs exploit their local structural preferences in their interactions with their partners to a great extent.

10.3 AMBIGUITY IN STRUCTURE

The concept of disorder is based on a binary classification (i.e., that order or disorder can be unequivocally assigned to a residue). In many cases, the difference is marginal, and slight environmental changes make the region of the protein cross the boundary or assume different structures.

10.3.1 Chameleon Sequences

The concept of chameleon sequences originated from the finding that identical short sequences may exist in completely different secondary structures in proteins (Kabsch and Sander 1984). Later, it was shown that such segments dubbed "chameleons" can be as long as 11 residues, suggesting that their conformational state depends on context (i.e., molecular environment) (Minor and Kim 1996). Chameleon sequences can change conformation within the same protein as a result of point mutations (Yang et al. 1998), ligand binding (Abel et al. 1996), or a change in pH (Carr and Kim 1993). Their structural plasticity may also be implicated in the pathogenesis of amyloidoses when helix-to-strand conversion is the key step in amyloid fiber formation (Kelly 1998).

In a database of 6,962 PDB structures, chameleon sequences were found to be rather frequent (Guo, Jaromczyk, and Xu 2007): 48,194 (4 residues in length), 24,144 (5), 1,519 (6), 56 (7), and 2 (8). Such sequences occur in different conformations, most often assuming alternative α-helix and β-sheet/strand structures. These latter type of chameleon sequences are enriched in Val, Leu, Ile, and Ala residues, which have nonpolar, aliphatic side-chains. Interestingly, Ala and Leu have strong helical propensity, whereas Val and Ile have strong β-strand propensity, which probably explains the structural ambiguity of these regions. Whereas the alternative conformational states of chameleon sequences are both structured, their adaptability is probably linked with that of dual-personality sequences.

10.3.2 Dual-Personality Sequences

Context-dependence of structure manifests itself in an even more subtle way in the case of protein fragments termed "dual personality" (DP) (Zhang, Stec, and Godzik 2007). Zhang and colleagues found that 3,412 of the 5,859 different structural groups (clusters) in the PDB have more than one structure with identical sequences and in 1,535 of them a segment occurs in both ordered and disordered conformational states. The vast majority of these DP segments (92.3%) are shorter than 10 amino acids, but occasionally they can be much longer, up to the limit of 71 amino acids. The occurrence of different secondary structural elements in the ordered form of DPs is significantly different from PDB averages (α-helix: 20% vs. 35%, β-strand: 7% vs. 24%, β-turn: 27% vs. 21%, and irregular: 46% vs. 20%), which is reminiscent of the values characteristic of IDPs. 50% of them are located immediately next to an IDR, and their amino acid frequencies fall between the composition of ordered and disordered proteins, with certain amino acids (e.g., Ala) being closer to that of ordered, others (e.g., mostly hydrophobic) closer to that of disordered regions, whereas some (e.g., Ser) are positioned halfway between the two. The prominence of six particular amino acids (Asp, Thr, Gly, Asn, Pro, and Arg) suggests that DP regions are different from both ordered and disordered proteins and define a distinct structural category.

70% of DP regions were found to be involved in post-translational modification (PTM), and a PTM site is 20% more likely to be within five residues from a DP segment than a disordered fragment and three times more likely to be in a DP segment than in a continuous ordered fragment. The structural and functional kinship of DPs to eukaryotic linear motifs (ELMs) and molecular recognition features (MoRFs) is emphasized by the ability of a DP to undergo disorder-to-order transition in an intramolecular fashion, a capacity reminiscent of what IDPs do in the presence of their cognate partner (Dunker 2007).

10.3.3 The Twilight Zone between Order and Disorder

A somewhat neglected point with respect to structural disorder is its context-dependence (i.e., that a correspondence between sequence and the lack of structure is not unequivocal due to the influence of other regions of the protein). The recognition of this fact motivated the development of predictors, such as VSL2 (Peng et al. 2006), which predict disorder of long and short segments separately (see Chapter 9, Section 9.5).

Definition of disorder of short IDRs is more limited, as demonstrated by analyzing the overlap in the distributions of ordered and disordered proteins/segments in the 20-D space of amino acid composition (Szilagyi, Gyorffy, and Zavodszky 2008). The overlap is much larger for short segments than for long ones. Because amino acid composition determines disorder primarily by virtue of simple physicochemical features (Uversky et al. 2000a; Weathers et al. 2004; Williams et al. 2001), this distinction is also apparent when the 20-D space is projected onto two dimensions of amino acid features: hydrophobicity and charge. The overlap is much reduced in the case of longer sequences. About 90% of sequences fall into the "twilight zone" of dubious identity for chains less than 50 amino acids in length, but only 25% for chains longer than 300 amino acids.

10.4 TERTIARY STRUCTURE: GLOBAL FEATURES OF IDP STRUCTURES

The global description of IDPs usually applies a single global parameter and leaves residue-level resolution out of consideration. Hydrodynamic and spectroscopic techniques can both provide such data. Our description of these states draws heavily on analogies with the unfolded states of globular proteins, as reviewed in (Bright, Woolf, and Hoh 2001) and detailed in Chapter 1, Section 1.7.

10.4.1 Hydrodynamic Description

The unusual hydrodynamic behavior of potential IDPs is often the first indication of their disorder. Several techniques can provide parameters such as R_G, R_S, and R_H (Chapter 4), which are often simply used to characterize the disordered character of the protein. Other times, they are converted to an apparent M_W. A few examples are discussed in some detail, and several more can be found in Chapter 4.

The observed R_G of the binding region D1–D4 of FnBP approached by pulsed-field gradient (PFG) NMR (Wilkins et al. 1999) is 26.2 Å, which is significantly larger than the R_G of a globular protein of similar number of residues, such as ribonuclease A (RNase A)(15.0 Å) or hen lysozyme (15.3 Å). On average, D1–D4 is almost 75% larger than expected for a protein of compact fold. A very similar approach was used to characterize the hydrodynamic radius of α-synuclein (Morar et al. 2001). The R_H determined (18.8 Å) is larger than that calculated for a globular state (17.7–18.3 Å), but smaller than that of the corresponding random-coil state (29.8–31.6 Å). Apparently, the protein is disordered but much more compact than a fully random state. GF and SAXS were combined for studying the hydrodynamic behavior of the segments corresponding to the carboxy-terminal 136 amino acids (CaD136) of caldesmon (Permyakov et al. 2003). Its R_S is much larger (R_S = 28.1 Å) than that expected for a globular protein (19.1 Å), and it slightly changes in the presence of 6M Gnd-HCl (35.3 Å). The R_S is closest to that expected for the pre-molten globule (PMG) state (27.4 Å), suggesting that CaD136 belongs to the class of native PMGs. These observations were corroborated by SAXS, suggesting an R_G value (40.8 Å) significantly smaller than that estimated for a random coil of a protein of this size (51.9 Å). These and other data suggest that typical hydrodynamic radii 1.5–2.0 times their expected value (4–6 times in terms of M_W, see Chapter 4, Section 4.1) are characteristic of fully disordered (random coil–like) IDPs, whereas values below this threshold are more indicative of molten globule (MG)–like states.

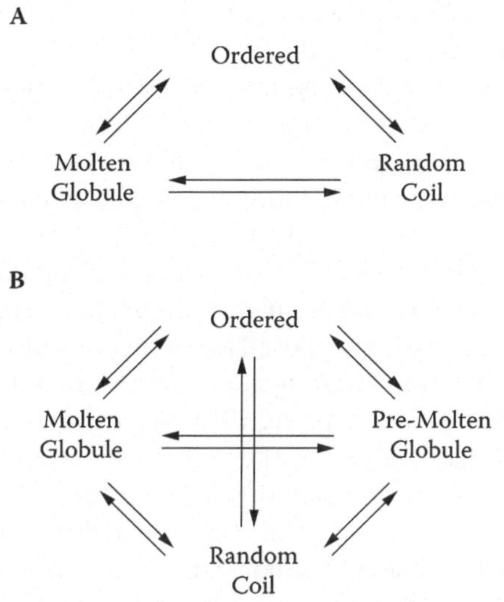

FIGURE 10.5 The protein trinity and protein quartet models of the structure-function relationship. These models provide a conceptual framework to extend the classical structure-function paradigm by suggesting that different proteins can exist in any of three (random coil, MG, and globular) or four (random coil, PMG, MG, and globular) structural states. Function can arise from any of the states or transitions between them. Reproduced with permission from Dunker et al. (2001), *J. Mol. Graphics Modelling* 19, 26–59, copyright by Elsevier Inc., and Uversky (2002) *Protein Sci.* 11, 739–56, copyright by the Protein Society.

The hydrodynamic behavior of IDPs was systematically analyzed on a collection of 91 experimentally verified IDPs, 35 of which has data on the hydrodynamic volume, V_H (Uversky 2002a). The data plotted as a function of the number of residues compared to similar data for various unfolded states (Native, MG, PMG, and random coil) of globular proteins show that IDPs tend to segregate into two categories, PMG-like and random coil-like, whereas MG-like IDPs are not represented in the data. Such data have led to the formulation of general models that attempt to provide a framework for the extension of the structure-function paradigm to disordered proteins (Figure 10.5). The first model of "protein trinity" (Dunker et al. 2001) suggests that proteins under native conditions can exist in three principal forms: folded globular, MG, and unfolded (random coil-like), and functions can arise from any of the three forms and from transitions between them. This concept was extended to the "protein quartet" model by suggesting the inclusion of a fourth separate thermodynamic state, PMG (Uversky 2002a).

10.4.2 Spectroscopic Approaches

Global characteristics of the structural ensemble of IDPs can also be addressed by spectroscopic techniques. An elegant example is tau protein (see Chapter 15, Section 15.3.1.2), for which a combination of fluorescence resonance energy transfer (FRET) and electron paramagnetic resonance (EPR) was used to elucidate the spatial relationship of domains and global folding state (Jeganathan et al. 2006). FRET pairs were created by engineered Trp residues (donors) and covalently attached IAEDANS molecules (acceptors). The observed FRET distances were found to significantly differ from the values expected for a random coil (Figure 10.6), suggesting that tau folds back so that its C-terminal end is in the vicinity of TBD, whereas the N-terminus remains outside the FRET distance of TBD, yet it approaches the other end of the molecule. The average distance between the C-terminal end and the TBD is about 19–23 Å, which is significantly shorter than the value predicted for the random coil ensemble (87.7–99.3 Å). The two ends of the molecule are about 20.8–24.2 Å apart, as opposed to the theoretical value of 170 Å. Gnd-HCl abolishes these interactions, corroborating non-random structural features in tau. FRET between green fluorescent protein pairs (GFPs)—CyPet and YPet—was also used to estimate the end-to-end distributions of IDPs of various length, such as the charged-plus-PQ domain of ZipA, the tail domain of α-adducin, and the C-terminal tail domain of FtsZ (Ohashi et al. 2007). Constructs of similar length give different FRET intensities. For example, the N-terminal 33 amino acids of the charged domain of ZipA gives strong FRET signals, whereas the C-terminal 33 amino acids of the PQ domain of ZipA gives only moderate FRET signals. Thus, variations of the donor–acceptor distance calculated from FRET efficiency suggest variable stiffness (persistence length) abolished by 6M urea (i.e., the structural ensemble is more compact than a random coil).

Hydrodynamic measures can also be obtained by MD simulations, which are constrained by distance restraints derived from long-range paramagnetic resonance enhancement (PRE) NOEs (Dedmon et al. 2005). This approach suggests that the R_G probability distribution of α-synuclein has a mean R_G of 24.7 Å and an average R_H of

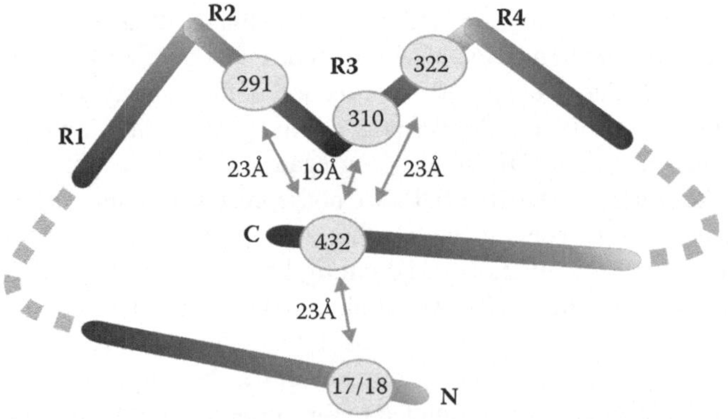

FIGURE 10.6 Global hairpin folding of tau in solution. Long-range intramolecular interactions within tau protein were assessed by FRET. The tubulin-binding repeats within TBD are marked R1 through R4. The positions of residues to which fluorescent labels were attached are indicated by the numbers in ovals. Major features of the global fold are that the C-terminus folds in the vicinity of the repeat domain, and it is also within FRET distance from the N-terminus, whereas this latter stays away from the repeats. Reproduced with permission from Jeganathan et al. (2006), *Biochemistry* 45, 2283–93. Copyright by the American Chemical Society.

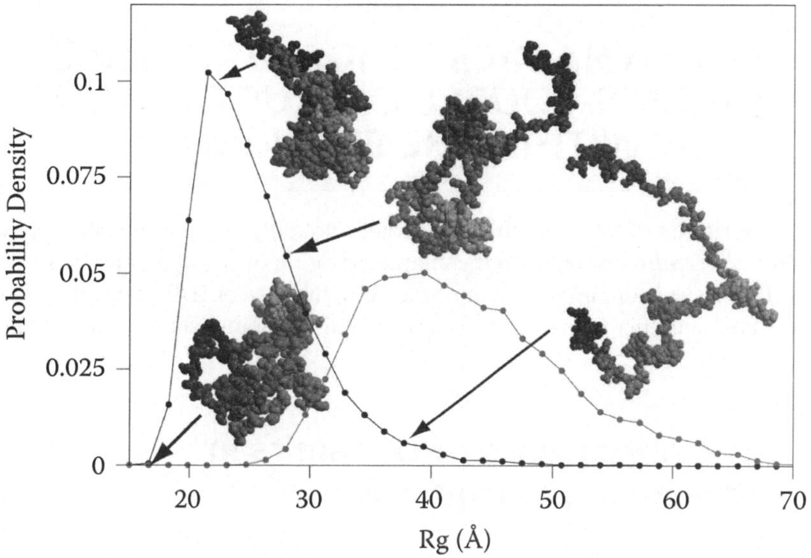

FIGURE 10.7 Structural ensemble of α-synuclein. Radius of gyration (R_G) probability distributions were calculated by MD simulations incorporating PRE-derived distance restraints for native (black trace) and random coil (gray trace) models of α-synuclein. Representative structures are indicated with arrows pointing to their corresponding R_G values. Reproduced with permission from Dedmon et al. (2005), *J. Am. Chem. Soc.* 127, 476–7. Copyright by the American Chemical Society.

27.2 Å, which closely matches the R_H value of 26.6 Å determined experimentally by PFG NMR (Morar et al. 2001). Thus, α-synuclein is represented by an ensemble more compact than expected for a random coil (Figure 10.7). A residual contact map generated from the simulated structures indicates that the primary cause of compactness is the interaction between the amyloidogenic NAC region and the acidic C-terminal region.

10.4.3 Global Structure: Is It Related to the Structure in the Bound State?

The test of the descriptive value of global models is the actual functional insight they provide. Whereas local structural preferences strongly correlate with regions of recognition function in IDPs (see Section 10.2.3 and Section 10.2.4), it is not clear if the same is true with respect to their global (tertiary) structure. So far, global topological similarities between the denatured states and folded structures of globular proteins have been reported, such as in the case of denatured *Staphylococcal* nuclease based on measuring NMR RDC values of oriented samples (Shortle and Ackerman 2001). The observed RDC values suggest a native-like topology (i.e., spatial positioning and orientation of chain segments), which persist in the protein denatured by 8M urea. Evidence is missing, however, that similar long-range topological order resembling the bound structure also applies to IDPs.

10.5 DYNAMICS OF IDP STRUCTURE: THE TIME-COURSE OF FLUCTUATIONS WITHIN THE ENSEMBLE

The full description of the structure of a protein not only involves a static representation of atomic coordinates but also the characterization of the dynamics of its internal motions. This issue is of paramount importance in the case of IDPs, because transitions between a large number of free conformations and the bound state is the very essence of their function.

10.5.1 The Importance of Dynamics in Structural Descriptions

As detailed in Chapter 5 and Chapter 6, many of the structural techniques applied for characterizing steady-state structures, most notably NMR, fluorescence polarization, fluorescence correlation spectroscopy (FCS), FRET, and EPR spectroscopy, also provide information on the dynamics of internal motions of IDPs. Basically, these motions manifest themselves at three different levels: dynamics of side-chains, local backbone motions, and tertiary rearrangements. Description of these motions extends the static structural picture in four different ways. First, increased flexibility may simply be taken as an indication of local disorder of the chain. Second, increased local dynamics in an otherwise ordered protein may mark the site of function. Third, locally decreased flexibility within a random coil-like region may point to a transient local structural element, which is also often implicated in function. Fourth, a significant decrease in dynamics in the presence of a partner may suggest a local transition to the bound state. Quantitative description of the timescale of motions is available only in a few cases, whereas a correlation of local structural state and dynamics is often seen in NMR (Table 10.1).

10.5.1.1 Local/segmental motions

Local motions at the residue/backbone level in several IDPs occur in the ns time regime. For example, fast local reorientation motions can be observed in the case of PKIα (Hauer et al. 1999b), where time-resolved fluorescence anisotropy decay measurements show local backbone fluctuations with time-constants in the range of 0.8–1.4 ns. Analysis of NMR relaxation data of FlgM by the model-free approach suggested an average local correlation time 3.6 ns within the fully disordered N-terminal half and about 6.3 ns in the C-terminal half, which has transient helical regions (Daughdrill et al. 1998). A combination of NMR and MD suggested local segmental motions in $p27^{Kip1}$ (Sivakolundu et al. 2005), which correlate with local transient structural elements (Figure 10.3) and occur on the high ps–low ns timescale. An EPR study of the internal dynamics of tau protein (Jeganathan et al. 2006) showed high local mobility, with effective rotational correlation times of 0.2–0.6 ns. In the case of an unfolded globular protein, denatured barstar (Saxena et al. 2006) fluorescence anisotropy decay analysis showed side-chain dynamics of 0.2–0.4 ns time-constants and somewhat slower local segmental motions on the order of 1–3 ns.

10.5.1.2 Restricted segmental motions

Large-scale segmental and/or tertiary-type of motions may occur on the ns-μs timescale. α-synuclein assumes conformations that are stabilized by long-range interactions between the central NAC region and the C-terminal end. PRE and NMR dipolar couplings show that these autoinhibitory conformations fluctuate on the ns–μs timescale, corresponding to that of secondary structure formation during folding (Bertoncini et al. 2005). Similar observations were made in the case of the NM region of the yeast prion Sup35 (Mukhopadhyay et al. 2007) (see also Chapter 5, Section 5.2.5). N is the amyloidogenic region of Sup35, whereas M is a charged region that keeps N in solution, and FCS analysis of the quenching by internal Tyr residues suggests long-range conformational fluctuations in the 150–250 ns and faster short-range fluctuations with typical time-constants of 20–40 ns. The chloride channel CIC-0 contains a 96-residue disordered linker region connecting its structured subdomains (Alioth et al. 2007). Large-scale interconversions in the structural ensemble of the protein occur in the ns-low μs range. Internal motions on the μs–ms timescale were also observed in the nuclear-receptor co-activator-binding domain (NCBD) of CBP (Ebert et al. 2008), which was taken to indicate the MG state of this disordered domain. Segmental motions were also suggested to fall on the μs–ms timescale by spectral density function analysis of NMR relaxation data of MBP, which suggests the formation of local secondary structural elements (Libich and Harauz 2008).

10.5.1.3 Reduced local motion signals transient structural elements

Local dynamics is very sensitive to the presence of a region of restricted mobility within an unrestricted random coil-like region, which often turns out to correspond to the binding site of IDPs (Table 10.1). For example, reduced relaxation rates and increased NOE values point to restricted local dynamics, which is indicative of local residual structure in region 18–34 of α-synuclein (Bussell and Eliezer 2001). Increased NOE and decreased R values indicate restricted local motions due to transient ordering in the N-terminal domain of histone messenger RNA (mRNA) stem-loop binding protein (SLBP) (Thapar et al. 2004). Transient helices in this region are probably involved in multiple interactions of the protein. Similar observations were made in the C-terminal half of the anti-sigma factor FlgM (Daughdrill et al. 1998). This region is known to assume transient helices that undergo further disorder-to-order transition upon the binding of σ^{28} (see Figure 8.4) (Sorenson et al. 2004). Reduced relaxation dynamics can also be seen in the N-terminal 20 amino acids of hRPA70, which assumes transient helical conformations (Olson et al. 2005).

10.5.2 A Reduction in Motility Signals Disorder-to-Order Transition

The local reduction of dynamics in the presence of a binding partner indicates folding induced upon binding. For example, a decrease in R_2 and an increase in NOE indicate reduction of internal mobility upon binding of activator for thyroid hormone and retinoid receptor (ACTR) to CBP (Ebert et al. 2008). The disordered loop region between helices II and III of lac repressor headpiece (HP) domain experiences a significant decrease in mobility upon DNA binding, as shown by relaxation and NOE measurements (Slijper et al. 1997). The large decrease in amide-N relaxation rates in the region 42–56 of the TAD of p53 indicate its binding to human replication protein A (Vise et al. 2005). A large positive shift in NOE values upon binding of Tβ4 to G-actin indicates that Tβ4 adopts a fully bound state, but somewhat smaller values in the center of the molecule (residues 20–25) indicate the retention of some internal flexibility in this region (Domanski et al. 2004).

10.6 A READOUT OF STRUCTURE: THE HYDRATE LAYER OF IDPS

The function of IDPs is often carried out by molecular recognition, in which changes in surface structure and hydration inevitably take place. In this sense, the hydrate layer is in intimate relationship with the structure and function of IDPs, and one may speak of, besides a structure-function paradigm, a structure-surface paradigm and an interface-function paradigm as well. In this spirit, systematic studies for quantifying the hydrate layer of several IDPs (Bokor et al. 2005; Csizmok et al. 2005; Kovacs et al. 2008; Tompa et al. 2006a) were carried out. Studies on calpastatin, MAP2c, and two plant dehydrins (ERD10 and 14) show that the hydrate layer of typical IDPs differs from that of globular proteins in several aspects. IDPs bind significantly more water (about 1.5–2 times) than globular proteins. Their hydrate layer is more heterogeneous from both the energetic and dynamic point of view, which suggests a greater variety of local chemical microenvironments on their surface. By their correlation times, water molecules bound to IDPs move faster, which suggests a faster rotational reorientation of their hydrate layer. Short- and/or long-range structural organization and deviation from a random-coil state are also signaled by suboptimal hydration of certain IDPs, such as calpastatin. The functional connections of these results are often not fully clear, but in the case of dehydrins it was suggested that their large hydration capacity may be critical in their function as a protective hydrate buffer shifting the osmotic balance of drying cells (Bokor et al. 2005; Kovacs et al. 2008; Tompa et al. 2006a).

11 Biological Processes Enriched in Disorder

This chapter presents an outline of the classification of the functional information available on intrinsically disordered proteins (IDPs) from the perspective of the Gene Ontology (GO) classification system. GO encompasses three separate ontologies: biological process (BP), molecular function (MF), and cellular localization (CL), which cover distinct aspects of functional information, such as the cellular process the protein takes part in (BP), the mode of its action at the molecular level (MF), and its ultrastructural localization within the cell (CL). Basically, disorder is related to almost all biological processes, of which the examples discussed in this chapter highlight the most notable generalities.

11.1 BIOLOGICAL FUNCTIONS ENRICHED IN DISORDER

The biological processes and molecular functions associated with protein disorder have been comprehensively addressed in several studies (Table 11.1). Ward and colleagues (Ward et al. 2004) applied DISOPRED2 to predict the frequency of proteins with intrinsically disordered regions (IDRs) ≥30 consecutive residues in 24 genomes from the three kingdoms of life and also the involvement of proteins in different GO annotations in yeast. Tompa and colleagues (Tompa, Dosztanyi, and Simon 2006b) combined the results of IUPred and PONDR® VSL1, and compared the proteomes of *E. coli* and yeast by several criteria, such as the percentage of all disordered residues, the percentage of fully disordered proteins, and proteins with IDRs ≥30 consecutive residues. Dunker and colleagues (Xie et al. 2007) predicted disorder in SwissProt entries and looked for keywords in their functional annotation record that showed statistically significant associations (238 out of 710 Swiss-Prot functional keywords). This study extended previous work, in which disorder in 12

TABLE 11.1 Biological process and molecular function ontologies enriched with disorder*

BP (JONES)	*BP (TOMPA)*	*SP FUNCTION (DUNKER)*	*MF (JONES)*
Ty transposition	Pseudohyphal growth	Differentiation	Transcription regulation
Development	Transcription	Transcription	Protein kinase
Morphogenesis	Morphogenesis	Transcription regulation	Transcription factor
Protein phosphorylation	Conjugation	Spermatogenesis	Binding
Regulation of transcription	Cell cycle/ cytokinesis	DNA condensation	DNA binding
Transcription, DNA-dependent	Meiosis	Cell cycle	Nucleic acid binding
DNA packaging	Signal transduction	mRNA processing	RNA polymerase II
Signal transduction	Ribosome biogenesis and assembly	mRNA splicing	Kinase activity
Actin cytoskeleton	Cytoskeleton o/b	Mitosis	Enzyme regulator
Pseudohyphal growth	Sporulation	Apoptosis	Cytoskeletal binding
Chromosome o/b	DNA metabolism	Protein transport	RNA binding
DNA recombination	Nuclear o/b	Meiosis	Signal transducer
Cytoskeleton o/b	Cell budding	Cell division	Intracellular transport
Epigenetic regulation of gene expression	Cell wall o/b	Ubl conjugation pathway	Carbohydrate transport
Gene silencing	RNA metabolism	Wnt signaling pathway	Nucleotide binding

* Top BP and MF categories in GO significantly enriched with protein disorder, as suggested in three bioinformatic studies. Jones and colleagues (Jones et al. 2004) used DISOPRED2, whereas Tompa and colleagues (Tompa et al. 2006b) used IUPred to estimate the frequency of disorder in proteins in various GO categories. Dunker and colleagues (Xie et al. 2007) used PONDR® VL3E to estimate long disorder in SwissProt (SP) proteins and looked for keywords in their functional annotation record. The categories are listed in the order of decreasing significance of the correlation. o/b stands for *organization* and *biogenesis*.

different functional categories, either related to BP or MF, was directly assessed (Iakoucheva et al. 2002). The bioinformatic analyses are complemented by statistically less rigorous studies, in which a census of functional annotations of IDPs (i.e., MF terms) (e.g., protein–protein binding, protein-DNA binding, metal binding, phosphorylation), and occasionally BP terms (regulation of proteolysis) is given (Dunker et al. 2002). Often, exclusive functional attributes of IDPs not yet incorporated into GO (e.g., flexible linker/spacer, entropic spring, protein detergent; see also Chapter 12) are included in these classification schemes.

These studies agree that disorder is significantly enriched in five large functional areas (see Table 11.1). The strongest correlations are found with the following:

1. Transcription and transcription regulation, also apparent in molecular functions such as deoxyribonucleic acid (DNA) binding and nucleotide binding
2. Signal transduction and the regulation of cell cycle
3. The biogenesis and functioning of nucleic acid containing organelles, such as the ribosome and the chromatin
4. Messenger ribonucleic acid (mRNA) processing, which includes splicing reactions
5. The organization and biogenesis of cytoskeleton, not independent of the execution of cell cycle

The studies also agree on the functional categories that show a characteristic depletion in disorder, these are usually the ones that require enzymatic and/or ligand-binding activities. BP categories such as biosynthesis, metabolism, respiration and energy pathways, and MF categories, such as oxidoreductase, catalytic, ligase, liase, and structural molecule, are noted most often.

11.2 DISORDER IN TRANSCRIPTION/ TRANSCRIPTION REGULATION

A high level of structural disorder in proteins involved in the regulation of transcription is consistently noted in the IDP field (Dunker et al. 2000; Iakoucheva et al. 2002; Sigler 1988; Ward et al. 2004). The correlation is apparent at three mechanistically intertwined functional levels, such as (1) transcription factors, (2) transcription co-activators, and (3) general transcription factors. Proteins involved in chromatin organization, although also directly involved in the regulation of transcription, are discussed in Section 11.4.2, among nucleic acid containing organelles.

11.2.1 Transcription Factors

Transcription factors activate or repress transcription by recognizing specific DNA sequences (enhancer/silencer regions and the promoter of the gene). They have modular architecture in which a DNA binding domain (DBD) binds DNA, whereas one or two trans-activator domains (TADs) engage in protein–protein interactions with co-activators and/or members of the general transcription machinery. These interactions result in the assembly of the pre-initiation complex (PIC). Transcription factors were among the first functional class of proteins where the functional role of structural disorder was noted. Many observations, such as swapping of TADs between transcription factors and/or their replacement with random sequences without significant loss of activity pointed to rather ill-defined structural requirements in initiating RNA polymerase II (RNAP II) transcription. These observations and the resistance to crystallization led Sigler to suggest that TADs function without the usual structural constraints of ordered proteins (i.e., as acid "blobs" or negative "noodles") (Sigler 1988) (see also Chapter 2). This bold suggestion has since received many lines of indirect evidence from the functional indifference of TADs to replacement, deletion or scrambling mutations (Gerber et al. 1994; Ng et al. 2007), and also direct structural verification, such as in the case of p53 (Bell et al. 2002; Dawson et al. 2003; Vise et al. 2005), c-Fos (Campbell et al. 2000), the androgen receptor (AR) (Kumar et al. 2004), cyclic-AMP response element-binding protein (CREB) (Richards et al. 1996; Zor et al. 2002), and nuclear receptors (McEwan et al. 2007), for example.

The prevalence of disorder in transcription factors is corroborated by bioinformatic analyses, which suggest that 82.6–94.1% of them possess extended regions of intrinsic disorder (Liu et al. 2006a). The level of predicted disorder is significantly higher in TADs (84.2%) than in DBDs (30.7%), with TADs always being very disordered (their disorder varies between 73.3% and 94.5%), whereas disorder in DBDs ranges from very low (around 10%, e.g., RING-type, PHD-type, fork-head, and T-box DBDs) through intermediate (around 50%, e.g., homeobox, high-mobility group (HMG) box, and AP2/ERF DBDs) to very high (above 80%, e.g., AT-hook and basic motifs) levels. The degree of disorder is significantly higher in eukaryotic transcription factors than in prokaryotic transcription factors. In agreement, transcriptional regulatory proteins are about two times longer than their prokaryotic counterparts, and only 31% of their sequences, as opposed to 72% in bacterial transcription factors, can be aligned to known domains (Minezaki et al. 2006). In addition, 49% of human transcription factor sequences are predicted to be disordered, and they often consist of a small DBD and long IDRs, frequently flanking unassigned regions.

11.2.2 Transcription Co-Activators

Complex modular proteins and multi-protein complexes communicate signals from transcription factors bound to enhancer and repressor sequences toward the core transcription machinery. Such co-activators have the capacity to interact with a variety of regulatory proteins, general transcription factors and RNAP II, and they are also involved in modifying chromatin structure. One of the best characterized co-activator is CBP/p300, which affects chromatin structure due to its intrinsic histone acetyltransferase (HAT) activity and serves as a scaffold for the assembly of the transcription machinery (Goodman and Smolik 2000). Approximately half of its 2,442 residues are found in disordered regions, including the nuclear coactivator binding domain (NCBD) domain and linkers between six folded domains (Dyson and Wright 2005). The six globular domains (Figure 11.1) serve as templates for the induced folding of disordered regions of many transcription factors, such as TAZ1 for HIF-1α (Dames et al. 2002), CITED2 for p53 C-terminal domain (CTD) (De Guzman et al. 2004), TAZ2 for E1A (Dyson and Wright 2005), KID-binding domain (KIX) for phosphorylated cyclic-AMP response element-binding protein kinase-inducible domain (CREB KID) (Chapter 6, Figure 6.3 [Radhakrishnan et al. 1997]), bromodomain for acetylated p53 (Figure 12.3 Chapter 12, [Mujtaba et al. 2004]). The disordered NCBD domain of CBP also provides an example of co-folding or mutual synergistic folding upon binding to the disordered activator for thyroid hormone and retinoid receptors

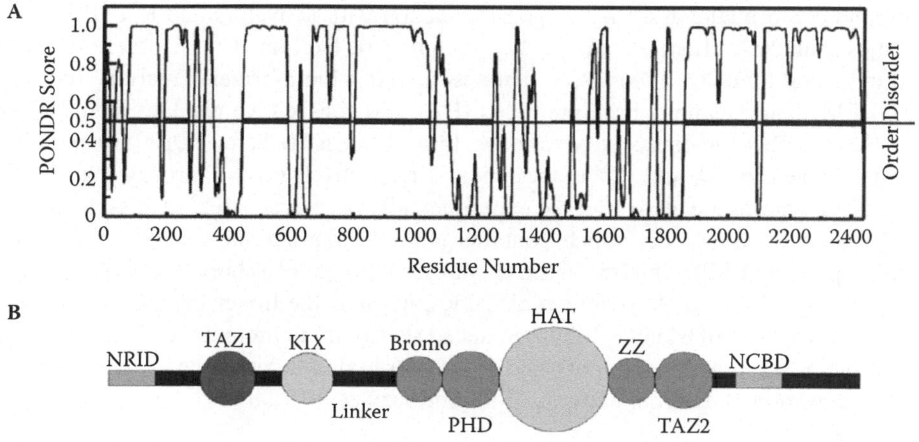

FIGURE 11.1 Domain structure and predicted disorder of CREB-binding protein. Predicted disorder (A) and schematic domain representation (B) of CREB-binding protein CBP. Known structured domains are represented by circles, whereas uncharacterized linker regions connecting domains and the disordered nuclear-receptor co-activator-binding domain (NCBD) and nuclear-receptor-interaction domain (NRID) are represented by connecting lines. Ordered transcriptional-adaptor zinc-finger-1/2 (TAZ1/2), KID-binding domain (KIX, see also Chapter 6, Figure 6.3), bromo domain (Bromo, see also Chapter 12, Figure 12.3), histone acetyltransferase domain (HAT), plant homeodomain (PHD), and zinc-binding domain (ZZ) are shown. Most regions between the ordered domains are predicted disordered. Reproduced with permission from Dyson and Wright (2005), *Nat. Rev. Mol. Cell Biol.* 6, 197–208. Copyright by the Nature Publishing Group.

(ACTR) domain of p160 co-activator (Chapter 14, Figure 14.3 (Demarest et al. 2002)). The long linker regions connecting ordered domains enable conformational flexibility required for CBP function, but they also serve as posttranslational modification sites (e.g., for SUMO-ylation (Girdwood and Specified 2003)) and harbor eukaryotic linear motifs (ELMs) binding regulatory proteins (e.g., three LXXLL motifs for steroid and retinoid receptors (Heery et al. 2001)). Due to the extreme complexity of signaling/binding functions, CBP/p300 is denoted as a "molecular interpreter" of the "words," "phrases," and "sentences" represented by different combinations of regulatory signals (Smith 2004), in which adaptability enabled by structural disorder might be indispensable.

The remodeling of chromatin and the assembly of PIC are also coordinated through the interplay between CBP/p300 and the Mediator complex (Black et al. 2006), a large multi-subunit assembly comprising about 25 components (Kornberg 2005). The two co-activators act synergistically, but CBP/p300 compete with the general transcription factor TFIID for binding to Mediator at the promoter. Mediator has three large modules, which appear to be specialized for certain functions (Asturias et al. 1999; Myers et al. 1999). The Head module mediates the interaction with RNAP II and other components of the basal transcription machinery (Takagi et al. 2006), the Middle module is involved in mediating repression signals and making contacts with the dissociable Cdk complex, whereas the Tail module is targeted by a number of regulatory proteins (Myers et al. 1999). Cryo-electron microscopy (EM) studies show that the Mediator undergoes profound conformational changes upon interaction with activators and RNAP II CTD (Taatjes et al. 2002; Taatjes et al. 2004), which may be facilitated by disordered regions within several Mediator subunits, as demonstrated in a bioinformatic analysis. In yeast, 80% of Mediator subunits have predicted IDRs ≥30 consecutive residues, and 24% of them have IDRs ≥100 consecutive residues. In the human Mediator, IDRs ≥30 and ≥100 residues appear in 75% and 32% of the subunits, respectively (Toth-Petroczy et al. 2008). Disordered regions are also observed experimentally in several cases, such as the Med8/Med18/Med20 submodule, which contains multiple binding sites for the TATA-box binding protein (TBP) complex. In the crystal structure, only a short α-helical region of Med8 can be observed (Lariviere et al. 2006), whereas the linker between the C- and N-terminal regions of Med8 exhibits enhanced sensitivity to proteolytic digestion in the free protein. The functional importance of disorder is also underscored by the evolutionary conservation of disorder patterns (Toth-Petroczy et al. 2008).

11.2.3 Disorder in the Core Apparatus

The transcription of protein-coding genes in eukaryotes is carried out by the multisubunit complex RNAP II. The CTD of its largest subunit contains 25–52 tandem repeats of variants of the sequence element YSPTSPS, which is missing from the X-ray structure of the complex (Figure 11.2, (Cramer, Bushnell, and Kornberg 2001)), and by several biophysical methods is highly disordered (Bienkiewicz et al. 2000). The CTD generated by repeat expansion in evolution (Chapter 13, Section 13.3.1.2) serves as a scaffold for the highly orchestrated assembly of a range of complexes involved in the initiation, elongation, and termination of transcription, linking these steps to mRNA maturation (Proudfoot, Furger, and Dye 2002). Central to its function is its structural malleability shown by its adaptability to multiple partners, such as the CID domain of the 3' RNA-processing factor Pcf11 (Meinhart and Cramer 2004), the nucleotidyl transferase domain of RNA guanylyltransferase Cgt1 (Fabrega et al. 2003), or the CTD phosphatase Scp1 (Zhang et al. 2006). The function of CTD is tightly regulated by phosphorylation (Proudfoot et al. 2002), which signals its transitions between states compatible with distinct phases of transcription. Biophyscial studies (Bienkiewicz, Woody, and Woody 2000) suggest the preponderance of PPII conformation in CTD, in line with its adaptability to a variety of partners.

Transcription by RNAP II is aided by a set of general transcription factors (GTFs TFIIA, B, D, E, F, and H), in which disordered regions have also been observed. For example, the globular CTD in TFIIB that contacts TBP at the TATA box is connected to the N-terminal RNAP II-interacting region by a linker that is disordered in solution. Part of this linker folds into a "B finger" upon interacting with RNAP II, reaching into the active site of the enzyme and playing a crucial role in determining the transcription start site (Bushnell et al. 2004). The rest of the linker remains disordered even in the presence of RNAP II, and might allow for different modes of interaction between the C-terminal portion of TFIIB and the polymerase (Chen and Hahn 2004). Cryo-EM studies of the complex of yeast RNAP II and TFIIF show scattered density of TFIIF around the RNAP II active site cleft, with a considerable fraction of the factor appearing disordered (Chung et al. 2003). The central region of the TFIIF subunit TFg1 is highly

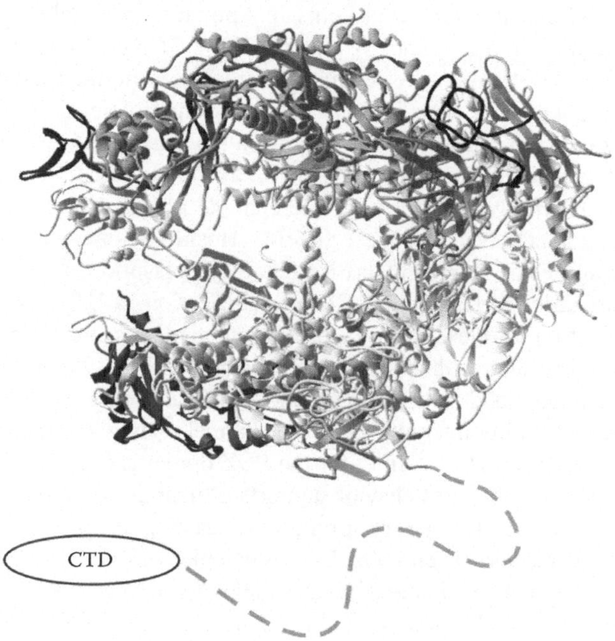

FIGURE 11.2 X-ray structure of eukaryotic RNA polymerase II complex. The structure of the 10-subunit yeast RNA polymerase II (pdb 1i50) has been solved at 2.8 Å resolution by X-ray crystallography (Cramer et al. 2001). The repetitive CTD of the largest subunit, Rpb1, is missing from the structure due to its structural disorder and is marked by a dashed line.

charged and is extremely sensitive to proteolysis, also suggesting an exposed and probably disordered structure (Yong et al. 1998).

11.3 DISORDER IN SIGNALING PROTEINS

Several bioinformatic studies (Iakoucheva et al. 2002; Ward et al. 2004; Xie et al. 2007) suggest very high levels of disorder in proteins of regulatory and signaling functions. In a general sense, signal transduction is part of the mechanism of communication between cells, which involves the binding of extracellular signaling molecules by membrane receptors, and downstream intracellular PTM cascades, which bring about changes in gene transcription, often leading to an altered state of cell cycle. Examples in three categories are discussed next.

11.3.1 Receptors and Membrane Proteins

Receptors and membrane proteins, critical elements of signaling cascades, show an elevated level of disorder. In a systematic study, the frequency and location of disorder in 2109 human plasma membrane proteins of known transmembrane topology was addressed (Minezaki, Homma, and Nishikawa 2007). IDRs ≥30 consecutive residues were found in 41.0% of them, far exceeding the frequency in inner membrane proteins of *E. coli* (4.7%). Long IDRs were found to have a strong preference for the cytoplasmic side. The functional consequences of disorder in a few cases are well-characterized.

The voltage-dependent potassium channel (Shaker-channel), is located in the plasma membrane of *D. melanogaster* neurons. The channel activates and inactivates rapidly when membrane potential becomes more positive (Hoshi, Zagotta, and Aldrich 1990), by virtue of a "ball-and-chain" action of its disordered cytoplasmic N-terminal tail (entropic clock mechanism, described in detail in Chapter 12, Section 12.1.2). The channel also has an IDR within its cytoplasmic C-terminal tail, which mediates interactions with intracellular scaffold proteins, such as the postsynaptic density 95 (PSD-95) protein (Magidovich et al. 2007). The unbound C-terminal segment is in a random coil state, with its length being critical in fine-tuning interactions of the channel. When it is made shorter or less flexible, the channel has higher affinity for the PDZ domains of PSD-95 than when it is made longer or more flexible. Its behavior supports a "fishing rod" molecular mechanism of binding, which is conceptually very similar to fly-casting and protein fishing (Dafforn and Smith 2004; Evans and Owen 2002; Levy, Onuchic, and Wolynes 2007). This binding mode may result in channel clustering at unique membrane sites, which is important for the proper assembly and functioning of the synapse.

Key components of cell-to-cell and cell-to-ECM communication in both vertebrates and invertebrates are calcium-dependent cell adhesion glycoproteins, cadherins (Alattia, Kurokawa, and Ikura 1999; Gooding, Yap, and Ikura 2004). Cadherins are single-pass transmembrane proteins involved in homotypic cell adhesion, in which they link the cytoskeletons of adjacent cells. In adherens junctions, their cytoplasmic domain binds β-catenin, which in turn binds to actin-associated α-catenin. The cytoplasmic domain of E-cadherin is about 70 amino acids in length, it is fully disordered (Huber et al. 2001), and adopts an extended structure upon binding to β-catenin (Figure 11.3). β-catenin is a signaling hub protein (see Chapter 12, Section 12.6.2) which can also bind several other partners, such as adenomatous polyposis coli (APC), Tcf and axin (Figure 11.3), proteins that either contribute to or compete with E-cadherin/catenin interaction. These interactions take part in two important developmental processes, cell–cell adhesion at adherens junctions, and the regulation of gene expression through Wnt signaling (Daniels, Eklof-Spink, and Weis 2001). In the resting state, β-catenin is either found in complex with E-cadherin and α-catenin, mediating the interaction with the actin cytoskeleton, or in direct contact with APC, axin, and glycogen synthase kinase 3β (GSK3β), functioning in Wnt signaling (Daniels et al. 2001; Gooding et al. 2004). Phosphorylation by GSK3β targets β-catenin for ubiquitin–proteasome-mediated degradation. If Wnt binds to its receptor Frizzled, it inhibits β-catenin phosphorylation and promotes is translocation to the nucleus where it serves as a co-activator to lymphocyte enhancer binding factor/T-cell factor (Lef/Tcf) family of transcription factors. They activate the transcription of developmentally important genes, as well as proto-oncogenes in humans. The extended binding mode of β-catenin partners (Figure 11.3) suggests their disorder in the free form, also directly shown in the case of E-cadherin (Huber et al. 2001), APC (Liu et al. 2006b), and Tcf/Lef transcription factors (Love et al. 2004). These catenin-binding proteins contain a homologous recognition motif, the disordered catenin-binding domain

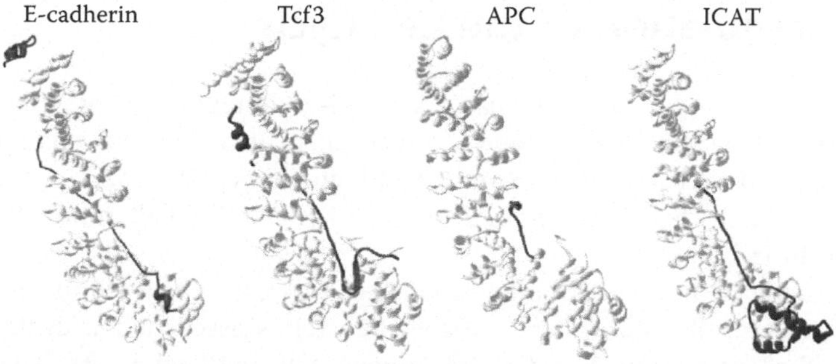

FIGURE 11.3 Many-to-one signaling involving β-catenin partners. β-catenin (light grey) has several binding partners (dark grey) such as E-cadherin (pdb 1i7w), Tcf3 (pdb 1g3j), APC (pdb ijpp), and ICAT (pdb 1luj). The β-catenin central domain (515 residues) contains 12 armadillo repeats, each consisting of 3 helices stacked to form a positively charged right-handed superhelix of helices, and serves as a scaffold for binding distinct partners. The interplay between various binding modes of β-catenin represents an example of many-to-one signaling enabled by disordered partners binding to the same ordered target. For further details, see Gooding et al. 2004.

(CBD) (see Chapter 14 Section 14.2.4), and their binding to the same partner represents a case of many-to-one signaling enabled by disorder.

A special example of disorder in receptor proteins is provided by the cytoplasmic domains of antigen receptors of immune cells, such as T cells, B cells, mast cells, and basophils (Sigalov 2004). In these cells, antigen recognition results in the initiation of immune response mediated by membrane-bound receptors termed multichain immune recognition receptors (MIRRs). MIRRs consist of multiple single-transmembrane subunits, each with extracellular ligand-binding domains and intracellular signaling domains, these latter containing one or more copies of an immunoreceptor tyrosine-based activation motif (ITAM), which gets Tyr-phosphorylated upon receptor clustering. The cytoplasmic domains of ITAM-containing signaling subunits were shown to be disordered, even in the homo-oligomeric (tetrameric) form, as explicitely demonstrated in the case of TCRzeta cytoplasmic domain (Sigalov, Aivazian, and Stern 2004; Sigalov et al. 2006). This seminal observation contributed to the development of the concept of fuzziness (Chapter 14, Section 14.8).

11.3.2 Scaffold Proteins and Hub Proteins

Specificity of signaling through cascades is often ensured by scaffold proteins, which can simultaneously bind several signaling proteins and determine the direction of flow of information in signaling. In fact, IDPs have the potential to bind more partners simultaneously than globular proteins (Gunasekaran et al. 2003), they are often found organizing large complexes (Hegyi, Schad, and Tompa 2007), and an increased level of disorder was predicted in functional categories specialized in organizing signaling complexes (i.e., scaffold proteins and hub proteins). These issues are covered in detail in Chapter 12, Section 12.6.2.

11.3.3 Regulation of the Cell Cycle

Often being the target of signaling cascades, cell-cycle regulatory proteins in general show a very high level of disorder, and several of them are discussed in detail elsewhere (e.g., Cip/Kip cell-cycle dependent kinase [Cdk] inhibitors $p21^{Cip1}$, $p27^{Kip1}$, and $p57^{Kip2}$ in Chapter 15, Section 15.1.3, securin in Section 15.1.5, and Sic1 Cdk inhibitor in Chapter 14, Section 14.11.1).

Cyclins, which are the activating subunits of Cdks, also have significant and functionally important disorder. There are specific cyclins associated with the G1 phase (Cyclin D), S phase (Cyclins E and A), and mitosis (Cyclin B and A, see also Chapter 15, Figure 15.3). The associated kinases are Cdk 4/6 in G1, Cdk1/2 in S phase, and Cdk1 in G2 and M. The kinases phosphorylate specific substrates required for a particular phase of the cycle, and their activity is also regulated by a variety of Cdk inhibitors, such as the Cip/Kip inhibitors and the INK4 gene family. Cyclins are targeted for degradation by ubiquitination carried out by specific ubiutin ligases, such as the SCF complex in the case of G1 cyclins and anaphase-promoting complex/cyclosome (APC/C) in the case of mitotic cyclins (Peters 2002). APC/C can ubiqutinate both securin and cyclin B, depending on its associated accessory protein (Cdc20 or Hct1) within specific sequence features termed destruction box (D-box). The N-terminal regions containing the D-box of both proteins are disordered (Cox et al. 2002).

11.4 NUCLEIC ACID-CONTAINING ORGANELLS

Large nucleic acid-protein complexes, such as the ribosome and chromatin are the most basic organelles of the cell. Whereas in a functional sense they are rather heterogeneous, a preponderance of disorder in proteins that contact the nucleic acid is a basic observation of the IDP field.

11.4.1 Ribosome

The ribosome performs protein synthesis in the cell by using mRNA as template. Eukaryotes have 80S ribosomes, composed of a small (40S) and large (60S) subunit (30S and 50S in prokaryotes). Their large subunit contains three structural elements—5S (120 nucleotides), 28S (4,700 nucleotides), and 5.8S (160 nucleotides)—RNA, and about 49 associated (L) proteins. The 40S subunit consists of a 18S (1,900 nucleotides) RNA and about 33 associated (S) proteins. Ribosomal proteins are consistently predicted to be among the most highly disordered proteins (Iakoucheva et al. 2002; Ward et al. 2004; Xie et al. 2007). Nearly 68% of them have predicted IDRs ≥30 consecutive residues, close to the value of regulatory and cancer-associated proteins in general (Iakoucheva et al. 2002). The structure of individual ribosomal proteins in complex with ribosomal

RNA is known at atomic resolution (Ban et al. 2000), but their structural disorder in isolation has not been studied in great detail.

Ribosomal proteins studied in isolation show clear signs of intrinsic disorder. The ratio of their CD ellipticities at 200 and 222 nm suggest that many of them exist in a PMG state in solution (Uversky 2002a). In certain cases, the structure solved by nuclear magnetic resonance (NMR) shows a globular domain and disordered N- or C-terminal tails, such as the N-terminal 22 amino acids of L18 (Turner and Moore 2004) or the N-terminal 25 amino acids and a 15 amino acid-long loop in L16 (Nishimura et al. 2004). In addition, certain ribosomal proteins, such as L7/L12 (Mulder et al. 2004), which constitutes a stalk-like extension on the 50S subunit, also show structural disorder in the ribosome-bound state. L7/L12 has two globular domains: the N-domain forms tetramers and connects to the body of the ribsosome, whereas the C-domain binds to GTPases in the course of translation (IF-2, EF-G, EF-Tu, RF3). They are flexibly tethered to the ribosome via a disordered linker, which can be replaced with an unrelated sequence without compromising ribosome function, whereas shortening or lengthening it seriously impairs translation (Bubunenko, Chuikov, and Gudkov 1992).

The disorder of ribosomal proteins is probably critical in ribosome assembly, which involves the sequential binding of numerous proteins via multiple pathways leading to large-scale changes in the conformation of both RNA and proteins (Xie et al. 2007). For example, L5 contributes to folding of rRNA in a mutual induced fit mechanism (DiNitto and Huber 2003). In addition, many ribosomal proteins appear to have extra-ribosomal functions (Wool 1996), which are often implicated in the regulation of transcription, RNA processing, DNA repair, and translation. These functions in general are closely associated with structural disorder, which suggests that disorder of ribosomal proteins is probably instrumental in these extra-ribosomal functions.

11.4.2 Disorder in Chromatin Organization

11.4.2.1 Histones

Genomic DNA in eukaryotes is extremely compacted in chromosomes. The primary level of compaction is binding by core (H2A, H2B, H3, and H4) and linker (H1 and H5) histones to form nucleosomes and chromatin fibers (Hansen 2002), to be further organized into higher order chromatin structures. This higher level of organization is primarily regulated through posttranslational modifications of the N-terminal tails of core histones, invisible in the crystal structure of the nucleosome (see Chapter 5, Figure 5.1, (Luger et al. 1997)). These regions also appear disordered in solution (Hansen, Tse, and Wolffe 1998), which might be important in their complicated patterns of protein-protein interactions, regulated by a bewildering variety of post-translational modifications (Ruthenburg, Allis, and Wysocka 2007; Shogren-Knaak et al. 2006). These modifications collectively represent a "histone code," an epigenetic regulatory feature of the accessibility of DNA for transcription (Jenuwein and Allis 2001). For example, the disordered NTD of core histone H4 functions as a platform for the assembly of chromatin-remodeling complexes, such as ISWI

(Clapier et al. 2001) and NURF (Xiao et al. 2001), and it can also interact with sequence-specific transcription factors resulting in both activation and repression (Fazzio, Gelbart, and Tsukiyama 2005).

Linker histones are nucleosome-binding proteins that stabilize condensed chromatin. They have a very simple domain organization, consisting of a central winged helix fold, a short N-terminal extension, and a long basic disordered CTD of about 100 residues in length (Hansen et al. 2006). The determinants required to condense chromatin fibers reside in the CTD, which binds to linker DNA (i.e., DNA between nucleosomes) and stabilizes nucleosome–nucleosome interactions. In addition, a 47 residue-long segment within linker histone H1 CTD binds and activates the DFF40/CAD apoptotic nuclease, irrespective of its primary sequence (Lu and Hansen 2004), which represents a case of sequence independence of recognition in fuzziness (see Chapter 14, Section 14.10).

11.4.2.2 Other chromatin organizing proteins

Structural disorder also plays important roles in the function of two classes of chromatin remodeling proteins, ATP-dependent complexes, such as ISWI, CHRAC, NURF, RSC, and SWI/SNF, as well as fully disordered ATP-independent architectural transcription factors (ATFs), such as high-mobility group (HMG) proteins, methyl CpG-binding protein 2 (MeCP2), silent information regulator protein 3 (Sir3), and decondensation factor 31 (Df31). By different mechanisms, both these classes regulate accessibility of the genomic DNA.

EM studies of the yeast SWI/SNF suggest that eight subunits are assembled into a modular and highly irregular structure (Smith et al. 2003), which utilizes IDRs for sliding along DNA (Hartlepp et al. 2005). This mobility is impaired upon removal of IDRs, which suggests that these mediate low affinity interactions of variable contact patterns with DNA. A significant content of disorder in Snf5 and Swi3 subunits is also demonstrated by gel mobility analysis. Structural disorder in these cases may be involved in assembly and recruitment, because Swi3 serves as an assembly scaffold that can bind histones, whereas Snf5 is involved in recruitment of SWI/SNF to specific regions of the genome (Wu et al. 2004).

The chromatin structure can also be remodeled by a different mechanism, through bending and distortion induced by ATFs, termed so because of their general effect on transcription due to affecting DNA structure and/or mediating interactions with transcription factors (Reeves 2001). Structural disorder of these factors enables their adaptable and simultaneous binding of several partners including DNA and proteins, cross-bridging nucleosomes and following unfolding transitions of DNA, without which they could not participate in higher-order chromatin organization. For example, HMG proteins contain multiple copies of a short basic sequence element AT-hook that tends to bind to the minor groove and significantly bend the DNA. One of them, HMGA (see Chapter 5, Figure 5.3), is described in detail in Chapter 12, Section 12.6.2.1 (Reeves 2001). Sir3p, which is involved in the initiation, propagation, and maintenance of transcriptionally silenced chromatin probably by nucleosome bridging, also has a high level of predicted and observed disorder (McBryant, Krause, and Hansen 2006). MeCP2, which functions as a methylation-dependent transcriptional repressor also involved in the maintenance of condensed chromosomal superstructures and regulation of mRNA splicing, is also highly disordered (Adams et al. 2007). Full structural disorder was demonstrated by several techniques for Df31, a Drosophila protein involved in chromatin decondensation and stabilization (Szollosi et al. 2008). Df31 can associate with poly-nucleosomes and participate in the higher-order folding of chromatin by several mechanisms, such as by uploading histones on DNA due to its histone chaperone activity.

11.5 DISORDER IN RNA-BINDING PROTEINS: TRANSCRIPTION AND RNA FOLDING

A wide range of mRNA-binding proteins also have a high level of disorder. These proteins regulate practically all stages of gene expression from transcription through mRNA processing and RNA folding to regulation of translation. The reason of this general preponderance of disorder is not readily obvious but may reside in the structural variability of the partner, RNA, itself.

Bacteriophage λ tightly regulates transcriptional termination in *E. coli* to control switching between its life cycles. It prevents termination of early gene expression by λN, a fully disordered phage-encoded protein (Greenblatt and Li 1982). By distinct and autonomous recognition elements, the protein binds several partners (i.e., a unique sequence (BoxB) within the untranslated region of mRNA, RNA polymerase (RNAP), the host-derived N-utilization substance (NusA), and additional factors) to form a ribonucleoprotein antitermination complex. Within the complex, RNAP is resistant to termination signals so that it reads through ρ-dependent and intrinsic termination sites (Van Gilst et al. 1997). Binding of the various partners occurs by local induced folding of independent binding elements, as shown in the case of NusA (Bonin et al. 2004) and RNA BoxB (Legault et al. 1998).

The initiation of translation is also precisely regulated by proteins, which have a high level of disorder. The process is mediated by the cap structure m7GpppN of mRNA, which is recognized by the eukaryotic initiation factor 4F (eIF4F) complex, composed of three subunits, the cap-binding protein eIF4E and the adaptor proteins eIF4G and eIF4A (von der Haar et al. 2006). Interaction of this complex with the cap structure recruits the 40S ribosomal subunit to the 5′ end of mRNA. The assembly of eIF4E and eIF4G is regulated by a competitive inhibitor, 4E binding protein (4E-BP) (Marcotrigiano et al. 1999). All the proteins involved are largely or fully disordered in the free state, and undergo mutual induced folding upon recognition (Figure 11.4A). eIF4E is an MG that folds upon binding the 5′ cap structure and/or the fully disordered eIF4G. 4E-BP is completely disordered (Fletcher and Wagner 1998), it undergoes local and predictable (see Chapter 9, Section 9.7) folding upon binding to eIF4E (Marcotrigiano et al. 1999) to a structure (Figure 11.4B) that mimics that of bound eIF4G (Figure 11.4C), which represents a prime case of molecular mimicry by a disordered protein (see Chapter 14, Section 14.14).

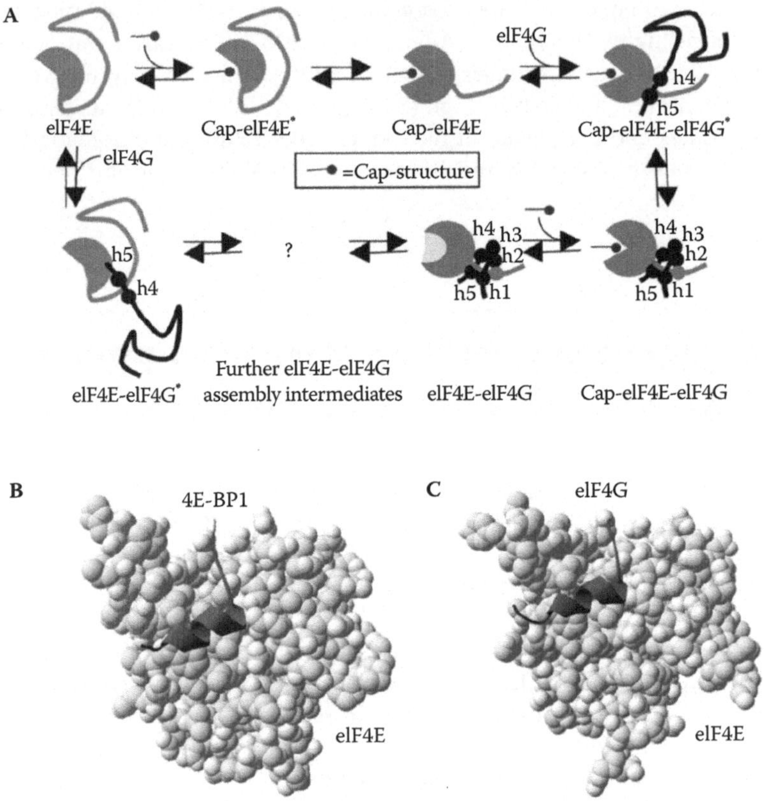

FIGURE 11.4 Disorder in eukaryotic translation initiation. (A) A model of structural transitions during the assembly of the cap-binding complex in translation initiation. Key features of the model are that elF4E is partially disordered, whereas elF4G is fully disordered before binding to the 5′ cap of mRNA or to each other. Reproduced with permission from von der Haar et al. (2006), *J. Mol. Biol.* 356, 982–92. Copyright by Elsevier Inc. (B) 4E-BP peptide bound to elF4E-7mGDP (pdb 1ej4), which is a molecular mimic of the binding of elF4G peptide (C, pdb 1ejh).

Disorder in RNA-binding proteins also has a general consequence of promoting proper folding of RNA, critical in many basic processes including splicing, transcription, biogenesis of the ribosome, and protein synthesis. RNA frequently misfolds into structurally stable but biologically inactive structures (Herschlag 1995), and many different RNA chaperone proteins have evolved to promote the formation and/or stabilization of the native RNA fold. Such RNA chaperones include heteronuclear ribonucleoprotein A1 (hnRNPA1) (see Chapter 12, Figure 12.5), prion protein, nucleocapsid proteins Ncp7 and Ncp9, and ribosomal proteins, among others (Tompa and Csermely 2004). As detailed in Chapter 12, Section 12.3.2, these proteins as a class are among the most disordered ones, with their structural disorder being critically involved in chaperone activity (Tompa and Csermely 2004).

11.6 CYTOSKELETAL PROTEINS

The cytoskeleton provides the internal scaffold of the cell and is composed of three basic components: microfilaments, intermediate filaments, and microtubules (MTs). Intriguingly, structural disorder is associated with all three, but in basically different ways. Microfilaments contain filamentous actin (F-actin) with significant disorder in proteins involved in regulating its assembly/disassembly, such as Tβ4 and Wiskott–Aldrich syndrome protein (WASP). Intermediate filaments (neurofilaments in neuronal cells) are extended coiled-coil structures of three principal components, IF-L, IF-M, and IF-H—all three with disordered tails that project away from the filament as side-arms. Microtubules are hollow tubes of protofilaments, which are polymers of tubulin α/β heterodimers. They are inherently unstable, with their stability depending critically on the presence of fully disordered accessory proteins, such as microtubule-associated protein 2 (MAP2), tau protein, and stathmin.

11.6.1 Microfilaments

Microfilaments are the most diverse and versatile elements of the cytoskeleton, contributing to cellular processes as diverse as muscle contraction, cell adhesion, cell migration, vesicle and organelle transport, signaling, cell division, and cytokinesis (Sparrow 1999). The elementary unit of the filaments is globular actin (G-actin), a 42 kDa conserved protein that polymerizes into F-actin. Microfilaments, which are composed of two intertwined fibers, are around 7 nm in diameter, and have two distinguishable ends termed barbed (+) and pointed (-) ends. They reach the highest density under the cell membrane, where they are responsible for resisting tension and maintaining cellular shape, and also for forming cytoplasmatic protuberances, such as pseudopodia and lamellipodia. When cells move, they form stress fibers, which are responsible for locomotion. They are the most dynamic among the three elements of the cytoskeleton and are under the influence of many regulatory inputs mediated by IDPs.

Tβ4 is an IDP of 45 amino acids in length, it binds and sequesters G-actin and inhibits actin polymerization (Hertzog et al. 2004). Tβ4 bound to G-actin (Figure 11.5) blocks both the barbed and pointed ends of G-actin, thus preventing its interaction with either end of the growing F-actin polymer (Irobi et al. 2004). Interestingly, a region homologous to Tβ4, named WASP homology domain 2 (WH2), can be found in many other proteins regulating the actin cytoskeleton (e.g., ciboulot, verprolin, spire, cordon bleu, WASP, WAVE, and WIP) (Paunola, Mattila, and Lappalainen 2002; Renault 2008). Its conserved features suggest that in all the homologs it functions in actin binding, but with context-dependent outcome. Whereas a single WH2 domain in Tβ4 sequesters G-actin, several tandem WH2 domains (e.g., in ciboulot) promote actin polymerization, probably due to tethering multiple actin monomers next to each other (Chereau et al. 2005).

The WH2 domain is also involved in regulating the actin cytoskeleton in a more sophisticated manner in the WASP, as also detailed in Chapter 14, Section 14.12.2.

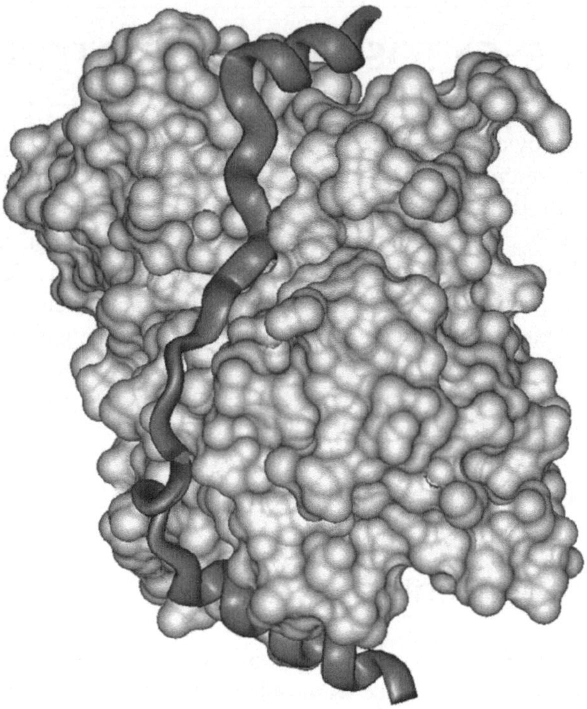

FIGURE 11.5 The model of thymosin-β4 bound to G-actin. The structure of Tβ4 (dark grey) bound to G-action (light gray) has been assembled by combining the structures of two fusion proteins containing either half of Tβ4 in complex with G-actin (pdb 1t44 and pdb 1sqk).

WASP mediates the effect of one of the Rho subfamily GTPases, Cdc42 (Caron 2002), which stimulates *de novo* actin polymerization. Mammalian WASP is a modular protein around 500 amino acids in length, with a WASP homology domain 1 (WH1), a basic region (BR), a GTPase-binding domain (GBD), a Pro-rich region, and a C-terminal VCA region composed of a verprolin-homology (V) or WASP homology domain 2 (WH2), a central hydrophobic region (C), and an acidic tail (A). The regions GBD and WH2 are disordered (Abdul-Manan et al. 1999; Kim et al. 2000a), and the inactive state of WASP is characterized by the interaction of its GBD and VCA regions, which occludes VCA (Kim et al. 2000a). Activation primarily results from Cdc42-binding at GBD, which releases VCA for interaction with the actin-related protein (Arp) 2/3 complex, as described in Chapter 14, Section 14.12.2 (see Chapter 14, Figure 14.3 and Figure 14.10) (Abdul-Manan et al. 1999; Panchal et al. 2003).

11.6.2 Intermediate Filaments

Intermediate filaments (neurofilaments) are the most abundant structural components in large-diameter myelinated axons (Lee and Cleveland 1996). They are obligate heteropolymers composed of three subunits, IF-L, IF-M, and IF-H, which differ in their M_W (68–70 kDa, 145–160 kDa and 200–220 kDa, respectively). The difference mostly comes from their sequentially diverged and disordered (Brown and Hoh 1997) C-terminal tail domains, which extend from the filament backbone and form lateral cross-bridges between adjacent filaments. The tail domain of the longest isoform, NF-H is highly repetitive containing more than 100 copies of a hexapeptide element, which harbors a characteristic KSP phosphorylation motif and has been generated by repeat expansion (see Chapter 13, Section 13.3.1). The importance of repeats derives from contributing multiple sites for phosphorylation, which determines interfilament spacing by virtue of tuning the entropic exclusion effect of tail domains by electrostatic repulsion (Brown and Hoh 1997).

11.6.3 Microtubules

MTs have the largest diameter (about 25 nm) of the three components of the cytoskeleton, they play a basic ultrastructural role in highly elongated cells, such as neurons, and also in cellular processes, such as mitosis, cytokinesis, and vesicular transport (Avila 1989). They are polymers of α/β-tubulin dimers, which polymerize end to end in protofilaments, 13 of which then bundle into hollow cylindrical filaments. MTs are nucleated and organized by microtubule organizing centers (MTOCs), such as centrosomes and basal bodies, and they form the mitotic spindle required for the segregation of chromosomes in mitosis. MT polymerization is driven by GTP-binding to tubulin. GTP hydrolysis at the tip of polymers may revert growing, and occasionally cause rapid depolymerization and shrinkage, termed a catastrophe. Due to this inherent instability, MT function depends critically on accessory proteins.

Microtubule-associated proteins MAP2 and tau proteins are fully disordered (Csizmok et al. 2005; Hernandez, Avila, and Andreu 1986; Schweers et al. 1994), they share a common tubulin-binding domain (TBD), and they have unrelated N-terminal projection domains. The projection domain of tau (Bodart et al. 2008) and probably also of MAP2 (Mukhopadhyay and Hoh 2001) remains disordered even in the bound state *in vivo*, and functions as an entropic spacer/bristle that provides proper spacing in the cytoskeleton. Due to its involvement in Alzheimer's disease, tau protein is among the best characterized IDPs (see Chapter 10, Section 10.2.3.3 and Chapter 15, Section 15.3.1.2). Stathmin is also fully disordered (Honnappa et al. 2006) but plays the opposite role as MAP2 and tau protein, because its interaction with tubulin destabilizes assembled MTs causing their catastrophic depolymerization (Gigant et al. 2000).

11.7 DISORDER IN STRESS PROTEINS

The correlation of disorder with both protein and RNA chaperone functions is briefly mentioned in Section 11.5. The broad class of stress-related LEA proteins of plants is discussed in some detail here. LEA proteins are expressed in late stages of seed maturation and they are also strongly associated with the toleration of abiotic stress conditions, such as dehydration caused by high salinity, high/low temperature or draught (Tunnacliffe and Wise 2007; Wise and Tunnacliffe 2004). Based on the presence of certain sequence motifs and function, LEA proteins are classified into three groups. Homologs of group 1 and 3 proteins are also found in bacteria and in certain invertebrates (Tunnacliffe and Wise 2007; Wise and Tunnacliffe 2004). Group 2 is termed dehydrins (DHNs). LEA proteins have high charge and are hydrophilic in character, and several of them, such as wheat EM (McCubbin, Kay, and Lane 1985), *A. avenae* LEA1 (Goyal et al. 2003), soybean DHN1 (Soulages et al. 2003), maize DHN1 (Koag et al. 2003), and *A. thaliana* ERD10/14 (Bokor et al. 2005; Tompa et al. 2006a), are fully disordered. Overall, it is reasonable to consider LEA proteins disordered in general (Goyal et al. 2003; Irar et al. 2006).

LEA proteins have several suggested functions, such as antioxidants, ion sinks, and membrane stabilizers (Tunnacliffe and Wise 2007; Wise and Tunnacliffe 2004), but results most consistently point to their stress-related function as chaperones. For two LEA proteins, *A. avenae* LEA1 and wheat EM, protection of citrate synthase from heat-induced aggregation and lactate dehydrogenase from cold-induced aggregation was demonstrated (Goyal, Walton, and Tunnacliffe 2005). A broad protein stabilization function of *A. avenae* LEA1 was also described, with potent inhibitory activity against polyQ aggregation *in vivo* (Chakrabortee et al. 2007). Cryoprotective activity was demonstrated for two soybean dehydrin-type proteins, Mat1 and Mat9 (Momma et al. 2003), and a similar effect was also shown for PCA60, a protein from winter bark tissues of peach (Wisniewskia et al. 1999). Potent chaperone activity of *A. thaliana* ERD10/14 was observed against heat-induced aggregation and/or denaturation of a range of substrates, such as lysozyme, alcohol dehydrogenase, firefly luciferase (Chapter 12, Figure 12.4), and citrate synthase (Kovacs et al. 2008). These results point to the role of disorder in stress-related proteins in general, and chaperones in particular (also discussed in detail in Chapter 12, Section 12.3 and Chapter 14, Section 14.15).

11.8 DISORDER AND METAL BINDING

Structural disorder is often implicated in metal-binding proteins. In a study on the functional implications of the pattern of disorder (Lobley et al. 2007), descriptors of IDRs ≤50 consecutive residues were found to correlate with metal binding function (see Section 11.10). Here a few prominent IDPs of metal binding function are discussed, also touched upon in Chapter 12, Section 12.5, on scavenger functions.

α-synuclein (NACP) is involved in Parkinson's disease and other neurodegenerative synucleinopathies (see Chapter 15, Section 15.3.2.1). The acidic CTD of the largely disordered protein of about 140 amino acids contains binding sites for Ca^{2+} ions and polyamines, which promote aggregation of the protein (Antony et al. 2003). Several other di- and trivalent metal ions (e.g., Cu^{2+}, Al^{3+}, Fe^{3+}, and Co^{3+}) also cause significant acceleration in the rate of fibrillation of the protein (Uversky, Lee, and Fink 2001c), which suggests that metal binding by its CTD is of rather broad specificity. The efficiency of different metal ions in stimulating fibrillation correlates with their ability to induce a conformational change in the IDP.

Prion protein (PrP) is also best known for its involvement in a range of fatal neurodegenerative diseases (Chapter 15, Section 15.3.4). As described in Chapter 13, Section 13.3.1.4, PrP has a disordered N-terminal half that contains a polymorphic octapeptide repeat region, which constitutes a high-affinity copper binding site capable of binding Cu^{2+} ion *in vitro* with a K_d of 10^{-14} M (Jackson et al. 2001). Mice in which the PrP gene is ablated exhibit severe reduction in the copper content of synaptosomal and endosome-enriched subcellular fractions of brain extracts, which suggests that copper binding is a function of the protein *in vivo* (Brown et al. 1997a). Because PrP null-mutant mice also have reduced copper/zinc superoxide dismutase activity, the prion protein might be a recycling transport protein for copper transport, and/or a superoxide dismutase enzyme itself.

Several IDPs have also been noted for their ability to bind metal ions with low affinity but high capacity. For example, (see Chapter 12, Section 12.5.3) calsequestrin can bind 40–50 Ca^{2+} ions per molecule, with an affinity of about 1 mM (He et al. 1993). The function of this protein is probably to store Ca^{2+} ions and regulate their traffic in the sarcoplasmic reticulum.

11.9 DISORDER AND ENZYME ACTIVITY

As suggested in Section 11.1, structural disorder is strongly anti-correlated with enzyme activity. Still, this topic deserves a special attention, both because of its evolutionary implications and its far-reaching consequences on extending the structure-function paradigm. Whereas enzymatic activity does require well-defined structure that ensures proper spatial positioning of catalytic residues (see Chapter 1, Section 1.11), in some cases it is claimed that either MG-type or random coil-type disorder is compatible with catalytic activity. In a strictly theoretical sense, these disordered enzymes probably do not violate the structure-function paradigm; they simply take the energy of substrate binding to complete folding and acquire a 3-D structure competent for catalysis.

The possibility of an enzymatic molten globule was demonstrated through an active, monomeric chorismate mutase generated from the wild-type dimeric helical bundle enzyme of *M. jannaschii* (Vamvaca et al. 2004). In the absence of the substrate, chorismate, the enzyme is monomeric and highly helical by GF and CD measurements, whereas NMR spectroscopy, ANS binding, DSC melting, and rapid H/D exchange suggest that it is highly flexible and probably has MG-type of disorder. This MG protein is enzymatically competent, catalyzing the rearrangement of chorismate into prephenate with the same k_{cat} ($3.2 s^{-1}$), and a K_M only three-fold elevated (170 μM) then the parent enzyme. Apparently, folding to the catalytically competent state is induced by substrate binding, as suggested by effective binding of a bicyclic dicarboxylic acid transition-state analogue inhibitor of the enzyme (K_i = 2.5 μM), which elicits a transition from a disordered to an ordered state.

UreG is one of four nickel chaperones (UreE, F, G, and D) involved in the assembly of active urease in bacteria. The CD spectrum of the protein indicates 15% α-helix and 29% β-strand structure, but NMR spectroscopy shows flexibility characteristic of a disordered protein (Zambelli et al. 2005). These observations are compatible with a MG state, even though the protein is a homo-dimer (Neyroz, Zambelli, and Ciurli 2006). UreG catalyzes the hydrolysis of GTP with a $k_{cat} = 0.04\ min^{-1}$, coupling energy requirement and nickel incorporation into the urease active site. The protein is specific for the metal ion and has been suggested by structure prediction (threading) to have a well-defined structure characteristic of GTPases. Apparently, UreG takes the energy from interaction with cofactors and/or other protein partners to complete folding and perform its catalytic function.

11.10 IS THERE A LINK BETWEEN THE PATTERN OF DISORDER AND FUNCTION?

The ultimate test of understanding the structure-function relationship of a protein is to predict its function from its sequence. Whereas this task is beyond our powers even in the case of ordered proteins, there appears to be some recognizable link between the pattern of disorder and protein function (Lobley et al. 2007). Pattern analysis of the distributions of disordered regions in human sequences suggests correlations between GO categories and length and/or position descriptors of predicted IDRs. Many useful generalizations arise, although they represent trends rather than strong correlations, and as yet lack real predictive power.

For example, transcription regulator-, DNA binding-, and RNAP II transcription factor functions are associated with IDRs in the protein interior, rather than toward the termini. A tendency of the IDR to fall toward the C-terminus is discernible in categories such as transcription factor activator, transcription factor repressor and transcription factor. Disordered residues are over-represented at the N-terminus of ion channels (potassium channels). Length descriptors show even more significant associations with function. IDRs >500 residues are over-represented in transcription-related functional categories. Shorter ones (≤50 residues) prevail in proteins of metal ion binding and ion channel functions, and GTPase regulatory functions. Proteins in Ser/Thr-kinase and phosphatase categories are over-represented with long IDRs on the order of 300–500 residues. The significance of these observations could be confirmed by incorporating the obtained feature vectors into an SVM, and observing the improvement in prediction accuracies in 26 GO categories related to signaling and molecular recognition (Lobley et al. 2007). The most significant improvements are observed for kinase, phosphorylation, growth factor, and helicase categories.

Molecular Functions of Disordered Proteins

12

This chapter classifies functional information on intrinsically disordered proteins (IDPs) with reference to the actual molecular modes of their action. In this respect, it is closely related to the molecular function (MF) classification of Gene Ontology (GO), but with categories that suit functions unique to IDPs, which cannot be described by the current MF ontology of GO. It complements Chapter 11 and extends GO to set the stage for a unified classification scheme that can handle both ordered and disordered proteins. The seven categories presented cover all the basic modes of IDP action and—alone or in combination—enable to describe the function of even complex proteins. Illustrative examples of the categories are given in Table 12.1.

12.1 ENTROPIC CHAIN FUNCTIONS

The unique structural feature of IDPs/intrinsically disordered regions (IDRs) manifests itself most clearly in entropic chain functions, which stem directly from the disordered state (Table 12.1). In these, functions result from the ability of the polypeptide chain to fluctuate between a large number of conformational states. The IDP/IDR may determine the distance distribution of functional elements, regulate the dynamics of their rearrangements, measure time, or respond to an ultrastructural change reducing its conformational freedom. The force generated against this insult may be largely or entirely entropic in origin.

12.1.1 Linkers and Spacers

Linkers and spacers are defined as IDRs of proteins that connect functional regions, whether they be ordered domains or disordered motifs. Their function is to provide appropriate spatial separation of the motifs, enable their spatially almost unrestricted

TABLE 12.1 Functional classification of IDPs

PROTEIN	*PARTNER*	*FUNCTION*
Entropic Chains		
Nup2p FG repeat region	n.a.	Gating in NPC
MAP2 projection domain	n.a.	Spacing in cytoskeleton
Titin PEVK domain	n.a.	Elasticity of muscle
K channel N-terminal region	n.a.	Timing of gate inactivation
Display Sites		
CREB KID	PKA	Site of phosphorylation
Cyclin B N-terminal domain	E3 ubiquitin ligase	Site of ubiquitination
Chaperones		
β-synuclein	e.g., α-synuclein	Prevention of aggregation
ERD 10/14	e.g., luciferase	Prevention of aggregation
Nucleocapsid protein 7/9	e.g., RNA	Trans-splicing
hnRNP A1	e.g., DNA	Strand re-annealing
Effectors		
4E-BP1	eIF4E	Inhibition of translation initiation
$p27^{Kip1}$	Cyclin A-Cdk2	Inhibition of cell-cycle
FlgM	σ^{28}	Inhibition of transcription
Securin	Separase	Inhibition of anaphase
Stathmin	Tubulin	Inhibition of tubulin polymerization
Assemblers		
RNAP II CTD	mRNA maturation factors	Regulation of mRNA maturation
SARA	Smad	Targeting TGFβ activity at Smad
Ciboulot	Actin	Promoting actin polymerization
$p21^{Cip1}$	Cyclin A-Cdk2	Assembly of cyclin-Cdk complex
CREB	p300/CBP	Initiation of transcription
Scavengers		
Casein	Calcium phosphate	Stabilization of calcium phosphate in milk
Salivary PRPs	Tannin	Neutralization of plant tannins

(Continued)

TABLE 12.1 (Continued)

PROTEIN	*PARTNER*	*FUNCTION*
ERD10/14	Water	Retention of water in dehydration stress
Calsequestrin	Calcium	Storage of calcium in sarcoplasmic reticulum
Prions		
Ure2p	Gln3p	Utilization of urea under conditions of growth on poor nitrogen source
Sup35p	NusA, mRNA	Suppression of translation termination, translation readthrough
CPEB	Cytoplasmic mRNA	Polyadenylation of dormant mRNA

search for binding partners, and/or increase binding affinity by increasing local concentration, by virtue of the entropic gain from the physical connection of two binding elements, also known as the chelate effect (Jencks 1981). Another result is processivity, which results from alternative binding interactions of the two connected recognition elements with multiple binding sites along an elongated partner, without full release at any point. This binding capacity results in rapid diffusive movements along the substrate, as observed in the case of bacterial cellulase, matrix metalloproteinase 9 (MMP-9), and myosin VI (discussed in detail in Chapter 14, Section 14.9).

An increase in binding strength and specificity is observed in the case of the transcription factor Oct-1, which regulates the expression of immunoglobulin genes at the Igκ promoter (Chang et al. 1999). Oct-1 has two globular deoxyribonucleic acid (DNA)-binding domains (POU homeodomain and POU-specific domain), each recognizing a 4–5 base pair sequence, connected by a 23 amino acid-long linker region. Upon interaction with the promoter region, the two domains connected by the linker target an octamer DNA sequence with high specificity. The linker is disordered in both the free and bound states, and it is rather resistant to deletions that shorten it down to about 10–14 amino acids; Oct-1 with an even shorter linker, 8 amino acids, has a high affinity for a promoter region in which the order of the two DNA recognition sequences is reversed (van Leeuwen et al. 1997). Interaction with a promoter in which the distance between the two sequences is increased by 3 base pairs requires the linker to be lengthened to 37 amino acids. Separation of the two domains (i.e., deletion of the linker) practically abolishes binding. Overall, flexibility and length of the linker region enable selective binding of differently spaced and oriented subsites of cognate DNA. Other notable linkers are discussed in relation with processivity (Chapter 14, Section 14.9), and the retention of linker function in spite of rapid evolutionary changes (Chapter 13, Section 13.4.1).

12.1.2 Entropic Clocks

A closely related function of IDPs is the entropic clock or timer function (Dunker et al. 2001), defined by the best studied example, the voltage-gated potassium channel

(Shaker channel) of nerve axons (Magidovich et al. 2006; Magidovich et al. 2007). The channel is activated by membrane depolarization, but within 1 ms it becomes inactivated even if membrane depolarization is maintained (Hoshi, Zagotta, and Aldrich 1990; Liebovitch, Selector, and Kline 1992). The molecular mechanism of inactivation (see Chapter 11, Section 11.3.1) can be accounted for by a ball and chain mechanism, in which a short helix (ball) is connected to the body of the channel by a disordered linker (chain). The linker enables the ball to freely move around and search in space for its cognate site. When bound, the ball sterically occludes the mouth of the channel and prevents ion translocation (Bentrop et al. 2001; Zagotta, Hoshi, and Aldrich 1990). Model calculations suggest that movement of the ball on the chain and inactivation kinetics of the channel can be described by a random spatial walk (Liebovitch et al. 1992). Entropic clock (timing) function results from the control of kinetics of channel inactivation by disorder of the chain, as substantiated by the dependence of channel kinetics on its length (Hoshi et al. 1990; Podlaha and Zhang 2003).

12.1.3 Entropic Springs

Entropic chains, which exert a force against physical extension by a mechanism analogous to that of rubber are defined as entropic springs (Dunker et al. 2002). In fact, the concept is borrowed from polymer chemistry, where elasticity of rubber is known to derive from the entropic force generated by stretching a polymer of random structure (Bright, Woolf, and Hoh 2001).

The prime example of this function is titin, the gigantic protein of striated muscle sarcomere (Granzier and Labeit 2002; Labeit and Kolmerer 1995). Titin is an extremely long protein of extensive internal sequence repetition (see also Chapter 13, Section 13.3.1.3), spanning half the 1 μm length of the sarcomere. It has three distinct regions, including a long disordered region (Pro, Glu, Val, Lys-rich (PEVK) domain). The function of the PEVK domain was probed by force–extension measurements, which show that the extensibility of single titin molecules (Kellermayer et al. 1997) or the isolated PEVK region itself (Watanabe et al. 2002) is best approximated by a worm-like chain behavior. Thus, the PEVK region behaves as an entropic spring and accounts for most of the elasticity of titin at low forces (see also Chapter 5, Section 5.7).

Another protein of similar structural properties is elastin, the primary function of which is to impart appropriate mechanical properties on soft tissues. This protein is the basic component of the fabric of skin, blood vessels, and elastic ligaments (Vrhovski and Weiss 1998). Essential to its function is its ability to contract reversibly after stretching, driven primarily by an increase in configurational entropy, as shown by a variety of techniques, such as nuclear magnetic resonance (NMR), Raman optical activity (ROA), and molecular dynamics (MD) simulations (Pometun, Chekmenev, and Wittebort 2004; Rauscher et al. 2006).

12.1.4 Entropic Bristles/Brushes

This function is closely related to that of springs but comes from the force generated against compression, usually exerted by another constituent of the cell. The ensuing effect was first described for cytoskeletal proteins (i.e., the side-arms of neurofilaments) (Brown and Hoh 1997) and projection domains of microtubule-associated proteins (MAPs) (Mukhopadhyay and Hoh 2001). In both cases, atomic force microscopy (AFM) measurements on proteins attached to a solid surface were performed to show that these regions exert a long-range repulsive effect on approaching macroscopic objects (the tip of AFM in this case). Typical distances on the order of 50 nm, as opposed to about 5 nm in the case of globular proteins of similar M_W, were observed. In both cases, it was found that the distance-force relationship of the protein and actual cytoskeletal spacing in cells are correlated, in agreement with the role of these disordered proteins/regions providing proper spacing in the cytoskeleton by entropic exclusion.

This molecular principle is also exploited for an entirely different purpose in the mechanism of gating within the nuclear pore complex (NPC). The nuclear pore is a huge assembly of approximately 50 MDa that selectively transports cargoes across the nuclear envelope (Alber et al. 2007). NPC in yeast is made up of about 450 copies of 30 different subunits, arranged as a large circle surrounding a central pore of about 9 nm (extensible to 30 nm, Figure 12.1). NPC has unusual size-selective filtering capacity as it lets molecules smaller than about 40 kDa freely through, and excludes everything above this threshold, unless it can bind to a specific carrier molecule termed karyopherin (Rout et al. 2000). A cargo bound to a karyopherin can translocate through the pore in either direction between the cytoplasm and the nucleus in an energy-dependent manner. This enigmatic molecular mechanism of NPC gating can be explained by the entropic effect of disordered NPC components (nucleoporins, Nups). 13 Nups in yeast contain long Phe-Gly repeats (thus termed FG Nups), which are intrinsically disordered both *in vitro*

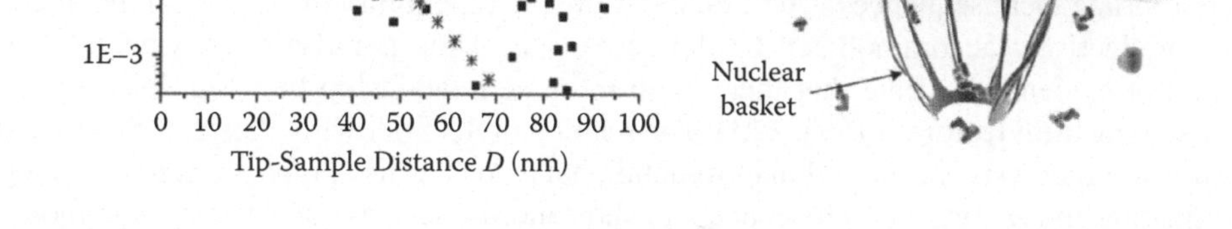

FIGURE 12.1 Entropic bristle function of FG Nups in the nuclear pore. (A) Compression of a nucleoporin (cNUP153) by the tip of AFM results in a force-distance curve which shows a long-range repulsion due to entropic exclusion by the disordered FG repeat region. (B) Artistic model of the gating device nuclear pore complex (NPC), with a ring scaffold made up of different Nup-s, having extensions forming cytoplasmic fibers, a meshwork of FG-domain filaments in its center, and the nuclear basket structure. A key element of the gating function of NPC is size-dependent filtering by entropic exclusion exerted by the disordered FG-domains. Reproduced with permission from Lim et al. (2006), *Proc. Natl. Acad. Sci. USA* 103, 9512–7, copyright by the National Academy of Sciences, and Patel et al. (2007) *Cell* 129, 83–96, copyright by Elsevier Inc.

and *in vivo* (Denning et al. 2003). These disordered appendages physically fill the central pore of NPC, provide multiple binding sites for karyopherins, and form a meshwork of random coil chains through which nuclear transport proceeds. Measurements of the associations of FG-domain coated beads (Patel et al. 2007), and AFM compressibility in a way similar to that applied in the case of MAPs and neurofilament side-arms, demonstrated long-range repulsive effects of entropic origin (Figure 12.1) (Lim et al. 2006). These and other observations on transient hydrophobic interactions suggest that FG Nups anchored at the NPC center form a cohesive meshwork of filaments primarily via hydrophobic interactions (Frey, Richter, and Gorlich 2006), whereas four peripherally anchored Nups are generally non-cohesive. The interplay of these two different behaviors results in a two-gate model of NPC featuring a central diffusion gate formed by a hydrophobic meshwork and a peripheral gate that principally operates by entropic exclusion (Patel et al. 2007).

The effect of entropic exclusion probably also constitutes a critical mechanistic element of the chaperone function of IDPs (Tompa and Csermely 2004). For example, in the case of late-embryogenesis abundant (LEA) proteins (Chapter 11, Section 11.7), part of their chaperone activity probably results from preventing the aggregation of their partners by serving as "space fillers" (detailed in Chapter 14, Section 14.15) (Chakrabortee et al. 2007; Tunnacliffe and Wise 2007). The entropic origin of this effect is underlined by its similarity to the function of caseins, which bind small calcium-phosphate seeds in milk and prevent their aggregation by an entropic exclusion/entropic brush mechanism (Holt and Sawyer 1993). This mechanism termed polymer brush has been known for a long time in polymer chemistry and colloid chemistry (Bright et al. 2001).

12.2 DISPLAY SITE FUNCTIONS

Posttranslational modification (PTM) of proteins has three structural requirements: an appropriate local sequence, structural exposure, and flexibility of the site so that it can be productively accommodated by the active site of the modifying enzyme. Several lines of evidence indicate that there is an intimate relationship between disorder and these structural features (Table 12.1). The relation of disorder with phosphorylation and limited proteolysis is explored in most detail, but evidence also points to its role in ubiquitination and acetylation. The concept of short motifs, such as eukaryotic linear motifs (ELMs)/short linear motifs (SLiMs) (see Chapter 14, Section 14.2), in IDP recognition is in close association with these concepts of PTM.

12.2.1 Phosphorylation Sites

Protein phosphorylation probably is the single most important and most frequently referred-to regulatory mechanism of the cell. Proteins are reversibly phosphorylated by protein kinases (Hunter 1987; Johnson and Hunter 2005; Manning 2005) on either Ser,

Thr, or Tyr residues, and the phosphate groups are removed by protein phosphatases (Cohen 1997; Cohen et al. 1996). Reversible phosphorylation is implicated in the regulation of practically all basic cellular processes, such as cell division (Murray 2004), differentiation (Frebel and Wiese 2006), migration (Panetti 2002; Xie and Tsai 2004), apoptosis (Ojala et al. 2000), and synaptic transmission (Chen and Roche 2007; Takahashi et al. 2003; Wang et al. 2005). By conservative estimates, one-third of eukaryotic proteins undergo reversible phosphorylation (Hunter 1987; Johnson and Hunter 2005; Manning 2005), and up to 2% of the genome encodes for kinases (kinome) (Manning 2005) and phosphatases (Cohen 1997; Cohen, Chen, and Armstrong 1996). The loss of control over the balance of phosphorylation/dephosphorylation is often implicated in cancer (Futreal et al. 2004).

Studies on individual proteins have shown that phosphorylation occurs in practically all known IDPs/IDRs, such as cyclic-AMP response element-binding protein (CREB) (Parker et al. 1996; Radhakrishnan et al. 1998), protein phosphatase 1 (PP1) I2 (Hurley et al. 2007; Park and DePaoli-Roach 1994), p53 (Chehab et al. 1999; Shieh, Taya, and Prives 1999), microtubule-associated protein 2 (MAP2) (Hernandez, Avila, and Andreu 1986; Sanchez, Diaz-Nido, and Avila 2000), tau protein (Mandelkow et al. 1996; Schweers et al. 1994; Uversky et al. 1998; Zheng-Fischhofer et al. 1998), $p27^{Kip1}$ (Galea et al. 2008a), the R domain of cystic fibrosis transmembrane conductance regulator (CFTR) (Baker et al. 2007; Cheng et al. 1991), stathmin (Honnappa et al. 2006; Wittmann, Bokoch, and Waterman-Storer 2004), DARPP-32 (Hemmings et al. 1990), osteopontin (Fisher et al. 2001; Singh, Devouge, and Mukherjee 1990), calpastatin (Averna et al. 2001; Salamino et al. 1994), the C-terminal domain (CTD) of ribonucleic acid polymerase II (RNAP II) (Fabrega et al. 2003; Meinhart and Cramer 2004; Zhang and Corden 1991), LEA proteins (Alsheikh, Heyen, and Randall 2003; Heyen et al. 2002; Irar et al. 2006), 4E-binding protein (4E-BP) (Marcotrigiano et al. 1999), the cytoplasmic domain (cytD) of E-cadherin (Huber and Weis 2001), securin (Agarwal and Cohen-Fix 2002), neurofilament sidearms (Aranda-Espinoza et al. 2002), histones (Bhaumik, Smith, and Shilatfard 2007; Hansen et al. 2006), and caldesmon (Hai and Gu 2006).

Systematic bioinformatic studies underline the general correlation of the site of phosphorylation and local disorder (Iakoucheva et al. 2004). By comparing a collection of more than 1,500 experimentally determined Ser (P_S), Thr (P_T), and Tyr (P_Y) phosphorylation sites to potential sites that are actually non-phosphorylated (N_S, N_T, and N_Y), it was found that segments surrounding phosphorylation sites are significantly enriched in amino acids of higher surface exposure, charge, and flexibility and lower hydrophobicity, reminiscent of the features of disorder-promoting amino acids (Dunker et al. 2001; Romero et al. 2001). By combining the sets of positive and negative examples and considering disorder, a predictor, DISPHOS (disorder-enhanced phosphorylation predictor), could be constructed. The predictor has an improved accuracy over other phosphorylation-site predictors, such as NetPhos (Blom, Gammeltoft, and Brunack 1999) and Scansite (Obenauer, Cantley, and Yaffe 2003), with accuracies of different sites being somewhat different, 76 % for Ser, 81% for Thr and 83% for Tyr residues. DISPHOS predictions suggest that phosphorylation sites primarily occur in regulatory, cancer-associated and cytoskeletal proteins, as opposed to proteins involved in degradation, biosynthesis, and metabolism (Figure 12.2).

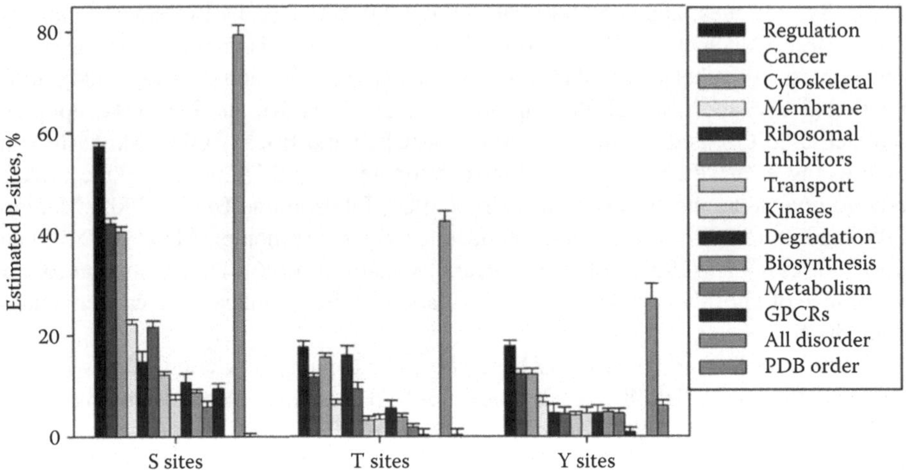

FIGURE 12.2 Level of phosphorylation in different functional classes of proteins. The percentage of actually phosphorylated potential Ser, Thr, and Tyr phosphorylation sites was estimated by DISPHOS in 12 functional protein categories in SwissProt and compared to disordered and ordered datasets. The datasets "all disorder" and "PDB order" were collected from the literature and from the Protein Data Bank. The error bars correspond to 1 SD. Reproduced with permission from Iakoucheva et al. (2004), *Nucleic Acids Res.* 32, 1037–49. Copyright by Oxford University Press.

12.2.2 Sites of Proteolytic Processing

Limited proteolysis that generates fragments of proteins with altered activity is an important irreversible PTM, not to be confused with degradation of the protein, to which IDPs in general are very sensitive. As detailed in Chapter 3, Section 3.4, preferential cleavage under limiting conditions occurs in exposed/flexible (disordered) regions of proteins, because the substrate has to engage in productive interaction with the active site of the protease along a stretch of about 12 residues (Hubbard, Eisenmenger, and Thornton 1994).

The importance of this effect is apparent in the case of calmodulin (CaM)-binding partners, for example, which bind CaM by virtue of locally disordered short recognition element (CaM-binding target (CaMBT); see Section 12.6.2) (Radivojac et al. 2006). These proteins are also stimulated by limited proteolytic digestion, as observed in the case of calcineurin (Manalan and Klee 1983) or cyclic nucleotide phosphodiesterase (Tucker, Robinson, and Stellwagen 1981). The protein thus undergoing limited digestion is no longer able to respond to CaM, or actually bind CaM.

A similar regulatory principle also seems to apply *in vivo*. The intracellular cysteine protease, calpain, has been implicated in the cleavage and *in vivo* activation of protein kinase C (PKC). Calpains constitute a family of calcium-activated cystein proteases, which have a strong preference for proteins over small peptides as substrates, and regulate a wide range of cellular processes by limited cleavage of substrate proteins (Suzuki et al. 2004). The most prominent feature around the scissile bond of these substrates is a strong preference for disorder-promoting amino acids, with a consensus pattern TPLKSPPPSPR (Tompa et al. 2004). Given that disorder was experimentally determined in several of the substrates, such as PKC, this finding suggests that disorder is a predominant recognition feature of limited cleavage by calpain *in vivo.*

The endoproteolytic activity of the proteasome also falls into this category (Liu et al. 2003). Thomas and colleagues showed that the proteasome can degrade intrinsically disordered substrates at internal peptide bonds even when they lack accessible termini (Liu et al. 2003). This limited endoproteolytic reaction suggests that disordered substrates promote gating of the proteasome, and provides a molecular mechanism for the regulated release of transcription factors from inactive precursors.

12.2.3 Ubiquitination Sites

Disorder may also be directly implicated in ubiquitination (the addition of a small conserved protein of about 50 amino acids), although the information on this relation is more limited. Its importance, however, is warranted by that targeted destruction of proteins is a critical regulatory mechanism of protein function. In the process three separate enzymatic systems take part. Ubiquitin-activating enzymes E1 activate ubiquitin in an ATP-dependent manner and transfer it to ubiquitin-conjugating enzymes, E2. Then, an E2 alone or in concert with a ubiquitin ligase (E3) binds the C-terminal carboxyl group of ubiquitin to the ε-amino group of a Lys residue in the target protein (Hershko and Ciechanover 1998). Addition of ubiquitin moieties usually continues to form a polyubiquitin chain, which targets the protein to the proteasome (see Chapter 8, Section 8.3.1). Importance of the ubiquitin/proteasome system is underscored by that it is involved in the regulation of key cellular process from cell-cycle control to inflammatory response (Hershko and Ciechanover 1998).

The involvement of structural disorder in ubiquitination was explicitly stated in the regulation of the cell cycle (Chapter 11, Section 11.3.3), in the mitotic destruction of securin and Cyclin B (Cox et al. 2002) by the E3 APC (Murray 2004). Securin (Chapter 15, Section 15.1.5) is the inhibitor of separase, the cysteine protease that initiates anaphase by cleaving the Scc1/Mcd1/Rad21 cohesin subunit, which holds sister chromatids together (Jallepalli et al. 2001; Waizenegger et al. 2002; Zou et al. 1999). Cyclin B is a mitosis-specific cyclin, the level of which rises during interphase and drops during mitosis. Securin has both D-box (RxxL) and KEN box motifs, whereas cyclin B only has a D-box. The N-terminal regions of cyclin B and yeast securin Pds1 encompassing the ubiqutination segments are intrinsically disordered (Cox et al. 2002).

Disorder of these regions sheds light on two intriguing experimental observations, multiple monoubiquitination and polyubiquitination. The N-terminal region of Cyclin B actually becomes ubiquitinated at several different Lys residues with no preference for a particular site (King, Glotzer, and Kirschner 1996). Such a mode of modification is most compatible with local disorder, which might enable several Lys residues to be brought into apposition to the active site. The mechanism of polyubiquitination (i.e., the formation of a chain of ubiquitin moieties) apparently also requires large conformational rearrangements following the addition of every ubiquitin molecule, enabled by disorder.

12.2.4 Acetylation Sites

Acetylation, in particular within the context of chromatin organization, also plays key regulatory roles (Kurdistani and Grunstein 2003; Yang 2005). An acetyl group is introduced by acetyltransferases (Yang 2004a), and is removed by deacetylases (Khan and Lewis 2005), whereas the signal itself, Ac-Lys, is recognized by specialized binding domains, such as bromo-domains (Pawson and Nash 2003; Seet et al. 2006) that generate downstream signals. Evidence of disorder has been presented on its role in histone acetylation, the recognition of the Ac-Lys moiety and deacetylation. As suggested in Chapter 11, Section 11.4.2, accessibility of genomial DNA is regulated by the epigenetic modification of core histone tails in nucleosomes. Modifications such as acetylation, phosphorylation, methylation, and ubiquitination lead to heritable changes in genome state (Pokholok et al. 2005; Shilatifard 2006; Yang 2004b). The importance of disorder in these post-translational modifications was explicitly stated (Hansen et al. 2006; Hansen, Tse, and Wolffe 1998).

The acetylation of p53 also involves disorder. The N-terminal trans-activator domain (TAD) and C-terminal regulatory domain of p53 are intrinsically disordered (Bell et al. 2002; Dawson et al. 2003), and contain numerous sites for regulatory post-translational modification, such as phosphorylation, acetylation, and ubiquitination (Alarcon-Vargas and Ronai 2002; Levine 1997). The interaction with CBP is partially mediated by the disordered C-terminal regulatory domain acetylated at Lys382 in response to DNA damage, which binds specifically to the bromo-domain of CBP (Mujtaba et al. 2004). The p53 peptide folds into a β-turn conformation upon binding (i.e., undergoes disorder-to-order transition in the presence of its partner) (Figure 12.3).

Disorder is also correlated with deacetylation, as demonstrated in the case of the Sir2 family of NAD-dependent deacetylases. These enzymes had been originally thought of as mediators of gene silencing through histone deacetylation, but several members of the family also have non-nuclear deacetylase activity against non-histone protein substrates (Blander and Guarente 2004). Their enigmatic substrate specificity was addressed in the case of the yeast homolog Hst2 by *in vitro* deacetylase assays and CD analysis (Khan and Lewis 2005), which suggested the lack of sequence specificity and ordered structure. It was concluded that Hst2 displays conformational rather than sequence specificity, preferentially deacetylating Ac-Lys within disordered regions of proteins.

12.3 CHAPERONE FUNCTIONS

In the most general sense, chaperones are protein machines that assist the folding of partner molecules by a combination of mechanisms, primarily by unfolding the misfolded substrate and preventing its aggregation, thus offering it another chance for folding attempts (Csermely 1999; Todd, Lorimer, and Thirumalai 1996). Because chaperones can assist the folding of a wide range of unrelated partner molecules in the extremely dense intracellular milieu of the cell (see Chapter 1, Section 1.6.4), they are considered

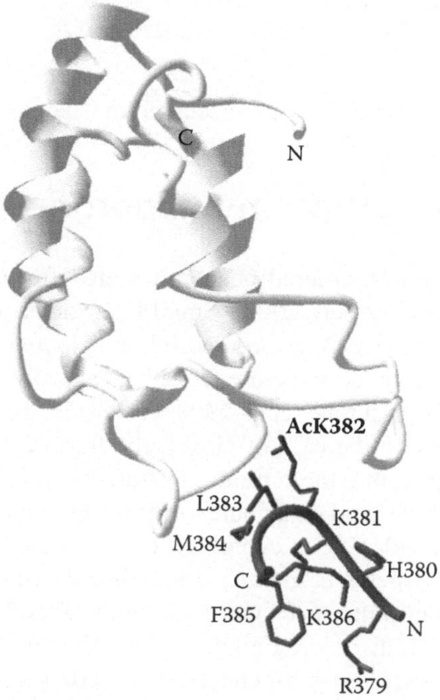

FIGURE 12.3 Structure of the CBP bromo-domain/p53 AcK382 peptide complex. Ribbon representation of the average minimized NMR structure of the CBP bromo-domain/acetylated p53 peptide complex (see Mujtaba et al. 2004). The peptide corresponds to residues Arg^{379}–Lys^{386} of p53 and encompasses acetylated Lys^{382}, the site of acetylation within the regulatory domain of the protein (pdb 1jsp).

to be highly sophisticated machines that use the energy of ATP hydrolysis to drive folding intermediates over the energy barrier of the folding trap (Csermely 1999; Todd et al. 1996). Due to their overall benefit to the cell and mechanistic demands of their action, their appearance is considered a critical early evolutionary invention (Csermely 1997). Due to the diverse mechanistic demands, chaperones are generally thought of as ordered proteins/complexes.

A bioinformatic analysis shows that chaperone action is compatible with structural disorder (Table 12.1). There is an elevated level of disorder in protein chaperones, and a very high level of disorder in RNA chaperones (see Chapter 11, Section 11.5), with 54.2% of their residues falling into disordered regions and 40% within IDRs ≥30 consecutive residues (Tompa and Csermely 2004). These numbers exceed even those of regulatory and signaling proteins, which are thought to be the most disordered functional classes (Iakoucheva et al. 2002; Ward et al. 2004), and strongly argue for the functional importance of disorder in RNA (and protein) chaperone functions. Whereas molecular details of their chaperone action are rather obscure, in principle they may act by

- preventing inactivation of a partner
- preventing the aggregation of a partner

- dispersing its aggregate, or
- helping actively refold it.

The possible mechanistic details are discussed in Chapter 14, Section 14.15.

12.3.1 Disorder in Protein Chaperones

Disordered regions of largely ordered chaperones are often associated with chaperone activity (Tompa and Csermely 2004). Small heat-shock proteins (sHsps), such as α-crystalline, Hsp16.9, and Hsp25, are composed of a globular crystalline domain and disordered N- and/or C-terminal tails, the latter being involved in the binding of substrates (Lindner et al. 2000; Lindner et al. 1998; Pasta et al. 2002; Smulders et al. 1996; van Montfort et al. 2001). In the case of GroEL, both the C-terminal and N-terminal disordered segments project into the central, substrate-binding cavity (Braig et al. 1994; Gorovits and Horowitz 1995) and contribute to chaperone action (Machida et al. 2008). The co-chaperone of Hsp90, p23, is composed of an ordered N-terminal part, which binds Hsp90 in an ATP-dependent manner, and a disordered CTD that is not required for Hsp90 binding but contributes to chaperone activity (Weikl et al. 1999). Hsp90 itself contains a highly charged, disordered hinge region (Kumar, Pavithra, and Tata 2007; Ali et al. 2006), which is necessary for chaperone function (Csermely et al. 1998).

Several reports have been published about fully disordered proteins displaying chaperone activity. β-synuclein prevents the formation of amyloid by α-synuclein, which may be relevant with respect to Parkinson's disease (Bertoncini et al. 2007). Intriguingly, the aggregation-prone α-synuclein itself has chaperone-like activity, as it can protect microbial esterases against heat, low pH, and organic solvents (Ahn et al. 2006; Park et al. 2002). Fully disordered α-casein was also described to prevent a variety of unrelated proteins/enzymes from thermally, or chemically induced aggregation (Bhattacharyya and Das 1999). A similar relation is also apparent between caseins themselves, as α- and β-casein are potent inhibitors of fibril formation by κ-casein (Thorn et al. 2005). MAP2 can prevent the DTE-induced aggregation of insulin and the thermal aggregation of alcohol dehydrogenase, whereas it can also reactivate enzymes, such as lactate dehydrogenase, malate dehydrogease, and α-glucosidase (Sarkar et al. 2004).

Whereas the physiological relevance of these *in vitro* observed chaperone effects is not clear, the situation is different in the case of LEA proteins. As detailed in Chapter 11, Section 11.7, disordered LEA proteins provide protection to seeds and mature plants under dehydration stress conditions (Goyal 2003; Irar 2006). *In vitro*, they can protect proteins against cold-, heat- (Figure 12.4), and dehydration-induced aggregation, which suggests that chaperone activity constitutes important part of their physiological functional repertoire.

12.3.2 Disorder in RNA Chaperones

In the RNA field "chaperone" proteins, which bind folding intermediates of RNA, are distinguished from "ligands," which stabilize the fold of RNA by specific binding

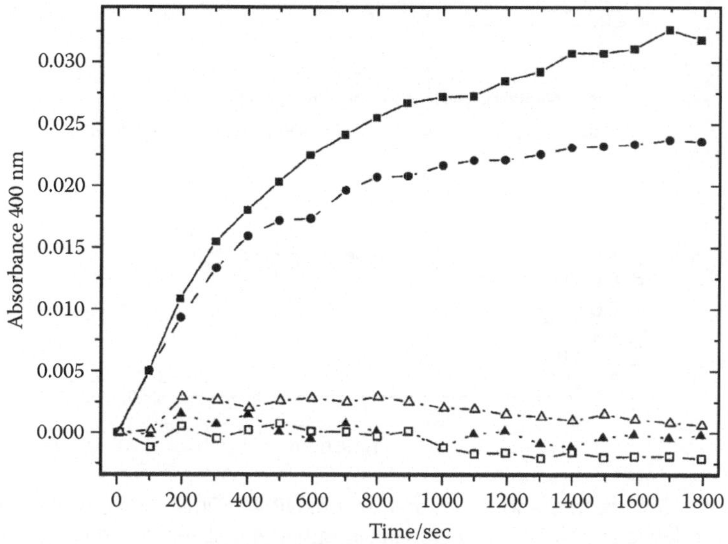

FIGURE 12.4 Chaperone effect of ERD10 and ERD14, two disordered LEA proteins. The effect of two plant LEA proteins, ERD10 and ERD14, on the heat-induced aggregation of firefly luciferase. Aggregation of 1.1 μM luciferase induced by heat (45°C) was followed without additions (■), or in the presence of 2 μM BSA (●), 2 μM HSP90 (▲), 2 μM ERD10 (□), or 2 μM ERD14 (△). Aggregation was measured by absorbance at 400 nm. Reproduced with permission from Kovacs et al. (2008), *Plant Physiol.* 147, 381–90. Copyright by the American Society of Plant Physiologists.

and incorporation into their permanent complex (Cristofari and Darlix 2002; Lorsch 2002). The distinction between the two categories is not always straightforward, but there are several cases when an elevated level of disorder in bona fide RNA chaperones is described. Probably the best characterized such protein is heteronuclear ribonucleoprotein A1 (hnRNP A1) protein, which is very effective in promoting renaturation of complementary nucleic acid strands (Figure 12.5). The disordered Gly-rich CTD of the protein promotes assembly of the protein–nucleic acid complex, and is involved in maximal renaturation activity of the protein (Pontius and Berg 1990). This observation led to the concept that nonspecific initial interactions of disordered regions of proteins can significantly accelerate macromolecular association reactions (Pontius 1993).

Nucleocapsid proteins are encoded by both the HIV virus (Ncp7) and the distantly related yeast Ty3 retrotransposon (Ncp9). These proteins have two zinc-finger motifs and disordered N-terminal and C-terminal segments, which facilitate strand transfer reactions during reverse transcription (Cristofari et al. 1999; Morellet et al. 1992). This was directly shown for nucleocapsid proteins of viruses of the Flaviviridae genera, such as GB virus B, West Nile virus, and bovine viral diarrhoea virus (Ivanyi-Nagy et al. 2007). A similar chaperone function was described and mapped into the disordered N-terminal half of the prion protein (Gabus et al. 2001). In a systematic *in vitro* trans-splicing assay of the RNA chaperone activity of ribosomal proteins of

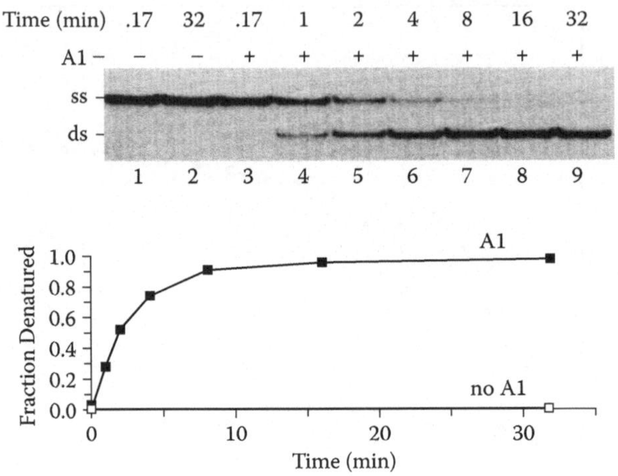

FIGURE 12.5 DNA renaturation facilitated by hnRNP A1. Time course of the renaturation of single-stranded (ss) DNA 124-nucleotide in length in the absence and presence of hnRNP A1. The time course was followed by 4.5 nM ss DNA. Lanes 1 and 2, no A1 added; further lanes, A1 at 32 nM for the time indicated. A1 under these conditions accelerates renaturation more than 3,000-fold (lower panel). Reproduced with permission from Pontius and Berg (1990), *Proc. Natl. Acad. Sci. USA* 87, 8403–7. Copyright by the National Academy of Sciences.

the large ribosomal subunit (Semrad, Green, and Schroeder 2004), it was found that several of them, such as L13, L15, L16, L18, and L19, are potent RNA chaperones. Some of these ribosomal proteins also possess protein-chaperone activity, which gave rise to the concept of "Janus" chaperones that can assist the folding of both RNA and protein partners (Kovacs et al. 2009). Another example is the fragile X mental retardation protein (FMRP), which possesses RNA-binding and chaperone activities *in vitro* under physiological conditions (Gabus et al. 2004). FMRP is a large and complex protein, and its RNA chaperone activity is thought to reside in its disordered region (Ivanyi-Nagy et al. 2005). A direct connection between the disordered region and RNA chaperone activity is shown when deletion of the underlying IDR abolishes activity of the protein. This was also observed in the case of the prion protein (Gabus et al. 2001), hnRNP A1 (Pontius and Berg 1990), and Ncp9 (Cristofari et al. 1999).

12.4 EFFECTOR FUNCTIONS

Effector functions of IDPs are probably the most straightforward to interpret in terms of the classical structure-function paradigm, because they result from binding to, and modification of, the activity of the partner (Table 12.1). Binding usually results in inhibition, but occasionally also in activation of the partner, and the resulting complexes are often found in the Protein Data Bank (PDB). When the effector has both activities, sometimes with the same partner, it is termed "multitasking" or "moonlighting."

12.4.1 Inhibitors

There are many examples of this function, a few of which are mentioned here. The archetypical inhibitor is one of the best characterized IDP, $p27^{Kip1}$ (see Chapter 14, Section 14.12.1, and Chapter 15, Section 15.1.3), which inhibits Cdk2 by binding to the Cyclin A-Cdk2 complex (Russo et al. 1996). Its close homolog, $p21^{Cip1}$, was the first IDP for which binding promiscuity was described, because it can inhibit distinct Cdks by binding to Cyclin A-Cdk2, Cyclin E-Cdk2, and Cyclin D-Cdk4 complexes (Kriwacki et al. 1996). In apparent contradiction with promiscuity, its inhibition is highly specific, as demonstrated by its inability to bind and inhibit non-cell-cycle dependent kinases (e.g., Cdk5 and Cdk7), due to the lack of specificity determinants on their cyclin partners, p35 and cyclin H (Lacy et al. 2004).

Further well-characterized IDP inhibitor-partner pairs are (see also Table 12.1) IA3-aspartic proteinase (Ganesh et al. 2006; Green et al. 2004), PKIα–cAMP-dependent protein kinase (Hauer et al. 1999a), I2-PP1 (Hurley et al. 2007; Park and DePaoli-Roach 1994), stathmin-tubulin (Honnappa et al. 2006; Wittmann et al. 2004), DARPP-32—PPI (Hemmings et al. 1990), 4E-BP1–eukaryotic translation initiation factor 4E (eIF4E) (Marcotrigiano et al. 1999), Tβ4-actin (Domanski et al. 2004; Hertzog et al. 2004), calpastatin–calpain (Kiss et al. 2008a; Moldoveanu et al. 2008), securin–separase (Jallepalli et al. 2001; Waizenegger et al. 2002), FlgM-σ^{28} (Daughdrill et al. 1997; Sorenson et al. 2004), and α-synuclein–phospholipase D2 (Jenco et al. 1998). High frequency of this functional relation is also indicated by the DisProt database (Sickmeier et al. 2007), which lists 22 IDP inhibitors.

12.4.2 Activators

Most effectors inhibit their partners, which probably follows from inhibition of activity of an enzyme being mechanistically less demanding than its activation. In fact, activation is always described for proteins that also have an inhibitory effect, suggesting multiple, often opposing functions for the same protein. To contain this kind of activity, the terms "moonlighting" or "multitasking" were suggested by Jeffery (Jeffery 1999; Jeffery 2003a) for ordered proteins. The effect is discussed in detail in Chapter 14, Section 14.6, where the most instructive examples are mentioned, such as $p21^{Cip1}/p27^{Kip1}$, which can both inhibit and activate cyclin-Cdk complexes (Bagui et al. 2003; Cheng et al. 1999); the random coil C fragment of dihydropyridine receptor (DHPR), which can interact with skeletal muscle ryanodine receptor (RyR) in two stochastically alternating modes, with one activating and the other inhibiting the partner (Haarmann et al. 2003); and Tβ4, which inhibits G-actin polymerization (Domanski et al. 2004; Hertzog et al. 2004) but can also activate integrin-linked kinase ILK (Bock-Marquette et al. 2004).

12.5 SCAVENGER FUNCTIONS

The open and extended structure of IDPs is particularly adapted to bind a large number of small ligands, such as ions and organic compounds. This may enable the ability to store and/or neturalize the compound, either for disposal or for later release upon the need of the organism (Table 12.1).

12.5.1 Salivary Proline-Rich Glycoproteins

In humans, other primates and herbivorous animals, disordered (Dunker et al. 2002; Uversky 2002a) salivary proline-rich glycoproteins (PRPs/PRGs) constitute about two-thirds of the total protein in saliva (Carlson 1993). Human PRPs are encoded by a family of six genes of significant sequence repetitions (Tompa 2003b), which show excessive length polymorphism, and high levels of substitution mutations, alternative splicing, posttranslational modification, and proteolytic processing events (Azen 1993; Carlson 1993). The key function of PRPs is to scavenge and neutralize polyphenolic plant compounds (i.e., tannins) (Baxter et al. 1997; Lu and Bennick 1998), which might cause growth retardation by inhibiting digestive enzymes and the absorption of minerals. This capacity of PRPs results from the formation of very stable complexes with tannins, via multi-dentate binding between multiple Pro side-chains of repeats and polyphenolic tannins (Baxter et al. 1997). Multi-dentate hydrophobic stacking and H-bonding with multiple Pro groups results in strong binding (Charlton et al. 1996; Hagerman and Butler 1981), due to which the tannin complex of full-length PRPs can withstand the harsh conditions within the digestive tract (Lu and Bennick 1998).

12.5.2 Caseins

Caseins constitute a family of proteins in the milk of mammals, traditionally thought to serve as nutrients for breast-fed newborns (Andrews et al. 1979; Creamer et al. 1981; Holt and Sawyer 1993). As discussed in the chapter on the history of disorder (Chapter 2, Section 2.2.4), structural disorder of caseins (i.e., rheomorphism) was among the first to be recognized (Holt and Sawyer 1993; McMeekin 1952). Perhaps as important as being nutrients in milk, caseins also function by binding and neutralizing calcium phosphate. Milk is a rich source of a great variety of nutrients, vitamins, and minerals, among which calcium and phosphate can reach concentrations as high as 20–30 mM. Calcium phosphate is not soluble in water at these concentrations, and its precipitation would have deleterious effects in the mammary gland. Caseins have binding sites for calcium phosphate seeds, and due to their open structure they can interact with small seeds with a large capacity and speed, with an apparent first-order rate constant rivaling the active-site activity of enzymes (Holt and Sawyer 1993; Holt, Wahlgren, and Drakenberg 1996).

12.5.3 Calsequestrin

Calsequestrin is a low-affinity, high-capacity calcium-binding protein, which can bind 40–50 Ca^{2+} ions per molecule, with an affinity of about 1 mM (He et al. 1993). The protein can be found in the terminal cisternae of the sarcoplasmic reticulum of muscle cells, where calcium concentrations reach millimolar levels. Thus, large storage capacity of a protein with a K_d value in the range of the concentration of the free ion enables calsequestrin to bind large amounts of Ca^{2+}, thus lowering the free Ca^{2+} concentration inside the sarcoplasmic reticulum and allowing the accumulation of Ca^{2+} via Ca^{2+}-ATPase. When the Ca^{2+}-release channel is stimulated to open, free Ca^{2+} at the terminal cisternae is increased due to dissociation of Ca^{2+} from calsequestrin (Ikemoto et al. 1991) localizing released Ca^{2+} directly at the release channel concomitant to its opening. Structurally, calsequestrin is an IDP that undergoes significant induced folding upon Ca^{2+} binding with an increase in α-helix content, compactness, and resistance to proteases (He et al. 1993), when its 3-D structure can be solved (Wang et al. 1998).

12.6 ASSEMBLER FUNCTIONS

Due to their open structure and frequent involvement in protein–protein interactions, some IDPs/IDRs function by binding and regulating the localization of other proteins relative to other constituents of the cell (Table 12.1). This localization effect may have two slightly different manifestations, targeting of activity and the assembly of large complexes, although in a strictly functional sense the two have very similar consequences. In principle, targeting can be defined as the event of binding that directs the activity of the rest of the protein onto a target. In a broader sense, the assembly of complexes also provides a proximity effect and has an element of targeting. A slight distinction between the two might come from that targeting may be provided by transient interactions, whereas assembly has a connotation of being involved in the formation of stable complexes.

12.6.1 Targeting Activity

One amply characterized example of IDP targeting is provided by RNAP II, the multi-protein complex that catalyzes the transcription of protein-coding genes in eukaryotes (Cramer, Bushnell, and Kornberg 2001; Proudfoot, Furger, and Dye 2002). As detailed in Chapter 11, Section 11.2.3, RNAP II is composed of 10 subunits (in yeast), the largest of which has a long, highly repetitive, and disordered (Bienkiewicz, Woody, and Woody 2000) CTD (Figure 11.2), which plays an essential physiological role demonstrated by deletion mutagenesis studies (Litingtung et al. 1999; Meininghaus et al. 2000). The importance of this region resides in its targeting activities, which result from the sequential, spatially, and temporarily highly coordinated binding (recruitment) of the enzymes/enzyme complexes involved in messenger RNA (mRNA) maturation (Orphanides and Reinberg 2002; Proudfoot et al. 2002).

A targeting interaction of slightly different molecular logic is represented by the binding of disordered Smad anchor for receptor activation (SARA) to Smads (Figure 12.6). The interaction is involved in transforming growth factor-β (TGFβ) signaling, which plays a central role in regulating cellular responses such as growth, differentiation, and decision on cell fate (Massague 1998). TGFβ is the ligand of a transmembrane Ser-Thr kinase receptor, the signaling of which to the nucleus is mediated by the Smad family of proteins. For specific signaling, receptor-regulated Smad 2 is recruited to the TGFβ receptor by SARA, it becomes phosphorylated by the receptor, and it hetero-oligomerizes and translocates into the nucleus (Massague 1998). SARA-Smad2 interaction is mediated by the MH2 domain of Smad2 and the 85-residue Smad-binding domain (SBD) of SARA (Wu et al. 2000). The actual process involves not only the direct interaction between SARA and Smad2 (Figure 12.6) but also between TGFβ receptor and both SARA and Smad2 (Tsukazaki et al. 1998). Targeting results from the specificity of interaction, because SARA does not interact with other Smads (1 or 5), which show 80% identity to Smad 2 in sequence.

12.6.2 Assembling Complexes

The large interaction capacity of IDPs also predisposes them to organizing the assembly of complexes. As discussed in detail in this section, various lines of observations provide evidence for this function. High-throughput screening (HTS) studies of protein–protein interactions generally apply two basic techniques—tandem affinity purification (TAP)-tag (Puig et al. 2001) and yeast two-hybrid (Y2H) (Fields 2005; Parrish, Gulyas, and Finley 2006)—to describe all protein–protein interactions in a

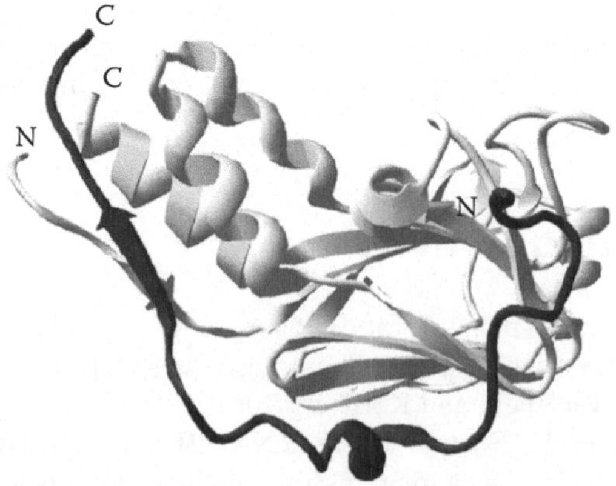

FIGURE 12.6 The complex of SARA and Smad2. Structure of the Smad-anchor for receptor activation Smad-binding domain (SARA SBD, dark grey) in complex with the MH2 domain of Smad2 (pdb 1dev). The interaction recruits Smad for phosphorylation by the transforming growth factor-β (TGFβ) transmembrane Ser-Thr kinase receptor (see Wu et al. 2000).

cell, denoted as the interactome (Aloy and Russell 2004; Arifuzzaman et al. 2006; Gavin et al. 2006). Considering the distribution of connectivities, the interactome is "scale-free" (Barabasi and Oltvai 2004) (i.e., the number of connections of proteins follows a power law). A few proteins in such a network possess a very large number of connections (hubs), whereas most others (ends) have very few, often only one, connections (Barabasi and Oltvai 2004). This arrangement suggests a functional specialization, in which hubs are preferentially involved in organizing the network, whereas ends are rather the executioners of specialized functions. The interactome shows an enhanced sensitivity to the removal of hubs (Jeong et al. 2001), which underscores the central role of proteins with multiple interactions. A range of bioinformatic studies suggest that hub proteins have an elevated level of disorder.

For example, the analysis of the Database of Interacting Proteins (DIP) suggests that predicted disorder is 21.7% for hubs and 17.2% for non-hubs. For hubs that can be found in PDB, the observed disorder is 41.2%, as opposed to 32.1% in non-hubs (Patil and Nakamura 2006). In a different approach comparing data in four interactomes (human, worm, fly, yeast) (Haynes et al. 2006), statistically significant differences were observed; for example in *C. elegans* (worm), the percentage of proteins with at least one IDR ≥40 consecutive residues is about 67% for hubs and 45% for non-hubs (Figure 12.7). By applying a dynamic threshold for hub proteins (Dosztanyi et al. 2006), proteins with the highest level of disorder were significantly enriched in hubs compared to non-hubs (32% vs. 16% in yeast, for example). When "party" hubs (which interact with most of their partners simultaneously) are compared to "date" hubs (which bind their different

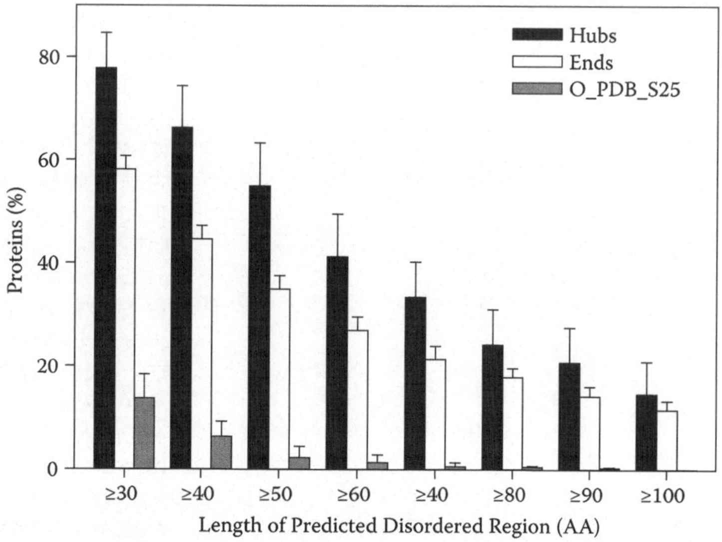

FIGURE 12.7 Distribution of predicted disorder in hubs and non-hubs. The percentages of hub (black), non-hub (end, white), and PDB (gray) proteins with at least one IDR ≥30 to ≥100 consecutive residues predicted by predictor of natural disordered regions (PONDR®) VL-XT for the *C. elegans* (worm) interactome. Reproduced from Haynes et al. (2006), *PLoS Comput. Biol.* 2, e100.

partners at different times or locations) (Han et al. 2004), 30.8% of date hubs but only 10.2% of party hubs were found to be mostly disordered by the charge-hydropathy analysis, and 20.4% of the residues in date hubs but only 7.8% of the residues in party hubs were found to fall into locally disordered regions.

Considering experimental data on hubs (Table 12.2), intrinsic disorder can apparently contribute to hub function in three different ways (Dunker et al. 2005). First, the disorder of a hub protein can provide the structural basis of binding promiscuity. Such hubs are exemplified by proteins which are fully disordered (α-synuclein, caldesmon, high-mobility group protein A (HMGA), and synaptobrevin) and some proteins, which are "mostly" disordered (i.e., have the majority of their residues in local disorder (BRCA1 and XPA)). Partially disordered hubs (p53 and murine-double minute 2 [MDM2]) have

TABLE 12.2 Hub proteins of different levels of disorder*

PROTEIN	*PONDR® %*	*STRING*	*ILLUSTRATIVE PARTNERS*
α-synuclein	100	27	Parkin, tau, CaM
Caldesmon	100	27	ERK, S100, myosin, actin, CAM
HMGA	100	18	AP1, NF-κB, C/EBPβ, Oct-1, Sp1
Synaptobrevin	100	8	Syntaxin 1, BAP31, VAMP-ass. prot., SNAP-25
BRCA1	79	119	p53, ATM, BRCA2, c-Myc, Chk1
XPA	63	41	RPA70, RPA34, ERCC1, TFIIH, XAB1
Estrogen receptor α	31	116	p53, BRCA1, CaM, c-Jun
p53	29	239	MDM2, ATM, ERK, p38, BCL-Xl
MDM2	26	72	p53, ARF, ATM, CK2, HIF-1α
Calcineurin, subunit A	16	31	NFAT, calcipressin, cabin1, SOCS-3, calsarcin
14-3-3	12	97	p53, Wee1, tau, Raf-1, Cdc25c, Bad
Cdk2	7	125	PP2A, CycE1, DNA Pol α, BRCA1, CycA
Actin	5	33	Profilin, RNase I, vit DBP, thymosin β4, cofilin
Calmodulin	3	9	Neurogranin, calcineurin, caldesmon, calponin, CaMK

* The table lists hub proteins, which are involved in multiple protein–protein interactions. The columns show the percent of disorder predicted by the PONDR® algorithm, the number of partners determined by the STRING search tool (Search Tool of the Retrieval of Interacting Genes/Proteins [von Mering et al. 2005]), and some illustrative partners. Adapted from (Dunker et al. 2005).

less residues in local disorder than in local order, and their disordered regions constitute domains/linkers next to, or between, ordered domains. Third, there are certain hubs (14-3-3 domain, actin, and CaM), which are well-structured and contain very little predicted disorder. All three types of behavior, however, are linked with protein disorder in one way or the other.

12.6.2.1 HMGA, a fully disordered hub protein

HMGA (actually a protein of two isoforms, HMGA1 and HMGA2, also termed as HMGI/Y) belongs to the high-mobility group family of nuclear proteins, which participate in a wide variety of processes through affecting chromatin dynamics and mechanics (Reeves 2001). HMGA regulates the availability of regulatory elements and structural genes in DNA, affects the expression of numerous genes *in vivo*, and influences a diverse array of physiological and pathological processes. Due to its action as a master regulator of transcription, HMGA acts as an architectural transcription factor (see Section 11.4.2) (Grosschedl, Giese, and Pagel 1994). HMGA is disordered by CD (see Figure 5.3B) (Lehn et al. 1988) and by NMR (Huth et al. 1997). The protein is extremely rich in charged residues and Pro, and contains three copies of a conserved DNA-binding peptide motif, AT-hook, of a consensus sequence A/T-x(1,2)-R/K(2)-G/P-R-G-R-P-R/K (Reeves 2001).

HMGA recognizes DNA structure, rather than nucleotide sequence, and its binding induces structural changes in DNA, such as bending, straightening, unwinding, and induction of loops. *In vivo*, the function of HMGA is carried out in interaction with a large number of other proteins, often transcription factors themselves, such as AP-1, ATF-2/c-Jun heterodimer, NF-Y, IRF-1, SRF, NF-κB p50/p65 heterodimer, C/EBPβ, Tst-1/Oct-6, HIPK-2, ELF-1, NF-AT, and PU-1 (Reeves 2001). These interactions and also post-translational modifications, including phosphorylation, acetylation, methylation, and possibly poly-ADP-ribosylation (Reeves and Beckerbauer 2001), regulate the output of HMGA action on chromatin structure and enhancosome assembly, owing to which HMGA affects the expression of more than 45 different eukaryotic and viral genes (Reeves 2001; Reeves and Beckerbauer 2001).

12.6.2.2 MDM2, a partially disordered hub protein

The oncoprotein murine double minute 2 (MDM2) is a cellular regulator of the p53 tumor suppressor (Brooks and Gu 2006; Iwakuma and Lozano 2003). MDM2 is an E3 ubiquitin ligase, which ubiquitinates p53 targeting it for proteasome-mediated degradation. This relation is the primary mechanism that regulates the transcriptional function of p53 (Kussie et al. 1996). Because p53 up-regulates MDM2 expression, the two proteins form a negative feedback loop that plays a critical role in regulating cell fate in cell division and cancer. MDM2 is 491 amino acids, many of which are in locally disordered regions (47.5%). Besides its N-terminal SWIB domain, it has a long acidic middle region followed by a Zn-finger domain of unknown function, and a RING domain at the C-terminus. The binding site for p53 is SWIB, whereas RING is common to ubiquitin ligases, and serves as the catalyst of p53 ubiquitination at multiple sites.

MDM2 is also involved in interactions with many other proteins. The partners are usually classified as effectors (i.e., upstream regulators of MDM2) and affectors (i.e., downstream proteins regulated by MDM2) (Iwakuma and Lozano 2003). Among the effectors, interaction with ARF blocks nucleocytoplasmic shuttling of MDM2 and thus enhances p53 function (Tao and Levine 1999). HIF-1α probably has a similar function, because direct interactions between HIF-1α and MDM2 modulate p53 function (Chen, Luo, and Gu 2003). MDM2 is also the target of several kinases, among which phosphorylation by ataxia-telangiectasia mutated (ATM) kinase and c-Abl interfere with the interaction of MDM2 with p53 and impair degradation of p53. Ribosomal proteins, such as L11 (Lohrum et al. 2003), L5, and L23 (Dai and Lu 2004), bind at the central acidic region, sequester MDM2 in the nucleolus, and/or directly interfere with p53 ubiquitination, thus stabilize p53. Interaction with p300/CBP, on the other hand, cooperates in the degradation of p53 (Grossman et al. 1998). MDM2 also affects the activity of several interacting proteins, such as retinoblastoma protein (RB), Sp1 transcriptional activator, E2F1, and p300/CBP. Some other proteins, such as the androgen receptor (AR) and Numb, are also targeted by the E3 ubiquitin ligase activity of MDM2 (Iwakuma and Lozano 2003).

12.6.2.3 Calmodulin, an ordered hub protein

Calmodulin (CaM) is a hub protein that can interact with a large number of partners but does not have a high level of disorder. CaM belongs to the major superfamily of Ca^{2+}-sensor proteins of nearly 600 members (Carafoli et al. 2001), it is 148 amino acids in length, and has a well-defined structure of two lobes each containing two EF-hands, simple helix-loop-helix motifs that can coordinate a single Ca^{2+} ion (Kretsinger and Nockolds 1973). The two stable lobes are connected by a flexible linker that enables a conformational change upon interaction with Ca^{2+}, which is characterized by the transition from a rather compact, inactive state to a dumbbell-shaped active species (Babu, Bugg, and Cook 1988).

CaM regulates more than 300 functionally and structurally diverse target proteins (Yap et al. 2000). CaM regulates the activity of kinases (e.g., myosin light chain kinases, CaM-dependent protein kinases, phosphorylase kinase), phosphatases (calcineurin), channels and receptors (e.g., G protein-coupled receptor kinases, plasma membrane Ca^{2+} ATPase pump, and inositol 1,4,5-trisphosphate receptors), and a bewildering variety of other enzymes and proteins (e.g., adenylate cyclases, glutamate decarboxylase, and nitric oxide synthases) (Ikura and Ames 2006; Yap et al. 2000).

Although CaM is considered an ordered protein, its interaction with its targets involves a significant element of flexibility on both sides. The classical mode of CaM interaction is that CaM wraps around a helical binding peptide/target (CaMBT) of about 20 amino acids in length in a Ca^{2+}-dependent manner, which enables CaM to bind to many different sequences with high affinities. Besides this wrapping-around mechanism, at least three other binding modes are known, in which different segments of CaM take part in the interaction with the partner (Ikura and Ames 2006). In these interactions both the plasticity of its backbone and flexibility of its side chains (nine Met residues in particular) are critical. As also discussed in Section 12.2.2, recognition by

CaM also involves the flexibility/disorder of the partner (Radivojac et al. 2006). A critical element of evidence is that often CaM-dependent enzymes are also stimulated by limited proteolytic digestion (e.g., calcineurin [Manalan and Klee 1983] or cyclic nucleotide phosphodiesterase [Tucker et al. 1981]), which brings them in a state where they can no longer responded to, or bind, CaM (see Section 12.2.2). Further, the wrapping of CaM around the binding peptide demands an open spatial location of the peptide, which is easiest to be reconciled with disorder, as also shown by the structural state of binding regions of CaM. Of 42 CaM-partner structures in PDB, CaMBT appears to be properly folded in 3 cases only, whereas it is missing in 4 cases (e.g., in the case of calcineurin [Kissinger et al. 1995]), it had been removed in 24 cases (as often done with disordered segments to help crystallization), and it is either in crystal contacts or in interchain contacts in a further 11 cases.

12.6.2.4 Disorder and complex size

The involvement of disorder in assembler functions is also manifested in its prevalence in large complexes. A systematic analysis of this feature is made possible by HTS TAP-tag/MS interaction studies, which provide data on actual multi-protein complexes (Arifuzzaman et al. 2006; Gavin et al. 2006; Gavin et al. 2002). Both prediction and actual structural data show statistically significant correlations between disorder and the number of protein subunits of complexes (Hegyi, Schad, and Tompa 2007). For example, the average predicted disorder in the yeast interactome is 6% for solitary proteins but 18.5% for components of complexes composed of 11–100 components (Figure 12.8). Proteins that are specific to large complexes (i.e., which do not occur in complexes of 10 proteins or less) are even more disordered than the average of the respective complexes (average disorder 21.0% vs. 18.0%).

12.6.2.5 Scaffold proteins

Scaffold proteins are defined by their ability to simultaneously bind several members of a signaling pathway. Their definition significantly overalps with three other closely related functional groups. *Adaptor* proteins, such as Grb2 and Nck, usually contain short globular modules (e.g., SH2 and SH3 domains) that link phosphorylated receptor tyrosine kinases to downstream effectors (Buday 1999). *Anchor* proteins are exemplified by A-kinase anchoring protein (AKAP), which localizes protein kinase A to different compartments within the cell (Pawson and Scott 1997). *Docking* proteins, such as insulin receptor substrate 1 (IRS1), are usually associated with an N-terminal membrane targeting element, such as a PH domain, a myristoylation site, or a short transmembrane domain. The docking protein becomes Tyr-phosphorylated on multiple sites by a tyrosine kinase and provides an interaction site for signaling proteins containing SH2 domains (Thirone, Huang, and Klip 2006). Bona fide *scaffold* proteins, such as Shank proteins, have the capacity to interact with several different proteins at the same time without the need of phosphorylation to create novel binding sites, and their primary function is to modulate signaling pathways (Sheng and Kim 2000). In general, these proteins usually have modular organization, with several ordered domains involved in protein–protein interactions

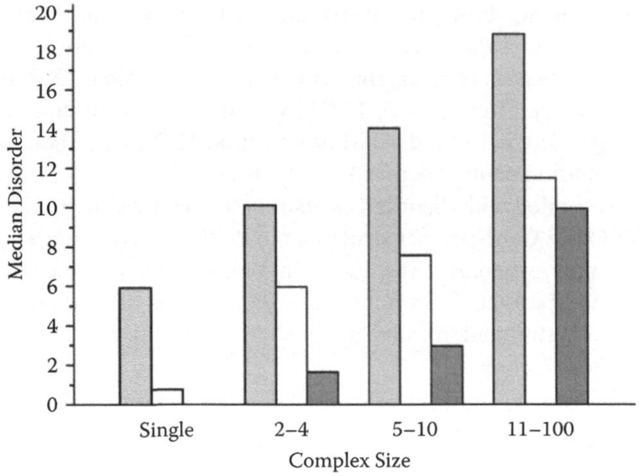

FIGURE 12.8 Predicted and observed disorder in complexes of various size. Average disorder of complexes of various numbers of subunits was either predicted by the IUPred algorithm or determined by examining their individual components in the PDB. The values thus determined for individual protein components are averaged within four groups (i.e., singular proteins and complexes of 2–4, 5–10, and 11–100 subunits) (light gray: yeast; white: *E. coli*, both predicted; dark grey: *E. coli*, observed). Reproduced with permission from Hegyi et al. (2007), *BMC Struct. Biol.* 7, 65. Copyright by BioMed Central Ltd.

connected by long uncharacterized regions. Several well-studied scaffold proteins have a high level of disorder.

The tumor suppressor gene breast-cancer 1, early onset (BRCA1) encodes a protein of 1,863 amino acids, with a central disordered region of about 1,500 amino acids in length (Mark et al. 2005). As detailed in Chapter 15, Section 15.1.4, it is implicated in a variety of cellular processes and cancer (Mark et al. 2005) and serves as a scaffold for a whole range of intermolecular interactions with p53, c-Myc, Rad50, BRCA2, and DNA, among others.

CBP is also a large, multifunctional protein that serves as a transcription co-activator of CREB in a variety of cellular functions (see Chapter 11, Section 11.2.2). It is 2,442 amino acids in length and contains several well-folded domains, but more than 50% of its sequence, including some functional domains, resides in intrinsically disordered regions (Chapter 11, Figure 11.1). The protein is involved in a complex array of interactions with various partners in regulating transcription.

Sterile 5 (Ste5) is a large scaffold protein required for signaling through the mating (or pheromone) response mitogen-activated protein kinase (MAPK) pathway in yeast (Elion 2001). It has separate binding sites for each member of the mating MAPK cascade, the MAPK Fus3, the MAPK kinase (MAPKK) Ste7, and the MAPKK kinase (MAPKKK) Ste11 (Bhattacharyya et al. 2006). Because several functionally distinct MAPK cascades use an overlapping set of kinase components (e.g., Ste11 is also a member of the osmoresponse and filamentation pathways, and Ste7 also functions in the filamentation pathway), this scaffold protein is particularly important for directing signals through the mating pathway. The protein is 917 amino acids in length, it contains only a single RING-type Zn-finger domain, and it has the capacity to tether and activate the respective pathway members

A scaffold protein in post-synaptic density (PSD) is the CASK-interactive protein Caskin (Tabuchi et al. 2002), a multi-domain protein of 1,430 amino acids, possessing 6 ankyrin repeats, 2 sterile-α motifs (SAM domains), and a single SH3 domain in the N-terminal part. There are no recognizable domains in its C-terminal 800 amino acids, which are dominated by a long, disordered Pro-rich region (Balázs et al. 2009). Caskin1 can bind the CASK adaptor protein (Tabuchi et al. 2002), the Abl-interactor-2 (Abi-2), and other nine proteins, and is presumably involved in the assembly of PSD and signaling related to Abl tyrosine kinases.

12.7 PRION FUNCTIONS

Prions were originally described as nonconventional infectious entities in mammals, which can exist in two completely different structural states, the cellular- and prion states (see Chapter 15, Section 15.3.4), the latter being implicated in a variety of deadly diseases collectively termed as prion diseases (Prusiner 1998). Propagation of the disease (i.e., the transmission of prions) results from the conversion of the cellular form to the scrapie state in a self-sustaining, autocatalytic reaction. Since the two forms are identical at the level of sequence and posttranslational modification, the sole difference between them is the conformation of the protein, and in this sense prion diseases are conformational disorders (Chien, Weissman, and Depace 2004). There are also prions that are not harmful (Table 12.1), as several proteins harness their capacity for self-sustaining conformational change for their normal, non-pathological functions (Fowler et al. 2007). There are about 10 such physiological prions known, which are unrelated but each contains similar Q/N-rich, disordered, portable prion domains (Wickner et al. 2000).

12.7.1 Sup35

Sup35 prion has been first described as the genetic element [PSI$^+$] in yeast, which causes translational read-through and is inherited in a non-Mendelian manner (Lindquist 1997). This unusual behavior can be ascribed to the altered conformation of a cellular protein, Sup35p, which is part of the translational termination complex. The protein is composed of a disordered, Q/N-rich N-terminal domain NTD or NM region of Chapter 5, Section 5.2.5.2 and Chapter 10, Section 10.5.1.2. (Mukhopadhyay et al. 2007) and a globular CTD that forms part of the complex. When the NTD undergoes self-sustaining transition to the prion (amyloid) state (Nelson et al. 2005), it occludes the globular domain, which can no longer take part in complex formation. This suppresses the termination of translation at stop codons, causing translational read-through, which may provide functional advantages under certain circumstances (Li and Lindquist 2000).

12.7.2 Cytoplasmic Polyadenylation Element Binding Protein

Arguably, the most intriguing example of disorder in a functional prion is cytoplasmic polyadenylation element binding protein (CPEB) of the marine snail *Aplysia californica*. This is a neuronal member of a larger family, which regulates mRNA translation by promoting the polyadenylation of cytoplasmic mRNA, thus activating "dormant" message and facilitating local protein synthesis at activated synapses (Si et al. 2003a; Si, Lindquist, and Kandel 2003b). Neuronal CPEB has a Q/N-rich NTD that resembles yeast prion-determinants with predicted conformational flexibility. Expressed as a fusion construct in yeast, this region brings about epigenetic changes of the cell, which is a hallmark of yeast prions (Li and Lindquist 2000; Wickner et al. 2004). In the synapses of the snail activated by repetitive neuronal stimuli, its expression is up-regulated, which promotes its transition to the prion state. This altered state of CPEB serves as a molecular marker that confers synapse specificity and promotes synaptic growth associated with the maintenance of long-term facilitation. Surprisingly, it is the dominant, self-perpetuating prion-like form that has an elevated capacity to stimulate translation of CPEB-regulated mRNA. By all criteria, CPEB is a prion with the physiological function of strengthening synaptic communication in memory formation (Si et al. 2003a; Si et al. 2003b).

Evolution and Prevalence of Disorder

13

The evolutionary history of disorder is of particular importance because disorder correlates with regulatory functions that have undergone an expansion in higher multicellular organisms. Such functions are often missing from bacteria, which raises several issues with respect to the generation and evolutionary modification of genes encoding for intrinsically disordered proteins (IDPs). In addition, tracking the evolutionary history of a protein is very closely related to understanding the molecular basis of its function, because selection among functional variants generated by mutations is intimately linked with their phenotypic effects (i.e., functional readout).

13.1 PHYLOGENETIC DISTRIBUTION OF DISORDER

The primary information pertaining to how disorder has evolved comes from comparing the level of disorder in various species, assessed by bioinformatic predictions. The level of disorder has been estimated for genomes, proteomes, and essential proteins, with similar conclusions that a sharp increase at the prokaryote/eukaryote boundary occurred.

13.1.1 Predicted Disorder in Genomes and Proteomes

Predictions of the level of disorder in entire genomes suggest that disorder is widespread. It is prevalent in most species analyzed, and it has a much higher frequency in eukaryotes than in prokaryotes (Table 13.1). The percentage of genes encoding for fully disordered proteins assessed by predictor of natural disordered regions (PONDR®) ranges from about 1–2% in bacterial genomes up to 17% in eukaryotes (i.e., in *D. melanogaster*) (Dunker et al. 2000; Romero et al. 1998). Proteins with partial disorder containing intrinsically disordered region (IDRs) ≥30 consecutive residues of

TABLE 13.1 Prevalence of disorder in whole genomes*

KINGDOM	SPECIES	DISORDER % (WARD)	PROTEINS % IDR≥30 (WARD)	CDF (DUNKER)	PROTEINS % IDR≥30 (DUNKER)
Archea	*Aeropyrum pernix*	4.7	2.1	18	57
	Archaeoglobus fulgidis	2.8	0.9	4	36
	Halobacterium sp.	6.2	5.0	11	53
	Methanococcus jannaschi	2.8	1.0	2	21
Bacteria	*Escherichia coli K12*	4.6	2.8	2	33
	Mycobacterium tuberculosis	9.1	7.0	7	51
	Staphylococcus aureus	6.2	4.5		
	Thermotoga maritima	3.3	1.8	3	36
	Bacillus subtilis			2	30
	Deinococcus radiodurans			8	52
Eukaryota	*Plasmodium falciparum*			3	48
	Saccharomyces cerevisiae	17.0	31.2	6	54
	Arabidopsis thaliana	16.8	33.8	8	57
	Caenorhabditis elegans	15.9	27.5	8	49
	Drosophila melanogaster	21.6	36.6	17	63
	Homo sapiens	21.6	35.2		
Bacteria		5.7	4.2		
Arhcaea		3.8	2.0		
Eukaryota		18.9	33.0		

* Data are taken from bioinformatics analyses of whole genomes. Ward and colleagues (Ward et al. 2004) used DISOPRED2, whereas Dunker and colleagues (Dunker et al. 2000) used PONDR® to estimate the percentage of proteins with at least one IDR 30 ≥ residues, or the percent of disordered residues in the whole genome (Ward), or the percent of fully disordered proteins by CDF analysis (Dunker). Please note that not all genomes were addressed in both analyses.

potential functional significance are even more prevalent, reaching 63% in Drosophila (Table 13.1). Predictions by DISOPRED2 corroborate these results, with somewhat different levels due to differences in the false-positive rates of the predictor (Ward et al. 2004). In this case, the frequency of proteins with at least one IDR ≥30 residues increases from 1–2% in bacteria to 36.6% in *D. melanogaster*, and 35.2% in *H. sapiens*. The sharp evolutionary increase is also apparent when averages calculated for the three kingdoms of life are compared.

These predictions have been carried out on putative protein-coding genes annotated in genomic sequences, which carries an element of uncertainty, because it is not clear what fraction of annotated genes actually encodes for expressed proteins. This point is illustrated, for example, by the steadily decreasing number of putative protein-coding genes in the human genome (Consortium 2004). Predictions of disorder for genes that are expressed (the proteome) show an even higher level of disorder than that inferred from the genome (Figure 13.1). For example, in *E. coli* the frequency of proteins with at least one long IDR ≥30 consecutive residues is 8.7% in the genome, but 13.7% in the actual proteome (40.8% and 49.1% in *S. cerevisiae*). The level of predicted disorder is even higher in essential proteins, deletion of the gene of which is lethal to the organism. Such proteins, identified in *E. coli* (Hashimoto et al. 2005) and *S. cerevisiae* (Giaever et al. 2002; Winzeler et al. 1999), have a level of disorder of 20.6% in *E. coli* and 52.4% in *S. cerevisiae* (Tompa, Dosztanyi, and Simon 2006b).

13.1.2 The Origin of Disordered Proteins in Eukaryotes

The sudden spread of disorder at the prokaryote/eukaryote boundary raises the question how genes encoding for IDPs have arisen. There are several possible mechanisms, such as *de novo* generation (Schmidt and Davies 2007; Sorek 2007), lateral gene transfer (LGT), and horizontal gene transfer (HGT) (Hotopp et al. 2007), but these have not

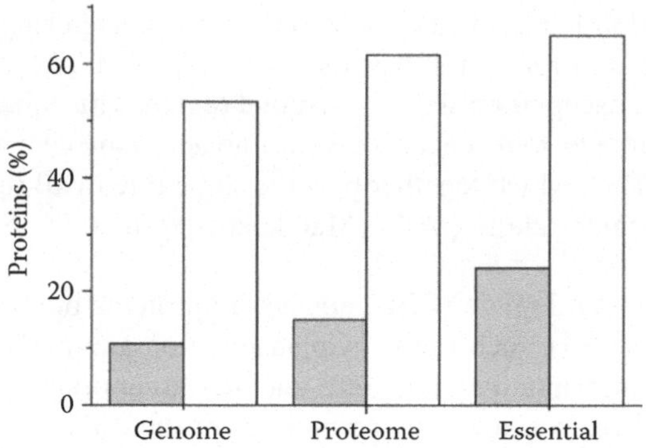

FIGURE 13.1 Structural disorder in *E. coli* and *S. cerevisiae* genomes and proteomes. Structural disorder was predicted by IUPred for the *E. coli* (gray columns) and *S. cerevisiae* (white columns) genomes, proteomes, and essential proteins. The percent of proteins with at least one IDR ≥30 consecutive residues are shown. Reproduced with permission from Tompa et al. (2006), *J. Proteome Res.* 5, 1996–2000. Copyright by Elsevier Inc.

been studied yet. The third possible mechanism is gene duplication, which is the leading mechanism of the generation of novel genes (Conrad and Antonarakis 2007). In terms of the generation of novel genes encoding for IDPs, this would assume that when genes of ordered proteins duplicated, one copy preserved its original structure, whereas the other has became an IDP by acquiring multiple mutations. This mechanism is somewhat unlikely, because it assumes a series of mutations that can lead from an ordered to a disordered state, preserving functionality without degenerating into a pseudogene (Chothia et al. 2003; Prince and Pickett 2002). There is more evidence for duplications and exchange at the domain level—attaching a disordered domain to an already existing protein. Such events allowed gradual evolutionary changes and experimentation with chimera constructs that preserved their original function, undergoing stepwise modifications.

13.1.3 The Generation of Disordered Domains by Gene Duplication and Module Exchange

There are several examples of protein families with common disordered domains, which have apparently arisen by domain duplication and subsequent relocation by gene rearrangements (Tompa et al. 2009). For example, the kinase inhibitory domain (KID) domain of Cdk inhibitors $p21^{Cip1}$, $p27^{Kip1}$, and $p57^{Kip2}$ (Chapter 15, Section 15.1.3) are homologous, which suggests their common evolutionary origin (Lacy et al. 2004). The rest of the molecules are extremely variable in length (the total length within humans changes from 164 amino acids for $p21^{Cip1}$ to 316 amino acids for $p57^{Kip2}$) and show no similarity in sequence either among themselves or with other proteins.

A common disordered domain is also apparent in the family of microtubule-associated proteins (MAPs) tau protein, MAP2 and MAP4 (Chapter 10, Section 10.2.3.3 and Chapter 11, Section 11.6.3) which are extremely variable in length (from 441 amino acids in the case of tau to 1,827 amino acids in the case of MAP2, both in humans) and display no similarity, except for a short C-terminal region. This tubulin-binding domain (TBD) consists of three or four repeats of 31 amino acid-long elementary microtubule-binding region (MTBR), which together play the same role in all these MAPs, binding and stabilization of microtubule (MTs) (Mandelkow et al. 1996; Sanchez, Diaz-Nido, and Avila 2000).

The catenin-binding domain (CBD) appears in proteins binding to β-catenin (see Chapter 11, Section 11.3.1), such as the cytoplasmic domain (cytD) of E-cadherin and the Lef family of transcription factors Tcf3 and Tcf4 (Gooding, Yap, and Ikura 2004; Poy et al. 2001). β-catenin is a crucial component of the cell adhesion machinery and is also an intracellular mediator of the Wnt signaling pathway, its various functions being enabled by its partners sharing the same disordered recognition domain (see Figure 11.3).

Wiskott–Aldrich syndrome protein (WASP) homology domain 2 (WH2) domain is about 40 amino acids in length, it is fully disordered (Domanski et al. 2004), and it appears in several actin-binding proteins, such as Tβ4, ciboulot, and WASP, among others (Paunola, Mattila, and Lappalainen 2002). The domain occurs in different sequence contexts, but almost always in a Pro-rich region and with conserved features, which suggest that in all homologs it functions in actin binding. The binding event has different functional outcomes, because a single WH2 domain in Tβ4 inhibits actin polymerization (Figure 11.5), whereas several tandem WH2 domains in other actin-binding proteins promote actin polymerization (Chereau et al. 2005) (see also Chapter 11, Section 11.6.1).

The possible generality of disorder spreading by domain duplications and exchange between genes is also underscored by the observation that about 14% of all Pfam domains are mostly disordered by prediction (for further examples, see Chapter 14, Section 14.2.4). Because Pfam domains are, by definition, homologous, they must have spread by duplications and module exchange, which suggests that these mechanisms contributed significantly to the spread of disorder in eukaryotes.

13.2 FAST EVOLUTION OF IDPS BY POINT MUTATIONS

A common observation in the IDP field is that homologs of IDPs/IDRs are more difficult to find by sequence similarity search than homologs of globular proteins, which infers the large variability of the former. The rate of this evolutionary change could be compared by aligning sequences in which a globular domain and a disordered region is also present (Brown et al. 2002). Homologs of 26 such families were aligned by the CLUSTALW algorithm (Thompson, Higgins, and Gibson 1994), and genetic distances between each pair of sequences were calculated by the Protdist algorithm from the PHYLIP package (Felsenstein 1997) separately for the ordered and disordered regions. A statistical measure of variability (average genetic distance) showed that the disordered region evolves faster than the ordered region in 19 families, at about the same rate in 5 families, and significantly more slowly only in the remaining 2 families (Figure 13.2). The functions of the faster-evolving disordered regions are diverse, including binding sites for proteins, deoxyribonucleic acid (DNA), and ribonucleic acid (RNA), as well as flexible linkers, whereas the more slowly evolving disordered regions are mostly involved in DNA binding. In general, disordered regions not only evolve faster but also tend to accept different amino acid replacements than ordered proteins. In the case of IDPs, tendencies for amino acid replacements are different from those of globular proteins, as demonstrated by developing a novel scoring matrix, DISORDER, for amino acid replacement statistics adjusted to suit disordered regions. This matrix results in a significant improvement in the ability to detect and discriminate related disordered proteins, the average sequence identity of which is below 50% (Radivojac et al. 2002). A critical feature of the matrix, for example, is that rare hydrophobic residues (Phe, Tyr, Trp) are very conserved in IDPs, which is probably a direct consequence of their involvement in recognition functions (see Chapter 14, Sections 14.2.2 and 14.2.5).

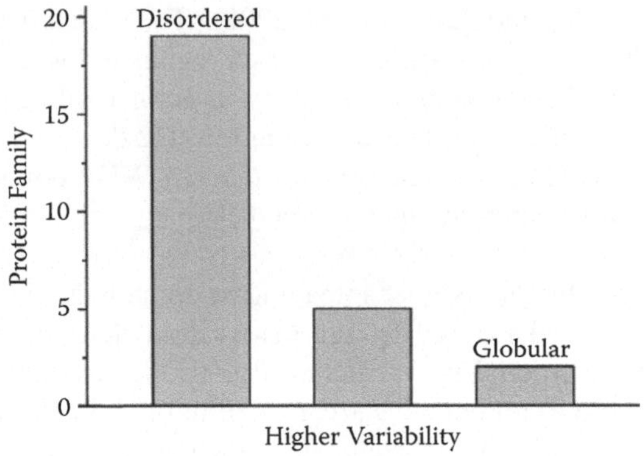

FIGURE 13.2 Evolutionary variability of disordered vs. globular proteins. Evolutionary variability of globular and disordered regions was calculated in 26 protein families, which contain separate globular and disordered regions. Average values were compared for each family to see which region evolves faster. (data from Brown et al. 2002).

13.2.1 Neutrality in the Evolution of IDPs

Fast evolution of IDPs can also be approached from the direction of the redundancy of the genetic code (see Chapter 1, Section 1.3). That is, certain base changes at the DNA level do not change the sense of the codon (i.e., the amino acid encoded (synonymous, sense, or silent mutations)), others change the amino acid (nonsynonymous or missense mutations), or occasionally they change the codon to a stop codon (nonsense mutations). As a consequence, changes at the DNA level are subject to different evolutionary selection forces, because nonsynonymous mutations affect the functionality of the protein and are less frequently retained than synonymous mutations. Synonymous mutations are considered neutral, not subject to selection forces at the protein level. The ratio of nonsynonymous over synonymous mutations (K_a/K_s) can be used to characterize whether a certain amino acid or region in a protein is subject to purifying selection forces ($K_a/K_s = 0.1$–0.2, the region is "functional," i.e., mutations are usually "disadvantageous"), or it is not ($K_a/K_s = 1.0$, the region is "nonfunctional," in which case the mutation is "neutral"), or maybe it is selected for ($K_a/K_s > 1$, in a region that undergoes "adaptive" evolution, in which case the mutation appears as "advantageous") (Hurst 2002).

This approach suggests close to neutral evolution in some IDPs/IDRs, such as the trans-activator domain (TAD) of transcription factor SRY, the sex-determining region of the Y chromosome, also termed testis-determining factor (Tucker and Lundrigan 1993; Whitfield, Lovell-Badge, and Goodfellow 1993). In the case of murine SRYs, the K_a/K_s ratio for the DNA-binding high-mobility group (HMG) domain is in the range of 0.1–0.2 in most cases, whereas for the TAD it falls in the range 0.4–0.6 (Tucker and Lundrigan 1993). In the case of primates, the results are similar, with $K_a/K_s = 0.1$–0.2 for the HMG domain, but much higher, 0.6–0.8, for the TAD (Whitfield et al. 1993). Given the general layout of transcription factors having a DNA-binding domain that tends to be ordered and a TAD(s) that tends to be disordered (see Chapter 11, Section 11.2.1), these results suggest that disordered TADs tolerate significantly greater amount of nonsynonymous mutations (i.e., they evolve in an almost neutral fashion).

The linker region (termed intrinsically unstructured linker domain, IULD), of RPA70, the 70-kDa subunit of replication protein A (RPA70), also appears to evolve neutrally. RPA70 plays a critical role in replication, recombination, and DNA repair and has an N-terminal DNA/protein-interaction domain (DBD F) connected by the linker to two tandem high-affinity, single-stranded, DNA-binding domains (DBD A and B). Sequences from distant species (animal, fungi, and plant) are too diverged to be aligned, and evolutionary variability of the linker region can only be approached by examining more closely related mammalian sequences (Daughdrill et al. 2007), very much like in the case SRY. Most sites in the linker region evolve nearly neutrally, with certain interspersed conserved sites, which happen to be mostly Gly residues (six are preserved in all nine mammalian homologs, and six are conserved in eight of them). Apparently, rapid neutral evolution is compatible with flexibility being the primary functional prerequisite of the linker (see Chapter 13, Section 13.4.1), which also explains the presence of conserved Gly residues critical for maintaining flexibility.

13.2.2 Disordered Regions May Also Be Conserved

Despite the general tendency of IDPs to evolve rapidly, certain disordered regions are rather resistant to evolutionary changes. In 10,802 domain families from the InterPro database, 30% (2,898 families) contain conserved IDRs ≥20 consecutive residues (Chen et al. 2006a, b). These regions termed conserved disorder prediction (CDP), are short; only 9% of them exceed 30 residues in length, and they usually cover less than 15% of the respective domain. The longest CDP is 171 amino acids (in dentin matrix protein); 8.7% of CDPs actually cover more than half of the respective domain, and 16 CDPs cover the entire domain. CDPs can be found in all kingdoms of life, but long ones (> 50 residues) are almost 10 times more frequent in viruses and eukaryotes than in bacteria and archaea. The functions of the domains harboring CDPs agree with the general functional classification of disorder (Chapter 11): most CDPs occur in proteins of DNA/RNA binding, ribosomal function, protein binding (both signaling/regulation and complex formation), and coat/capsid formation functions.

13.3 FAST EVOLUTION OF IDPS BY REPEAT EXPANSION

IDPs not only evolve rapidly by point mutations, but also by repeat expansion enabled by their repetitive nature. Tandem repeats of short motifs can be found in many IDPs, and they often are directly involved in the function of the protein.

13.3.1 Micro- and Minisatellites in Protein Evolution

Repetitive regions are highly prevalent in genomes. Repeats are classified roughly according to the length of the repeat unit as satellites (several thousand bp), minisatellites (10–100 bp), and microsatellites (1–10 bp), also called variable number tandem repeats (VNTRs) (Bois and Jeffreys 1999; Vergnaud and Denoeud 2000). About half of the genome is made up of repetitive elements; most conspicious is the satellite nature of the centromere (typically 25–170 bp repeat units) and telomere (typically 6–8 bp repeat units) region of chromosomes. Repeats appear to be most frequent in non-coding regions, but there are also many examples that they occur in coding regions (i.e., within proteins themselves) (Bjorklund, Ekman, and Elofsson 2006; Heringa 1998; Karlin et al. 2002; Marcotte et al. 1999). In fact, an influential theory on the evolution of proteins assumes that primordial protein-coding genes have all been generated by internal duplications of small sequence units (Ohno 1984; 1987). This mechanism may have had many evolutionary advantages, such as the increased chance of encoding for an open reading frame, the high probability of generating super-secondary structural motifs (domains), the resistance to randomly sustained mutations, and self-templating in replication processes. In accord, many repeats can be seen even in present-day domains/proteins (Marcotte et al. 1999; Soding and Lupas 2003). Repetitive elements may correspond to whole domains spread out in different genes of the genome, they may be repeated in tandem within the same gene, or they may form supersecondary structural elements of domains (Patthy 1996; 1999; Soding and Lupas 2003). Often, they appear as internal tandem repeats within disordered proteins.

A related comparative proteome analysis of five complete eukaryotic genomes (*H. sapiens*, *D. melanogaster*, *C. elegans*, *S. cerevisiae*, and *A. thaliana*) shows that runs of amino acid (≥5 identical amino acids) are also frequent in proteomes (Karlin et al. 2002; Karlin and Burge 1996). Each proteome contains numerous runs, with the percentage of proteins with at least one run being 13% in worm, 15% in yeast, 20% in human and weed, and 27% in the fly. Ser, Ala, and Gln account for a significant proportion of hits in each species. Proteins with Ser runs range from 13.7% (human) to 33.4% (weed), with Ala runs from 4.7% (yeast) to 26.3% (fly) and Gln runs from 5.8% (weed) to 33.9% (fly). Amino acid runs usually correspond to small polar or acidic residues, and they avoid aliphatic and aromatic (Ile, Val, Met, Tyr, Phe, and Trp), plus Arg and Cys residues (Kreil and Kreil 2000).

Homopolymeric runs tend to be locally disordered, as indicated for stretches of Gln (Chen, Luo, and Gu 2003; Vitalis, Wang, and Pappu 2007), Ala (Chen, Liu, and Kallenbach 2004), Ser (Howard et al. 2004), and Pro (Bochicchio and Tamburro 2002). By looking for sequence patterns in IDPs, Lise and Jones (Lise and Jones 2005) also found that the most significant patterns are invariably repeats (Chapter 10, Section 10.1.2). The functional significance of these regions in IDPs is rather enigmatic, but they may be involved in protein–protein interactions, as demonstrated by raplacing Pro- or Gln-rich TADs of transcription factors by homopolymeric stretches of these amino acids (Gerber et al. 1994).

The analysis of a collection of 126 IDPs showed directly that the percentage of proteins with tandemly repeated segments is much higher in IDPs (39%) than in SwissProt (14%), yeast (18%), or human (28%) proteins (Tompa 2003b). Repeat regions make up a very large fraction, about 34%, of all IDPs, as opposed to about 7% of SwissProt proteins. Microsatellite and minisatellite sequences are about equally represented, and they are often essential to the function of the protein. In addition, these regions often show an exceptional evolutionary activity (i.e., repeat length variation). The possible mechanisms are discussed next.

13.3.1.1 Mechanisms of repeat expansion

The replication of repetitive DNA is often accompanied by changes in the number of repeat units. Such events that increase (expand) or decrease (contract) the length of the repeat region occur about six orders of magnitude more often than point mutations and make repetitive regions the evolutionary hot spots of functional changes. The exact genetic mechanism depends on the length of repeat units. In the case of microsatellites, the preferred mechanism results from the occasional stalling of RNA polymerase during DNA replication. Stalling may bring about the dissociation and off-register reannealing of the leading and lagging strands of DNA, causing the length of the DNA copied become different from that of the template. This mechanism is termed "polymerase slippage" (Wells 1996). In the case of longer repeats encoded by minisatellites, the primary mechanism is nonhomologous meiotic recombination by unequal crossing over or gene conversion, in which repetitive regions align in the wrong register (Buard and Jeffreys 1997; Vergnaud and Denoeud 2000). This causes one allele to become longer, and the other shorter, by the same number of repeat units.

Due to these effective mechanisms, repeat regions are extremely mutagenic, and often show allelic variations within a species (polymorphism) or differences between orthologues of different species. Because of the frequent involvement of repeat regions in the functions of IDPs (Tompa 2003b), the ensuing evolutionary activity confers on them an exceptional functional variability and may also partially explain the generation of long repeat regions in IDPs. This may be one underlying mechanism of the spectacular evolutionary spread of disorder at the prokaryote/eukaryote boundary and afterward.

It should be noted that expansion of homopolymeric repeats may also be of catastrophic functional consequences, as seen in the case of polyQ regions (see Chapter 15, Section 15.3.3). In a range of neurodegenerative diseases, such as Huntington's disease or Kennedy's disease, pathogenic expansion of a normal polyQ region less than 40 Gln residues to above 40 residues results in the deposition of the protein in the form of amyloid that causes neuronal death (Perutz 1999; Schaffar et al. 2004). Because of the underlying dependence of the mutation rate of the repeat region on length, repeat regions are also termed hypermutable sites or mutable mutators (Bois 2003). As shown next, repeat expansion generates viable functional variants in many proteins.

13.3.1.2 Tandem repeats in the CTD of RNA polymerase II

RNA polymerase II (RNAP II) catalyzes the transcription of protein-coding genes in eukaryotes (Cramer, Bushnell, and Kornberg 2001; Proudfoot, Furger, and Dye 2002). This protein actually contains 10 subunits in yeast, the largest of which (RPB1) has a long, disordered (Bienkiewicz, Woody, and Woody 2000) C-terminal domain (CTD) missing from the structure, as determined by X-ray crystallography (Cramer et al. 2001). As detailed in Chapter 11, Section 11.2.3 and Chapter 12, Section 12.6.1 (see also Figure 11.2), the CTD is involved in the assembly of transcription pre-initiation complex, whereas changes in its phosphorylation pattern signal various phases of the transcription process (Orphanides and Reinberg 2002; Proudfoot et al. 2002). A large part of the CTD is composed of heptapeptide units of a consensus sequence YSPTSPS, and it has undergone a stepwise expansion in evolution due to which its present length correlates with the complexity of the organism. The number of repeats is 11 or 17 in malarial parasites (Giesecke et al. 1991), 26 or 27 in *S. cerevisiae*, 40 in *A. thaliana*, 42 in *C. elegans*, 45 in *D. melanogaster*, and 52 in mammals (Young 1991). Removal of more than half of the repeats is lethal in mammalian cells (Meininghaus et al. 2000), whereas deletion of about half of them results in various viability phenotypes, such as heat- and cold-sensitivity in yeast (Nonet, Sweetser, and Young 1987). Even removal of a shorter region, such as 13 repeats, causes dwarfism in the mouse (Litingtung et al. 1999).

13.3.1.3 Tandem repeats in the PEVK region of titin

Titin is the largest protein (about 36,000 amino acids) encoded in the human genome. This protein spans half the length of the vertebrate striated muscle sarcomere (Granzier and Labeit 2002; Labeit and Kolmerer 1995). Due to its elastomeric property, a primary function of titin is to provide the resting tension of muscle (i.e., to generate passive force when a relaxed muscle is stretched) (see Chapter 5, Section 5.7 and Chapter 12, Section 12.1.3). There are two regions of completely different nature that combine to impart unique physical elasticity on titin—a highly repetitive I-band region composed of tandem Ig domains, and a long repetitive disordered region (Kellermayer et al. 1997; Ma, Kan, and Wang 2001; Ma and Wang 2003), known as the Pro, Glu, Val, Lys-rich (PEVK) domain, which is mainly composed of 28-mer Pro-rich units interspersed with homopolymeric runs of Glu residues (Gutierrez-Cruz, Van Heerden, and Wang 2001; Labeit and Kolmerer 1995). The PEVK domain shows large tissue-specific differences due to alternative splicing. In humans, its length varies from 163 residues in cardiac muscle to 2,174 residues in soleus muscle (Labeit and Kolmerer 1995).

The connection of length with function is demonstrated by the fact that tissue-specific combinations of the various forms have basically different mechanical properties: short isoforms tend to have a high resting tension, whereas longer isoforms provide a much lower tension (Watanabe et al. 2002). Large variations in length derive from differences in the number of repeat units, from 5 in the cardiac N2B isoform up to 60 in soleus muscle (Greaser 2001; Gutierrez-Cruz et al. 2001). The evolutionary plasticity of the PEVK region can be inferred from comparing human titin with its *D. melanogaster* homolog, D-titin, which actually contains two PEVK regions in the place of one in the human protein, both being very different in length (1,240 and 5,065 residues (Machado and Andrew 2000)).

13.3.1.4 Tandem repeats in prion protein

The NMR structure of murine (Riek et al. 1997), bovine (Lopez-Garcia et al. 2000), and human (Zahn et al. 2000) prion protein (PrP), involved in prion disease (Chapter 15, Section 15.3.4), shows that the protein can be divided into a globular C-terminal half and a disordered N-terminal half. Within the disordered half, there is a characteristic octapeptide repeat region, which most often consists of five variant repeats and binds a copper ion with high affinity *in vitro* (Jackson et al. 2001) and probably *in vivo* as well (Brown et al. 1997a) (see Chapter 11, Section 11.8). The genetic instability of this region results in significant variations in repeat number, which is characterized by 3–6 octarepeats in mammals and 6.5–9 hexarepeats in birds, for example (Wopfner et al. 1999). Deviations from the most frequent allele (5 repeats) have been described in the form of repeat-number polymorphism (–1) or insertion/deletion mutations resulting in 2–9 repeats in individuals or families (Goldfarb et al. 1991; Palmer and Collinge 1993). The potential functional significance of these differences may come from high-affinity copper binding, which cannot be evolutionarily explored because of the extreme amyloidogeneicity of the longer alleles (Goldfarb et al. 1991; Palmer and Collinge 1993).

13.3.2 A Functional Model of Repeat Expansion in IDPs

These and other examples of the functions of repeats (Tompa 2003b) demonstrate different possible functional consequences of repeat expansion in the evolution of IDPs (Figure 13.3). From a purely functional perspective, repeat expansion of RNAP II resulted in a gradual change in function, which yielded functionally non-interchangeable variants, as demonstrated by deletion of large parts of the CTD of mammalian RNAP II. Repeat units resulting from the expansion became functionally nonequivalent, because

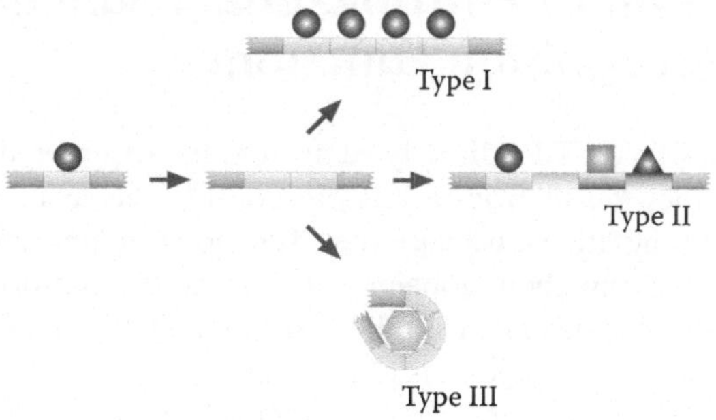

FIGURE 13.3 Repeat expansion in the evolution of IDPs. IDPs/IDRs are often made up of internal repeats, which may follow three evolutionary routes of expansion. Type I denotes regions in which repeats generated by tandem duplication(s) remain functionally equivalent. Repeat units in type II regions diversify due to mutations leading to changes in sequence. A type III repeat region is envisaged to acquire a novel function as a consequence of expansion. Reproduced with permission from Tompa (2003), *BioEssays* 25, 847–55. Copyright by Wiley Periodicals.

of some changes in repeat sequences and/or different distances from the catalytic unit of the polymerase. Titin followed a different evolutionary path, in the sense that repeat units of its PEVK domain remained functionally equivalent in terms of the physical elasticity they provide. Copper binding by the prion octarepeat represents still another evolutionary alternative, because it indicates a possible sudden functional change when the repeat region acquired physiologically significant affinity upon extending to four repeats. These three alternative mechanisms represent three different types of logic in IDP function and evolution by repeat expansion (Tompa 2003b).

13.4 FAST EVOLUTION AND FUNCTIONALITY OF DISORDERED PROTEINS

IDPs evolve much faster than globular proteins by point mutations and/or repeat expansion. Apparently, this observation raises a very serious issue concerning the functionality of these proteins, because by definition structural and/or functional constraints should manifest themselves in evolutionary restraints. Some evolutionary variability is advantageous because it provides the raw material for selection among functional variants, but too much variability is disadvantageous, because it works against retention of function. What is the solution to this dilemma in the case of IDPs? As discussed throughout the book (see Chapter 11 and Chapter 12), the molecular logic of IDP functions differs from that of ordered proteins, and thus they can retain function despite respectable changes in sequence. Within IDPs, different functional categories display different behaviors.

13.4.1 Retention of Entropic-Chain Functions and Recognition Functions

Entropic chain functions of IDPs directly stem from the disordered state. These functions, especially if they result from a random coil-like behavior of the polypeptide chain, are resistant to mutations, because their function remains unchanged as long as the mutation(s) do not bring about a major transition in the conformational ensemble. This indifference was demonstrated by the resistance of the binding function of the transcription factor Oct1 to mutagenesis of its linker region (van Leeuwen et al. 1997), or the resistance of entropic gating function to rapid changes in the FG repeat region of nuclear pore complex (NPC) component Nups (Denning and Rexach 2007).

The interplay of linker function and sequence variations was directly addressed in RPA70, a protein involved in replication, recombination, and DNA repair (Daughdrill et al. 2007; Olson et al. 2005). As shown in Section 13.2.1, the linker (IULD) evolves almost neutrally. Its disorder is essential for the mobility of DBD F, so that it can interact with ssDNA, other proteins, or the other subunits of the replication protein A heterotrimer. Sequentially, five entirely different linker domains of RPA70 homologs from the three kingdoms of life—two from animals, one from fungi, and two from plants—show very similar backbone flexibilities by NMR. Thus, the entropic-chain function of an IDP can be retained in the face of negligible sequence conservation.

The conservation of recognition functions poses an even more serious challenge in light of the range of sequence variations commonly observed in IDPs (Brown et al. 2002). The solution may reside in the special binding mode of IDPs (i.e., that they often recognize their partners by virtue of short recognition motifs) (see Chapter 14, Section 14.2). These elements are often constructed from a few specificity determinant residues interspersed in highly variable and disordered regions (Fuxreiter, Tompa, and Simon et al. 2007). Apparently, a large fraction of these recognition sequences function as a linker that is rather free to mutate, and only the very little fraction of direct recognition residues are subject to evolutionary constraints, practically falling into the level of noise when considering the variability of the entire domain/protein.

Calpastatin, the inhibitor of calpain, demonstrates this situation. Calpastatin is composed of four equivalent inhibitory domains of about 140 amino acids, each capable of very tight and specific inhibition of the enzyme with inhibitory constants ranging from 4.5 pM to 4 nM (Hanna, Garcia-Diaz, and Davies 2007). Because the inhibitor has co-evolved with its cognate enzyme and each domain can inhibit the same enzyme species, the absolute conservation of a recognition function is assured. Alignment of the domains (Figure 13.4) shows the presence of short, conserved segments within each domain (termed subdomains, marked A through D). Subdomains A, B, and C (Ma et al. 1994; Ma et al. 1993; Takano et al. 1995), and probably also subdomain D (Kiss et al. 2008b) serve as the recognition determinants of the inhibitor, and are in direct contact with the enzyme, whereas the linker regions connecting them remain free even in the state bound to the enzyme (Kiss et al. 2008b; Moldoveanu, Gehring, and Green 2008). Binding occurs through a few specificity-determinant residues only, but the intervening

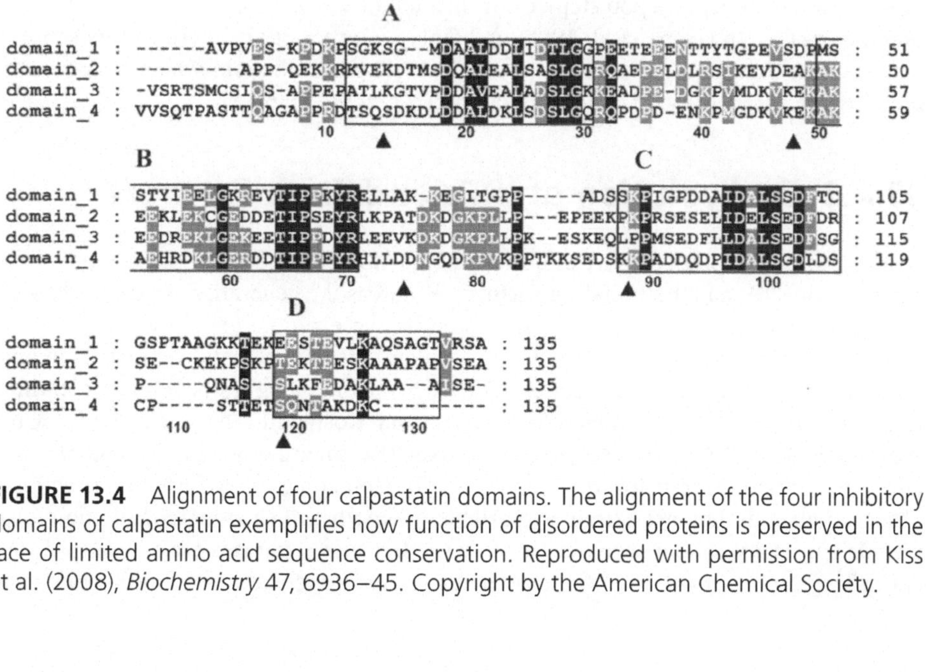

FIGURE 13.4 Alignment of four calpastatin domains. The alignment of the four inhibitory domains of calpastatin exemplifies how function of disordered proteins is preserved in the face of limited amino acid sequence conservation. Reproduced with permission from Kiss et al. (2008), *Biochemistry* 47, 6936–45. Copyright by the American Chemical Society.

regions are rather insensitive to the identity of the actual residues (Betts et al. 2003). The flexible linkers separating subdomains are largely variable, because they only have to ensure a range of distances and relative orientations of the subdomains for effective recognition. Thus, binding results from the combination of very short subsites connected by flexible linkers, which overall provides respectable binding strength and specificity, yet it enables large evolutionary variability (Hanna et al. 2007).

13.4.2 Recognition Another Way: The Lessons from Fuzziness

Fuzziness is a concept of disorder in the bound state of IDPs, discussed in detail in Chapter 14, Section 14.8, which is also relevant with respect to the apparent contradiction between rapid evolution of IDPs and their retention of function. In certain cases, molecular recognition and partner binding occurs without ordering of the IDP, as described for T-cell receptor ζ-chains (Sigalov, Aivazian, and Stern 2004; Sigalov 2004) and the product of the umuD gene (Simon et al. 2008). In other cases, sequence-independence of recognition was described (cf. Chapter 14, Section 14.10), when recognition is apparently resistant to the scrambling of sequence, as in the case of transcription factors (Hope, Mahadevan, and Struhl 1988; Sigler 1988), linker histones (Hansen et al. 2006), or prions (Ross et al. 2005; Wickner et al. 1999). Probably all these cases are associated with a distributed array of rather loosely defined and transient contacts with the partner, which do not bring about a well-defined ordered structure even in the complexed state. The underlying lack of strict geometric complementarity in binding (Sigler 1988) might permit a rapid evolution, because most residues are not critical for contacting the partner. A manifestation of this unorthodox mode of binding is the gradual, as opposed to sudden, loss of function upon stepwise truncation of IDRs, as in the case of transcription factors Gcn4p (Hope et al. 1988) and Gal4p (Gerber et al. 1994). In this sense, it may be said that globular proteins die suddenly upon mutation, whereas disordered proteins dye "gracefully" (i.e., lose function in a gradual manner).

13.4.3 Co-Evolution of IDPs and Their Partners

A recognizable evolutionary relationship might also be eroded by the (adaptive) co-evolution of IDPs and their binding partners. Whereas the generality of this mechanism has not been addressed, it may be illustrated by the example of an enzyme-inhibitor pair, the separase–securin system, which regulates sister chromatid separation in anaphase (details in Chapter 15, Section 15.1.5). Securin analogues have been identified in *S. cerevisiae* (Pdslp (Yamamoto, Guacci, and Koshland 1996)), *S. pombe* (Cut2 (Hirano et al. 1986)), *D. melanogaster* (pimples (Leismann et al. 2000)), and *H. sapiens* (pituitary tumor transforming gene (PTTG) (Zou et al. 1999)) based on functional observations, but they are extremely variable in length and show practically no recognizable similarity in sequence (i.e., homology). Because separases are also very different in length (*H. sapiens*: 1,600 amino acids, *D. melanogaster*: 600 amino acids) and show extreme variability in sequence, it follows that co-evolutionary adaptive changes have occurred in this system, and a large part of the variations are accounted for by the partner, not the IDP itself.

13.5 STRUCTURAL VARIABILITY AND EVOLVABILITY OF NEW FUNCTIONS

On the flip side of the coin, evolutionary variability and structural adaptability (see also Chapter 14, Section 14.6) of IDPs presents ample opportunity for developing new functions (James and Tawfik 2003). Although the concept primarily draws on examples of ordered proteins, the underlying ideas are relevant with respect to protein disorder, due to the critical element of flexibility/adaptability. Based on the development of new specificities of antibodies, it was suggested that conformational diversity and functional promiscuity are evolvability traits that enable existing proteins to rapidly acquire new activities (James and Tawfik 2003). The basic idea is that the predominant conformation of an existing protein carries its "main" function, whereas an alternative, scarcely

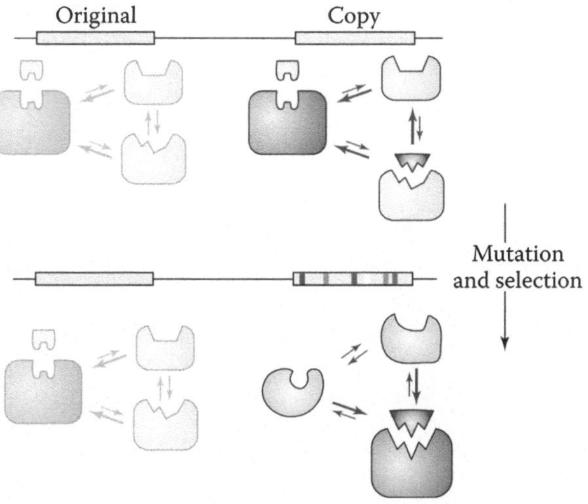

FIGURE 13.5 Structural variability increases functional evolvability. Functional diversity enabled by structural variability may form the basis of rapid evolution of new functions (see James and Tawfik 2003 for details). The protein is in equilibrium between different conformations, of which the native substrate (left side) selects the dominant conformer. An alternative conformation can bind a second substrate (right side), but this secondary activity confers only limited selective advantage due to the presence of the natural partner. Gene duplication enables one copy to evolve improved activity with the promiscuous substrate, while the original gene maintains its original function. Reproduced with permission from James and Tawfik (2003), *Trends Biochem. Sci.* 28, 361–8. Copyright by Elsevier Trends Journals.

populated conformation has more potential for another (binding) function. Initially, this secondary activity provides only a limited fitness advantage, because binding of the primary substrate will sequester most of the protein. Improvement through mutation is only possible to a limited extent because such mutations might decrease the primary activity. Following gene duplication, however, one gene copy becomes free to evolve without compromising the original activity, and its mutations could improve the secondary activity very rapidly (Figure 13.5). After successive rounds of mutation and selection, primary activity of the second copy may completely differ. Because the ensemble of structures of IDPs already harbors the capacity to manifest different functions, as formulated in the concepts of binding promiscuity (Kriwacki et al. 1996) and moonlighting (Tompa, Szasz, and Buday 2005), this evolutionary scenario may be of prime importance in the case of IDPs.

14 Extension of the Structure-Function Paradigm

This chapter discusses how the rapidly accumulating structural and functional information on intrinsically disordered proteins (IDPs) appears to solidify into a consistent framework of functional modes (i.e., how function can be interpreted in terms of the structural features of IDPs). This information is closely related to the functional classification scheme outlined in Chapter 12, but with a different emphasis. Rather than trying to classify IDPs by function, here special mechanistic features of their action are considered, which are often thought of as imparting "functional advantages" on IDPs. The two dominant elements of these are molecular recognition accompanied by local induced folding and entropic-chain-type functions that directly stem from disorder. These principles appear in various combinations in actual situations, and together they contribute toward formulating an extended structure-function paradigm that can encompass both ordered and disordered proteins.

14.1 FUNCTIONS THAT STEM DIRECTLY FROM THE DISORDERED STATE

As discussed extensively in Chapter 12, the function of IDPs may occasionally stem directly from disorder. In these entropic chain functions, the lack of a stable structure of the polypeptide chain per se forms the basis of function, either due to the freedom in structural search it experiences or the force it can generate against effects that would reduce its conformational freedom. Here, we present a census of the four basic varieties of entropic chain functions. These are:

1. Linkers and spacers, which provide appropriate spatial separation and search of binding/catalytic domains or elements (e.g., linker region of cellulase E (von Ossowski et al. 2005))

2. Entropic bristles/brushes, which generate force against compression (e.g., microtubule-associated proteins (MAPs) (Mukhopadhyay and Hoh 2001))
3. Entropic springs, which generate force against physical extension (e.g., titin Pro, Glu, Val, Lys-rich (PEVK) domain (Trombitas et al. 1998))
4. Entropic clocks, which provide a timing function (voltage-sensitive K channel (Bentrop et al. 2001)).

With respect to the issue of extending the structure-function paradigm, it only needs to be made clear that these entropic chain functions represent the most radical deviation from the classic view of protein structure and function.

14.2 RECOGNITION FUNCTIONS: RECOGNITION BY SHORT MOTIFS

IDPs most often function by molecular recognition, when they bind a partner and undergo induced folding or disorder-to-order transition (Dunker et al. 2002; Dyson and Wright 2002a; 2005; Gunasekaran et al. 2003). Many examples of complexes involving IDPs can be described (Fuxreiter et al. 2004; Gunasekaran, Tsai, and Nussinov 2004). In six out of seven functional categories, IDP function is associated with protein-partner interaction (Fuxreiter et al. 2004; Gunasekaran et al. 2004; Le Gall et al. 2007; Vacic et al. 2007), the level of disorder is increased in hub proteins that are specialized to have multiple interactions (Dosztanyi et al. 2006; Dunker et al. 2005; Ekman et al. 2006; Haynes et al. 2006; Patil and Nakamura 2006; Singh, Ganapathi, and Dash 2006), and average disorder increases with the number of subunits of complexes (Hegyi, Schad, and Tompa 2007). Thus, the issue of the molecular mechanism of partner binding is of paramount importance for understanding practically all aspects of protein disorder. IDPs use a recognition strategy basically different from that of ordered proteins. The latter have evolved a great variety of domains often specialized in molecular recognition (Copley et al. 2002; Pawson and Nash 2003; Ponting et al. 2000; Seet et al. 2006), whereas the counterparts of these domains, such as SH3 (Ferreon and Hilser 2004; Hiroaki et al. 2001; Yu et al. 1994), 14-3-3 (Bustos and Iglesias 2006), or phospho-tyrosine-binding domain (PTB) (Obenauer, Cantley, and Yaffe 2003), tend not to be domains but short motifs of rather flexible nature. Due to the involvement of disorder, domain–motif interactions represent a different evolutionary strategy from domain–domain recognition (Neduva and Russell 2005). Whereas the distinction between domains and motifs is clear, there are at least four different concepts of approaching and describing short recognition elements in protein-protein interactions.

14.2.1 Preformed Structural Elements

In many cases, the structure of the recognition element of an IDP in complex with its partner is known, which enables the analysis of the physical nature of their molecular interfaces (Gunasekaran et al. 2004; Meszaros et al. 2007) and also the probable structural preferences of IDPs in the unbound state. Such studies have led to the concept of preformed structural elements (PSEs) (Fuxreiter et al. 2004) and intrinsically folded structural units (IFSUs) (Sivakolundu, Bashford, and Kriwacki 2005).

The analysis of 26 such intrinsically disordered region (IDR)/partner complex structures (Fuxreiter et al. 2004) showed that the accuracy of predicting secondary structural elements in IDPs in the bound state is higher than that of their partner proteins and is significantly higher than the corresponding values for random sequences (Figure 14.1). This observation suggests that IDPs have rather strong intrinsic preferences for the conformation they attain when bound to their partners, which may be interpreted in terms of the partial preformation of their recognition segments in the free state. The relationship is strongest for helices and weakest for coils. Although these results are not conclusive with respect to the mechanism of binding (i.e., whether these elements are truly preformed), often a similar structure in the unbound and bound states is observed when the IDP is characterized by nuclear magnetic resonance (NMR) (see Chapter 10, Sections 10.2.3 and 10.2.4, and Table 10.1). For example, this correlation was observed in the case of the kinase inhibitory domain (KID) of cyclic-AMP response element-binding protein (CREB) (Parker et al. 1999; Radhakrishnan et al. 1998), $p21^{Cip1}$/$p27^{Kip1}$ (Kriwacki et al. 1996; Lacy et al. 2004; Sivakolundu et al. 2005), p53 (Lee et al. 2000), FlgM (Daughdrill, Hanely, and Dahlquist 1998; Dedmon et al. 2002; Sorenson, Ray, and Darst 2004), PKI alpha (Hauer et al. 1999a), Tβ4 (Domanski et al. 2004), and measles virus nucleoprotein (Longhi et al. 2003). Whether such preformed elements

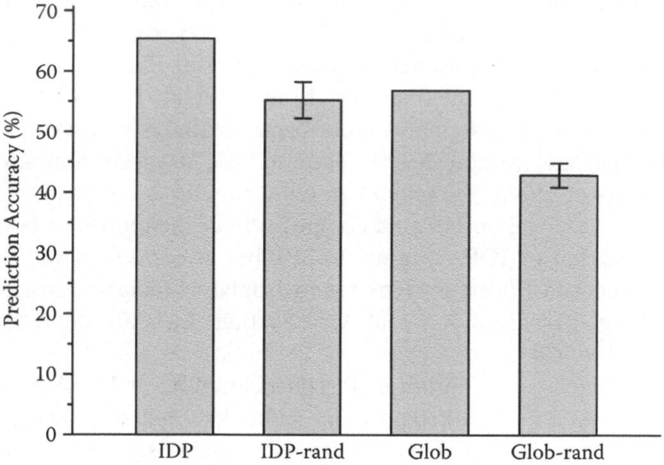

FIGURE 14.1 Predictability of the secondary structure of IDPs in the bound state. A selection of 26 IDP structures in complex with their partners was analyzed for the predictability of the secondary structure attained in the complex by ALB for IDPs (IDP), randomized sequences of IDPs (IDP-rand), sequences of globular partners (Glob) and randomized sequences of partners (Glob-rand). Intrinsic structural preferences of IDPs are strongly correlated with their conformation attained in the bound form. (data from Fuxreiter et al. 2004).

also serve as initial contact points of interaction is a matter of speculation, but they probably limit the entropic penalty of the induced folding process.

14.2.2 Linear Motifs

The concept of linear motifs (LMs, also denoted as eukaryotic linear motifs (ELMs) and short linear motifs (SLiMs)) derives from analyzing the sequences involved interactions. In certain proteins, the element of recognition is a short motif of discernible conservation, often denoted as a "consensus" sequence, such as modification sites of kinases or binding sites of SH3 domains (Neduva and Russell 2005). LMs are usually constructed as a few conserved specificity determinant residues interspersed with residues hardly constrained, with a typical length between 5 and 25 residues. They are usually described as a short sequence pattern, in which certain sites are restricted (RSs, e.g., P in Pxx for the SH3 binding sites), whereas others are rather freely exchangeable (i.e., non-restricted sites) (NRSs, marked with "x" in the above pattern). The first set of residues serve as specificity determinants, whereas the second set likely act as spacers. Due to their limited information content, LMs are much more difficult to identify by sequence comparisons than domains. A traditional Basic Local Alignment Search Tool (BLAST) search cannot positively identify LMs, and special algorithms that combine functional/structural clues with sequence analysis of non-globular regions had to be developed for this purpose (e.g., DILIMOT (Neduva and Russell 2006) and SLiMDisc (Davey, Shields, and Edwards 2006)).

LMs described in the literature have been collected in the eukaryotic linear motif (ELM) database available via the ELM server (Puntervoll et al. 2003), which contains about 800 examples of more than 100 ELMs. LMs are generally thought to correlate with local disorder (Linding et al. 2003b; Puntervoll et al. 2003), as confirmed in a systematic bioinformatic analysis of the ELM database (Fuxreiter, Tompa, and Simon 2007). The analysis suggests that LMs and their flanking segments of about 20 residues in both directions tend to be locally disordered (Figure 14.2A). The amino acid composition of the motifs resemble the characteristic composition of IDPs (Figure 14.2B), but at certain points the similarity breaks down, because LMs are enriched in hydrophobic residues Trp, Leu, Cys, and Tyr and the charged residues Arg and Asp. Further, LMs are depleted in Gly and Ala and enriched in Pro.

Marked differences in the amino acid frequencies of RS and NRS positions explain these propensities. At the conserved positions, either hydrophobic and rigid, or charged and flexible residues are preferred, whereas in NRS positions, excessive flexibility, very similar to that of IDPs, can be observed. The only exception is Pro, which is in excess in both RS positions and LM flanking regions, indicating its dual role as a contact residue within LMs and promoter of an open structure outside LMs. Overall, the unique amino acid composition suggests a mixed nature of LMs, with a few specificity-determinant residues strongly favoring order, grafted on a completely disordered carrier sequence flanking and intervening the region critical for interaction.

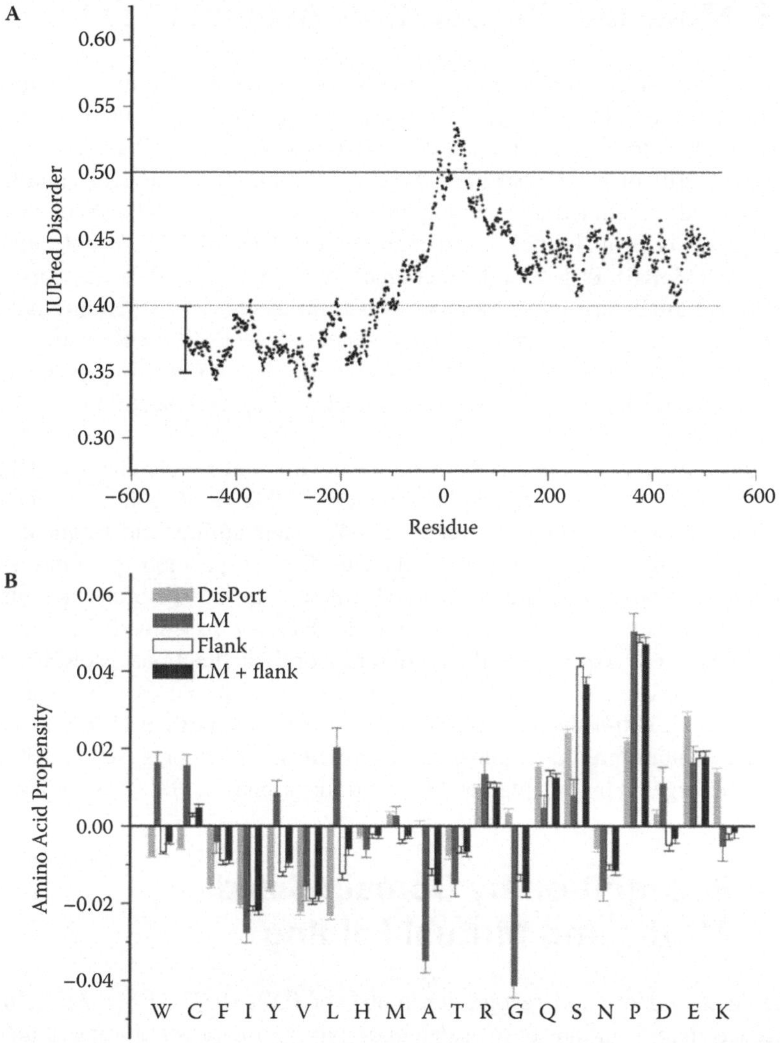

FIGURE 14.2 Linear motifs tend to fall into local disorder and are enriched in a special set of amino acids. Short recognition elements (linear motifs, LM) have been collected from the ELM database (Puntervoll et al. 2003). (A) Disorder profiles by the IUPred algorithm were computed and averaged. A thin horizontal line at 0.5 shows the threshold of disorder, whereas a dotted line at 0.4 shows the average score for experimentally verified disordered proteins in DisProt (Sickmeier et al. 2007). Standard error of the mean (SEM) values are displayed by an error bar. (B) Amino acid propensities of LMs and their flanking regions were also calculated and are shown as the difference between that of LMs and globular proteins. IDPs of the DisProt database (light gray), LMs (dark gray), 20-residue-long LM flanking segments (white), and LMs plus 20-residue-flanking segments (black) are shown. Reproduced with permission from Fuxreiter et al. (2007), *Bioinformatics* 23, 950–6. Copyright by Oxford University Press.

14.2.3 Molecular Recognition Elements/Features

The idea of short motifs in recognition can also be approached from a structural point of view. In the Protein Data Bank (PDB) there are 372 complexes in which one partner is shorter, whereas the other is longer, than 30 amino acids (see Chapter 9, Figure 9.4) (Oldfield et al. 2005b), or one partner is between 10 and 70 amino acids, and the other is a well-ordered protein (Mohan et al. 2006; Vacic et al. 2007). The shorter partner in these complexes is termed molecular recognition element (MoRE) or molecular recognition feature (MoRF). Based on the dictionary of protein secondary structure (DSSP) classification, MoRFs fall into four basic structural categories, depending on the secondary structure dominating in their bound state. α-MoRFs, β-MoRFs, and ι-MoRFs, the latter with residues mostly in coil conformation, have been distinguished, with a fourth type, mixed MoRFs, also described (see also Chapter 9, Section 9.7 and Chapter 10, Section 10.2.4).

The local structural preferences of αMoRFs are well predictable (see Chapter 9, Section 9.7), exceeding that of globular proteins (Mohan et al. 2006), suggesting that they are related to PSEs (Fuxreiter et al. 2004). Their amino acid frequencies show similarities to those of IDPs, being enriched in disorder-promoting amino acids and depleted in order-promoting amino acids (Dunker et al. 2001). Some notable deviations occur, such as the relative enrichment of MoRFs compared with IDPs in Cys and Phe, and depletion in Asp, Glu, and Lys. In terms of their functions, MoRF-containing proteins are correlated with signaling and regulation. The application of DISPHOS (Iakoucheva et al. 2004) shows that 159 of the MoRFs longer than 11 amino acids (out of 305 total) contain predicted phosphorylation sites, which suggests that they might frequently be targeted by this regulatory post-translational modification (PTM).

14.2.4 Recognition by Domain-Sized Motifs and Mutual Folding

The three concepts of short recognition motifs of IDPs (PSEs, LMs, MoRFs) can be considered as manifestations of the same underlying principle of binding of an ordered partner by a short segment within a disordered region, which undergoes disorder-to-order transition or folding induced upon binding (Dyson and Wright 2002a). The implicit assumption in this view is that motifs are always short, on the order of 5 to 25 residues in the case of LMs (Fuxreiter et al. 2007) and less than 30 residues in the case of MoREs (Oldfield et al. 2005b). The definition of MoRFs allows somewhat longer disordered binding motifs that fall between 10 and 70 residues (Mohan et al. 2006), which raises the possibility that interactions of IDPs might actually conform to two principally different concepts.

There are several examples in the literature that binding of an IDP involves a domain-sized segment, far too long to be considered as a short motif (see Chapter 13, Section 13.1.3). For example, such a binding mode has been described in the case of the disordered Smad-binding domain (SBD) of the Smad anchor for receptor activation (SARA) binding to the MH2 domain of Smad (Chapter 12, Figure 12.6) in transforming growth factor beta (TGF-β) signaling (Wu et al. 2000), the KID domain of the cyclin-dependent kinase (Cdk) inhibitor p27^{Kip1} binding to the Cyclin A-Cdk2 complex (Chapter 10, Figure 10.3) (Russo et al. 1996), botulinum neurotoxin serotype A (BoNT/A) binding to SNAP-25 (Breidenbach and Brunger 2004; Brunger et al. 2007), the Wiskott–Aldrich syndrome protein (WASP) homology domain 2 (WH2) domain of Tβ4 binding to G-actin (Chapter 11, Figure 11.5) (Irobi et al. 2004), E-cadherin cytoplasmic domain (cytD) (Huber and Weis 2001) or the catenin binding domain (CBD) of T-cell factor 3 (Tcf3) binding to β-catenin (Chapter 11, Figure 11.3) (Graham et al. 2000), and the GTPase-binding domain of WASP binding to the small GTPase Cdc42 (Figure 14.3A) (Abdul-Manan et al. 1999). Although in principle these recognition events can be visualized as binding by several short neighboring or even overlapping motifs, often there are no apparent motifs within these domain-sized regions, and their binding is better described as recognition by a disordered domain. In fact, about 14% of Pfam domains are mostly disordered (see Chapter 13, Section 13.1.3), which by definition arose by evolutionary divergence (unlike motifs, which arise by convergece). which have led to the extension of the domain concept to the disordered state (Tompa et al. 2009).

A further apparently closely related deviation from the simple picture of recognition by a short disordered segment is the mutual recognition of two IDPs in a process of mutual induced folding, also termed as "co-folding" or "synergistic folding," as reported in the case of the interaction between Bob1 and Oct1 trans-activator domain (TAD) (Lee et al. 2001), multiple vesicle-associated proteins (Dafforn and Smith 2004; Williamson 1994), and CBP/p300 and p160 nuclear receptor co-activators (Demarest et al. 2004; Demarest et al. 2002). Whereas in most cases these inferences rely only on biochemical and functional studies, in the case of the mutual binding of the activator for thyroid hormone and retinoid receptors (ACTR) domain of p160 and the MG-like nuclear-receptor co-activator-binding domain (NCBD) of CBP, the structure of the resulting complex is known in atomic detail (Figure 14.3B). The two domains completely wrap

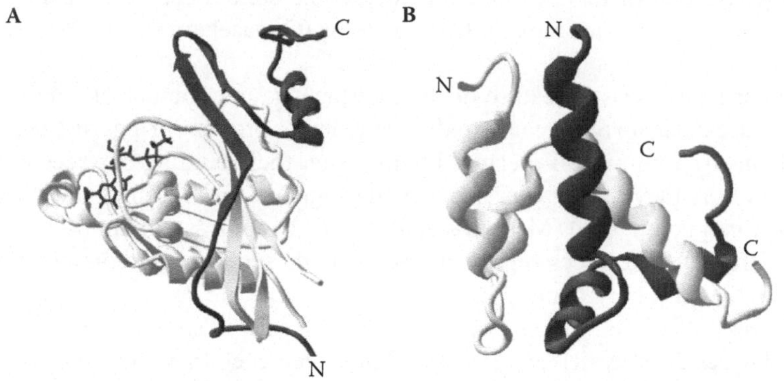

FIGURE 14.3 Binding of IDPs by disordered domains and mutual induced folding of two disordered recognition segments. (A) The disordered GTPase binding domain (GBD) of WASP (dark gray) bound to the small GTPase Cdc42 (light gray, pdb 1cee). (B) The structure of the complex that results from the mutual folding of disordered NCBD of CBP (light gray) and disordered ACTR domain of p160 (dark gray, pdb 1kbh).

around each other and bury a surface area of 1,500 Å^2 of primarily hydrophobic nature. The interaction is rather tight ($K_d = 3.4 \times 10^{-8}$ M) and is driven by enthalpy ($\Delta H° = -31.7$ kcal mol^{-1}), which compensates for the high entropic cost ($T\Delta S° = -21.3$ kcal mol^{-1}) associated with the induced folding of both ACTR and NCBD domains.

14.2.5 Recognition Interfaces

Notwithstanding the length of the region involved, a special area of interest is the chemical nature of the interface of IDP complexes. The interfaces of globular proteins usually fall on the order of 1,600 ± 400 Å^2 in area (Lo Conte, Chothia, and Janin 1999), and they are distinguished from the average surface of proteins by an elevated hydrophobicity, evolutionary conservation of certain anchoring residues, and characteristic shapes that differ for various classes of complexes, such as homodimers, heterodimers, or enzyme-inhibitor complexes (Jones and Thornton 1996; Keskin, Ma, and Nussinov 2005; Lo Conte et al. 1999; Nooren and Thornton 2003). Three studies of the interfaces of IDPs agree that they significantly differ from those of globular proteins. In one study, 10 two-state complexes, 44 ribosomal proteins, and 5 complexes of bona fide disordered proteins were analyzed (Gunasekaran et al. 2004), whereas in another, 39 complexes of experimentally verified IDPs (Meszaros et al. 2007) were analyzed. In the third study, a detailed analysis of 258 MoRFs, selected on the criterion of length and shown to correlate reasonably with disorder (Vacic et al. 2007), was carried out.

In terms of size, the interfaces of IDPs are slightly smaller than those of ordered complexes (1141 ± 110 Å^2 [Vacic et al. 2007]), with some bias against very large interfaces (>3,000 Å^2) (Gunasekaran et al. 2003; Meszaros et al. 2007; Vacic et al. 2007). When surfaces and interfaces are normalized to chain length, IDPs apparently have a much larger per-residue surface area and a much larger per-residue interface area than globular proteins (Figure 14.4A). As indicated by the ratio of the two values, IDPs also use a larger portion of their surface for interaction, sometimes 50% of the whole, as opposed to only 5–15% for most ordered proteins (Gunasekaran et al. 2004; Meszaros et al. 2007).

Another characteristic feature of the interfaces is the number of continuous segments they are made up of. Because folding of globular proteins brings distinct segments of the chain in proximity to create a binding site, their binding surfaces are usually fragmented (i.e., they hardly ever use a single segment for binding), and occasionally the number can go up to 10 (Meszaros et al. 2007). In contrast, in almost two-thirds of the cases, the IDP interfaces represent only a single sequentially continuous segment (Figure 14.4B), and they contain more than three separate segments in a single case only. The ratio of buried-to-exposed area of IDPs is much smaller for both polar and hydrophobic residues, which suggests that IDPs keep even their very few hydrophobic residues exposed for contact with the partner, instead of generating a hydrophobic core (Meszaros et al. 2007). In fact, IDPs keep a larger fraction of their hydrophobic residues exposed than ordered proteins (IDPs: 40–90%, ordered proteins: 15–50% [Meszaros et al. 2007]), and their interface is more hydrophobic than the buried regions of the protein (Gunasekaran et al. 2003; Meszaros et al. 2007; Vacic et al. 2007). In this sense, the hydrophobic core of IDPs is in the interface and not in the core of the partner-bound

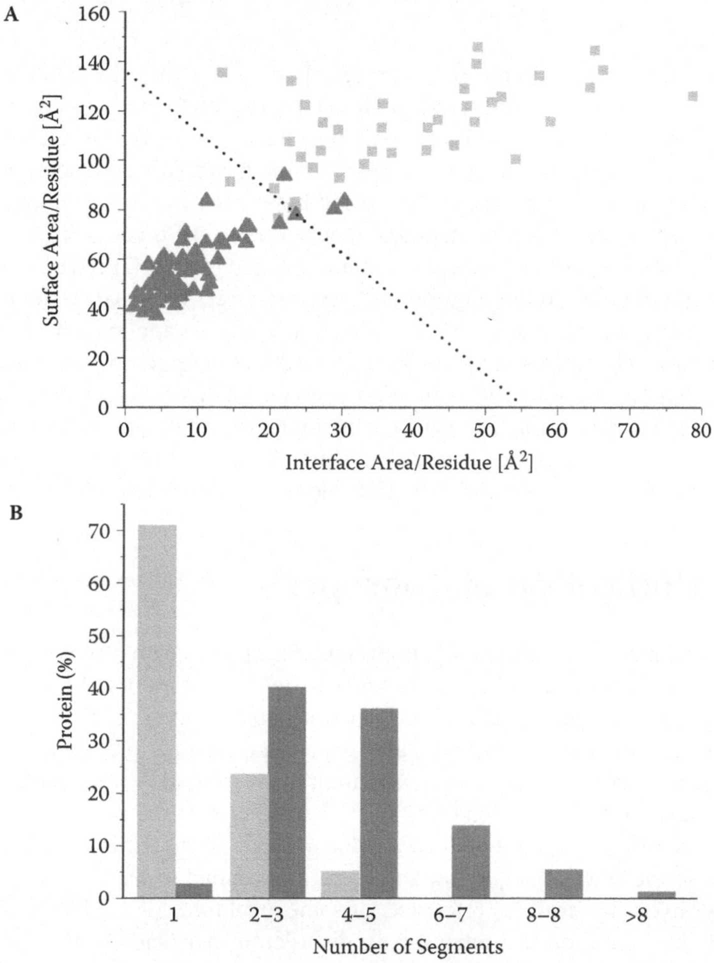

FIGURE 14.4 Surface and interface area and segmentation of interfaces of IDPs. (A) IDPs use a large fraction of their surface for binding. The total surface area per residue is given as a function of the interface area per residue for the smaller chain of ordered complexes (dark gray triangles) and for disordered proteins in complex with an ordered protein (light gray squares). (B) IDPs (light gray) tend to use fewer segments to make up the binding site than globular proteins (dark gray). The distribution of interfaces with the given number of non-continuous sequence segments is shown. Reproduced with permission from Meszaros et al. (2007), *J. Mol. Biol.* 372, 549–61. Copyright by Elsevier Inc.

state. Exposure of hydrophobic amino acids and/or a compositional bias favoring them at the interface also follows from the analysis of ELMs (Fuxreiter et al. 2007), two-state complexes (Gunasekaran et al. 2004), and MoRFs (Vacic et al. 2007).

In line with these characteristic composition values, IDP interfaces make much more hydrophobic–hydrophobic contacts (IDPs: 33%, ordered proteins: 22%), whereas ordered proteins make significantly more polar–polar contacts (IDPs: 27%, ordered proteins: 33%) (Gunasekaran et al. 2004; Meszaros et al. 2007). The probable reason for these distinctions is that IDPs require more enthalpic stabilization to counteract their decrease in configurational entropy, but probably also that they are less able to shield interactions of polar residues from hydrate water. In relation to this difference, IDP interfaces are tighter (i.e., structurally more complementary), probably due to a better adaptation to the structure of the partner enabled by their induced folding. Structural adaptation of ordered proteins is limited due to their much lower level of conformational freedom. These observed differences also manifest themselves in differences in the interaction energies of the two types of complexes, as demonstrated by the IUPred (Dosztanyi et al. 2005a; Dosztanyi et al. 2005b) algorithm developed to estimate pair-wise inter-residue interaction energy of IDPs (see Chapter 9, Section 9.4.2). Its application toward analyzing the interfaces (Meszaros et al. 2007) suggested that ordered proteins tend to realize more stabilizing interactions within their polypeptide chains, whereas IDPs derive more stabilization from the interaction with the partner than from interactions within their own chain. The overall balance is therefore shifted towards the folded state only in the presence of the partner, explaining why IDPs do not fold in isolation.

14.2.6 Unification of Concepts?

The different concepts of short recognition elements are based on different premises and do not necessarily correspond to the same structural and functional feature, although some unifying themes do appear. PSEs by definition exist in a similar conformational state in solution than in the bound state, which corresponds primarily to α-helices. In this respect, their overlap is most apparent with α-MoRFs and maybe also with β-MoRFs. In accord, the predictability of PSE structures in the bound state (Fuxreiter et al. 2004) is also observed for MoRFs (Mohan et al. 2006). LMs, on the other hand, are defined at the level of sequence, and could correspond to all three MoRF classes, and, if they have a discernible preference for some local fold, also to PSEs. On the other hand, PSEs by definition are intrinsically disordered in isolation, whereas LMs and MoRFs can also occur in ordered regions of the proteins. Thus, in many instances, a short recognition element conforms to all three definitions, but further work is needed to arrive at a unified concept.

14.3 DISORDER-TO-ORDER TRANSITION IN RECOGNITION: MECHANISTIC AND THERMODYNAMIC ASPECTS

The crux of the binding of an IDP, whether mediated by short recognition element or a longer domain-sized region, is folding induced upon binding (also termed disorder-to-order transition). Whereas induced folding has important functional consequences, such as uncoupling specificity from binding strength and adaptability of recognition (Dyson and Wright 2002a), the exact mechanism and thermodynamic consequences of the process are often rather obscure. One key issue is whether folding occurs before, after, or concomitant to binding. Experimental evidence seems to support all these varieties, and a certain level of unification is feasible, especially if the strong mechanistic parallels of induced folding with the process of protein folding (Daggett and Fersht 2003; Gianni et al. 2003) are taken into account (see Chapter 1, Section 1.6).

The implicit assumption in the PSE, and maybe also in the MoRF concept, is that the polypeptide chain preferentially samples the local conformation attained in the complex. Due to the inherent stability of α-helix conformation, this correlation usually corresponds to a transient helical structure (see Section 14.2.1 and Chapter 10, Section 10.2.3 and Table 10.1). In some cases, an extended conformation may also appear locally, such as a β-strand in the case of fibronectin binding protein (FnBP) (Penkett et al. 1998) and polyproline II (PPII) conformation in the case of partners of the Pro-rich peptide binding GYF domain (Gu et al. 2005). To uncover the actual mechanism of binding at atomistic detail, the process of induced folding has been approached by site-directed mutagenesis, molecular dynamics simulation, and NMR spectroscopy. The results appear to depend very much on the actual system studied, with no apparent generalizations.

14.3.1 Site-Directed Mutagenesis Studies of Induced Folding

Site-directed mutagenesis can be primarily used to stabilize or destabilize recognition helices in IDPs, to address if the change inflicted upon the unbound state affects the process of binding. One of the best characterized systems is the binding of $p27^{Kip1}$ to the Cyclin A-Cdk2 complex, mediated by the KID domain (see Chapter 3, Section 3.7.2; Chapter 10, Section 10.2.3.1; and Chapter 15, Section 15.1.3).

As by NMR and MD, the structure of KID in solution has a significant preference to locally populate the conformational elements of the bound state (see Chapter 10, Figure 10.3), and thus binding may proceed from recognition by the PSE α-helix (linker helix LH, also termed an IFSU ((Sivakolundu et al. 2005))). Analysis of the kinetics and thermodynamics of binding of various truncated constructs, however, shows that binding is initiated by the N-terminal coil segment (domain 1), followed by wrapping around in a staple-like fashion, binding at the active site of the kinase (domain 2), and finished off by stabilization of LH (details in Chapter 3, Section 3.7.2). The helix defines the geometry of binding and may function as a PSE or IFSU initiating the recognition process. This is not the case, however.

Mutagenesis studies showed that the final formation of the helix only occurs after the transition state of binding (i.e., binding is initiated by a non-structured state (segment) of the protein) (Bienkiewicz, Adkins, and Lumb 2002). Stabilization of LH helix by a triple-Ala mutation (E40A/D44A/K47A) or its destabilization by single-Pro mutations (L41P, K47P, and A55P) hardly affect the equilibrium K_d characteristic of the formation of the p27-KID–Cyclin A-Cdk2 complex (8 ± 2 nM). The single mutant L41P, which has half the preference for the helix in the unbound state, binds with about the same affinity (9 ± 1 nM), whereas the other mutants K47P and A55P, in which helix formation is practically abolished, bind with only slightly lower affinities (16 ± 3 nM and 13 ± 2 nM, respectively). In addition, stabilization by a triple-Ala mutation does not lead to a corresponding increase in stability, rather to a slight destabilization of the complex (12 ± 3 nM). In kinetic experiments, the binding of the helix-stabilized E40A/D44A/K47A mutant is actually three times slower than that of the wild-type protein or a single-Pro mutant. Thus, stabilization of the helix results in a kinetic impediment to binding, suggesting that binding is initiated by a locally unfolded or partially folded state of $p27^{Kip1}$ KID, and it proceeds through a rather disordered transition state.

Similar helix-stabilizing and destabilizing mutations unveiled a different binding mechanism in the case of the transcription factor Gcn4p. Gcn4p contains a dimeric Leu-zipper deoxyribonucleic acid (DNA)-binding motif, the coiled-coil element of which has approximately 70% helix content in the absence of DNA, implying only partial preformation of the zipper (Weiss et al. 1990). In the presence of DNA, α-helix content increases to at least 95%. To probe into the importance of the formation of helical segments in the transition state, a series of quadruple amino acid replacements spanning the entire helix propensity scale were generated at positions that do not directly interfere with DNA binding (Zitzewitz et al. 2000). Binding strength of DNA was found to correlate with helical propensity, which suggests that preformed elements of secondary structure play the key role in recognition. This scenario was also corroborated by mutations of the noncontacting Asp-Pro residues at the N-terminal end of the helical region. Such N-capping motifs, which can stabilize α-helical structure, contribute significantly to the stability of the complex of Gcn4p with cognate DNA (Hollenbeck, McClain, and Oakley 2002).

14.3.2 Molecular Dynamics Simulations of Induced Folding

Molecular dynamics (MD) simulations of the molecular path of dissociation of the IDP from its partner provide additional insight into the binding mechanism. Two systems, $p27^{Kip1}$ KID–Cyclin A-Cdk2 and CREB KID–CBP KID-binding domain (KIX), have been studied in most detail.

The hierarchy of loss of structure of $p27^{Kip1}$ KID upon unbinding from Cyclin A-Cdk2 was studied by high-temperature Monte Carlo simulations (Verkhivker 2004, 2005; Verkhivker et al. 2003). The unbinding/unfolding trajectories confirm the mechanism of binding suggested by experimental studies (Bienkiewicz et al. 2002; Lacy et al. 2004) (i.e., the recruitment of the inhibitor by the hydrophobic docking site of coil conformation) and local disorder of the LH helix in the unbound form. Rapid formation of the docking interaction dictates the folding mechanism by reducing dimensionality of search and overwhelms the local folding preferences for creating a stable α-helix.

A similar approach was applied in the case of CREB KID binding to the KIX domain of the co-activator CBP (see Chapter 6, Section 6.3.1 and Chapter 10, Section 10.2.3). The structure of the complex containing phosphorylated KID (pKID) solved by

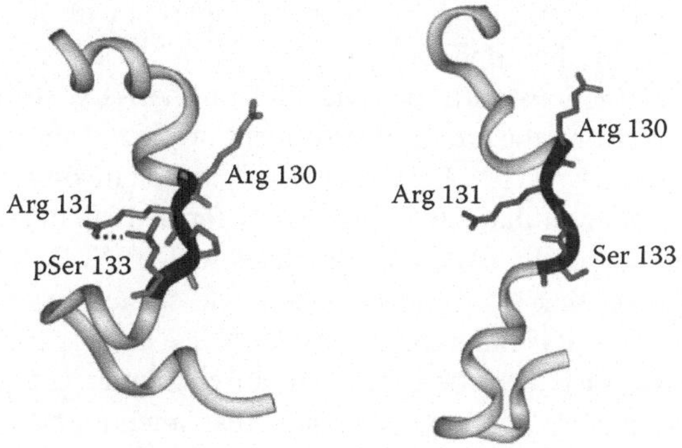

FIGURE 14.5 Change in the turn conformation of CREB KID upon phosphorylation of Ser^{133}. The conformation of CREB KID was assessed by MD simulations in the Ser^{133}-phosphorylated state (pKID, left) and nonphosphorylated state (KID, right). Representative configurations are shown, with the turn region Arg^{130}–Ser^{133} displayed in dark gray. In pKID, a hydrogen bond is maintained between pSer^{133} and Arg^{131} (marked by a dotted line), which stabilizes the region in a binding-competent closed conformation. Reproduced with permission from Solt et al. (2006), *Proteins* 64, 749–57. Copyright by Wiley-Liss.

NMR (Radhakrishnan et al. 1997) shows that CREB KID residues 120–144 undergo induced folding, which results in two perpendicular α-helices (αA: Asp^{120}–Ser^{129}, αB: pSer^{133}–Asp^{144}), connected by a short turn-like segment that harbors the phosphorylation site Ser^{133} (Chapter 6, Figure 6.3), which plays a critical role in the function of CREB (Parker et al. 1996; Zor et al. 2002). Because the two helices are transiently populated in the solution state of CREB, (αA 50% of the time, αB 10% of the time, see Chapter 6, Section 6.3.1), the question whether helices form prior to or after the transition state of the binding reaction was addressed by MD simulations (Solt et al. 2006). It was found that helical populations are hardly affected by phosphorylation that initiates the interaction, whereas a subtle change in the turn region that connects the two helices occurs (Figure 14.5). Here, phosphorylation induces a transient structural element that resembles the bound conformation of the molecule, stabilized by the pSer^{133}–Arg^{131} interaction. Its formation may limit the conformational search of the flanking helices and initiate binding, thus serving as a PSE (and/or a primary contact site [PCS], see Section 14.5.1). In the context of the role of local structure in the transition state of binding, this MD study suggests that less importance be assigned to preformed helices, and more importance to the turn region that connects them.

14.3.3 NMR Studies of the Mechanism of Induced Folding

NMR studies canvas a somewhat different scenario for the binding of CREB KID to CBP KIX (Sugase, Dyson, and Wright 2007), whereas they also suggest that preformed helices are not the primary determinants of the transition state (see Chapter 6, Figure 6.3). Based on NMR titrations and ^{15}N relaxation dispersion experiments, pKID forms an ensemble of transient encounter complexes with KIX, and is stabilized primarily by non-specific hydrophobic contacts. In this complex, pKID explores an ensemble of weak interactions with multiple sites on the KIX surface. The encounter complex is characterized by the strong involvement of $pSer^{133}$ and has further stabilizing hydrophobic contacts by Tyr^{134}, Ile^{137}, and Leu^{138}, which lie on the contacting face of helix αB. CSI values suggest that αB is only partly formed (up to 30%), whereas αA is almost fully formed, but hardly makes any contacts with KIX. R_2 relaxation dispersion experiments corroborate that αA behaves as a single cluster, whereas αB behaves as two separate clusters (i.e., it is incompletely folded).

Overall, the process is best described by an encounter complex dominated by $pSer^{133}$ interactions and hydrophobic contacts that anchor the pKID αB helix to the hydrophobic groove of KIX in a partially formed state. Within this encounter complex, there is a continuing conformational search for the favorable intermolecular interaction, without pKID dissociating from KIX. This analysis and the MD study (Solt et al. 2006) agree that the $pSer^{133}$ region makes contacts critical for the formation of the encounter complex, and formation of the helices is not essential for the recognition step.

14.3.4 The Analogy of Folding and Induced Folding

The results of mutagenesis, MD, and NMR studies agree that there is no general mechanism for the binding-induced folding of IDPs, but each protein follows a different and individual path. This situation is reminiscent of the mechanism of protein folding. As detailed in Chapter 1, Section 1.6 (see also Chapter 1, Figure 1.7), the actual mechanism of protein folding lies between two extremes described by the *framework* model, which assumes that secondary structural elements form in the unfolded state and tertiary structure forms by their diffusive collisions, and the *hydrophobic collapse* model, in which a rapid compaction driven by hydrophobic interactions is followed by the formation of secondary structural elements. The best description of the transition state of folding can be achieved by the intermediary scenario *nucleation condensation*, in which formation of secondary structural elements and hydrophobic compaction run in parallel (Daggett and Fersht 2003).

Because of the conceptual and mechanistic parallels of the folding of globular proteins and the induced folding of IDPs, and mixed results on the induced folding process, it may be suggested that induced folding also proceeds by a sort of "nucleation condensation" mechanism. In this, preformed secondary structural elements play a role, but they are not stable enough to dominate the folding pathway. On the other hand, tertiary-type of interactions are important, but can only have an effect in the context of the ensuing stabilization of the secondary structural elements. One might visualize the process as the parallel and reinforcing formation of intermolecular (tertiary) and intramolecular (secondary) contacts, both contributing to the transition state. Because induced folding is a bimolecular reaction, it is conceivable that in this case the emphasis is shifted toward local interactions (i.e., preformed elements).

14.4 RECOGNITION FUNCTIONS: UNCOUPLING SPECIFICITY FROM BINDING STRENGTH

It is generally held that the primary thermodynamic consequence of induced folding is that it uncouples binding strength from specificity to enable weak and reversible interactions of proteins. Here, both aspects (i.e., specificity and binding strength) are looked at in some detail. At the outset, it needs to be clarified that specificity is an often used and actually abused concept, which lacks a clear and generally acceptable definition.

In our view, specificity of an interaction is primarily a biological and not a physical concept, and thus specificity of a given interaction cannot be decided by the K_d of the interaction of two isolated proteins *in vitro*. This is clearly demonstrated by the biological relevance of an ultra-weak interaction between the SH3 domain of Nck-2 and the LIM4 domain of PINCH-1. The interaction has a K_d of 3 mM; still, it is indispensable in regulating focal adhesion dynamics during integrin signaling, as shown by genetic evidence (Vaynberg et al. 2005). Thus, specificity of binding means that a given interaction is realized at a given location and time in the cell, at the expense of competing interactions. Besides the strength and speed of formation, it may have many components, such as co-expression and co-localization, local concentration, the presence of assisting/targeting factors, and maybe many more. Disorder is most apparently related to the appropriate tuning of strength and speed of interaction.

14.4.1 Disorder May Contribute to Recognition of Specific Sites

The influential analysis of Spolar and Record (Spolar and Record 1994) concentrated mainly on the role of disorder-to-order transition in protein-DNA interactions. Specificity in DNA recognition is defined as the ratio of binding strength of a "specific" site corresponding to the consensus sequence, to that of another "nonspecific" site. Specific sites have a typical binding constant 10^9 to 10^{12} M^{-1}, some 10^3 to 10^7 times stronger than that of nonspecific ones. In many cases, the protein undergoes significant conformational changes upon binding, including ordering of disordered loops or formation of helices from disordered regions (see Chapter 11, Section 11.2.1 for disorder in DNA binding domains (DBDs)). The underlying entropy changes can be dissected into various and sometime opposing terms, such as burial of nonpolar surface area, changes in chain conformation, and loss of translational and rotational entropy. At temperature T_S, where the net entropy change of association is zero:

$$\Delta S_{assoc} = 0 = \Delta S_{HE}\,(T_S) + \Delta S_{rt} + \Delta S_{other} \tag{14.1}$$

where ΔS_{HE} is the entropy change that results from the burial of hydrophobic surface and ΔS_{rt} is the entropy change that results from the loss of rotational-translational freedom. Often, ΔS_{HE} is much larger than that expected assuming a rigid-body association and much larger than the magnitude of ΔS_{rt}, which suggests a large value for ΔS_{other} associated with a change in conformational entropy (in the protein, DNA or both) upon binding. In general, specific DNA sequences serve as better templates for folding of the protein, and local or even global folding transitions are coupled to DNA binding at specific sites. Frequently and often unjustifiably, this analysis is generalized to all protein–protein interactions of IDPs, and is thought to suggest that structural disorder confers the ability of specific binding on IDPs, which may turn out to be generally true. In terms of the original issue (i.e., the question of uncoupling specificity from binding strength), it actually provides evidence for the opposite, showing that the increased conformational freedom enables IDPs to actually realize a stronger binding with the specific partner.

14.4.2 Disorder May Make Interactions Weaker

The most often asked question with respect to IDP binding is whether the decrease in configurational entropy upon binding has a discernible unfavorable effect on binding strength, which might be directly incorporated into the regulation of protein function. The comparison of the binding of an inducible (CREB) and a constitutive (c-Myb) transcription factor to the same co-activator—the KIX domain of CBP (Parker et al. 1999)—may offer some insight. As detailed in Sections 14.3.2 and 14.3.3 (see also Chapter 6, Figure 6.3), the KID domain of CREB binds KIX in a conformation of a helix followed by an extended turn and a second helix (Radhakrishnan et al. 1997). Phosphorylation of Ser^{133} within the turn region is critical for binding and biological function (Parker et al. 1996), which makes the interaction inducible. Binding of c-Myb, on the other hand, is constitutive and does not require any signal or modification (Dai et al. 1996). Because both transcription factors stimulate protein expression via binding by an amphipathic helix at the same shallow hydrophobic groove of KIX, the difference in their behavior may reside in their level of disorder in the unbound state.

As determined by CD, the helix content of the binding region of c-Myb is about 30%, but it is only 1% in the case of CREB (10% by NMR (Radhakrishnan et al. 1998)). ΔG values of binding do not critically differ for the two transcription factors (–8.8 kcal/mol for pKID-KIX complex, and –6.0 kcal/mol for the c-Myb-KIX complex), but the binding of pKID entails a large entropic penalty ($\Delta S = -6.0$ kcal/mol K, $\Delta H = -10.6$ kcal/mol), which is basically different from the respective values of cMyb ($\Delta S = +7.5$ kcal/mol K, $\Delta H = -4.1$ kcal/mol). Thus, the formation of the complex between pKID and KIX is driven by enthalpy, which compensates for the unfavorable change in entropy. In the case of cMyb, both terms are favorable, and the absence of induced folding makes the interaction overall stronger. Thus, disorder-to-order transition lowers the binding strength of CREB KID to CBP KIX and brings it into a region where regulation may be effective. In this sense, disorder of CREB uncouples binding strength from specificity.

14.4.3 Strong Multivalent Binding and Weak Aspecific Binding

As opposed to weak but specific binding enabled by structural disorder, some IDPs engage in strong multivalent and very specific binding. Multivalent binding exploits

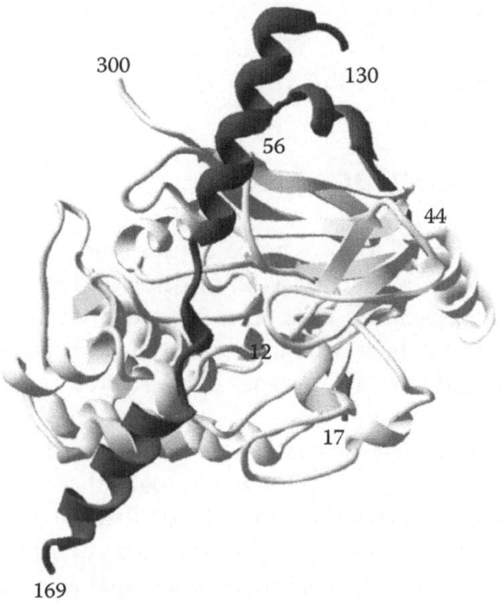

FIGURE 14.6 The structure of the tripartite I-2–PP-1 complex. Inhibitor-2 wraps around protein phosphatase 1cγ and contacts the enzyme by three isolated binding segments (encompassing residues 12–17, 44–56, and 130–169). The structure has been solved by X-ray crystallography (Hurley et al. 2007). The light gray ribbon represents the structure of PP1cγ, and the inhibitor is shown in dark gray (pdb 2o86).

the chelate effect, in which binding by multiple weak subsites of the same molecule results in a strong and probably very specific binding. For example, PRPs/PRGs in saliva function by the strong multidentate binding of tannins (Charlton et al. 1996; Hagerman and Butler 1981) by their tandemly repeated Pro-rich sequences (see Chapter 12, Section 12.5.1). Inhibitor-2 (I2) binds protein phosphatase 1 (PP1) with a K_d of 2 nM, wraps around, and contacts the enzyme via three separate binding regions (Figure 14.6, (Hurley et al. 2007)). Calpastatin, the strong (K_d = 4.5 pM (Hanna, Garcia-Diaz, and Davies 2007)) and specific inhibitor of calpain, also wraps around its partner and binds it via three separate binding regions, termed subdomains (Kiss et al. 2008a; Moldoveanu, Gehring, and Green 2008).

14.5 IMPLICATIONS OF DISORDER FOR THE KINETICS OF INTERACTIONS

Protein disorder is often mentioned in connection with an increased speed of interaction, which is largely based on the classical observations that non-specific initial interactions characteristic of flexible or disordered regions of proteins, or even simple polyamines speed up DNA renaturation (Chapter 12, Figure 12.5) (Pontius 1993; Pontius and Berg 1990, 1991). This issue can be approached from both mechanistic and thermodynamic points of view.

14.5.1 Primary Contact Sites

The concept of PCSs is related to recognition by short motifs, but it principally differs from PSEs, LMs, and MoRFs (Section 14.2) in being a kinetic, rather than structural or thermodynamic, concept. PCSs have been derived from the observation that IDPs can often attain the bound state very rapidly, which suggests that certain regions within their fluctuating structural ensemble might be exposed on time average for initiating productive interaction with the partner.

Such transient exposure was tested experimentally in the case of two IDPs: calpastatin and MAP2 (Csizmok et al. 2005). It was found that proteases of either narrow (trypsin, chymotrypsin, and plasmin) or broad (subtilisin and proteinase K) substrate specificity preferentially cleave both proteins in regions destined to form contact with the partner. It was found that in calpastatin, subdomains A, B, and C, whereas in MAP2c the central Pro-rich region (PRR) is spatially exposed. The structural basis of this not fully random behavior is long-range tertiary interactions in calpastatin and extended PPII helix conformation in MAP2. Provided the protease is thought of as a structural probe for local exposure and flexibility of the chain (see Chapter 3, Section 3.4 and Section 3.5, and Chapter 12, Section 12.2.2), these results are demonstrative of the transient exposure of sites destined for initiating interaction (i.e., for making the primary contact with the partner).

Endocytosis also provides evidence for the PCS concept. Endocytosis is driven by the assembly of large membrane-bound complexes of highly repetitive and disordered membrane-associated proteins, such as AP180, epsin1, and auxilin (Kalthoff et al. 2002; Scheele et al. 2003). Rapid assembly of the complexes is essential for the proper execution of endocytosis, and the large capture radius of specific, exposed recognition elements made possible by the disordered nature of these proteins is suggested to provide the key ingredient of this mechanism (Dafforn and Smith 2004), termed "protein fishing" (Evans and Owen 2002).

14.5.2 Fly-Casting in Recognition

The PCS concept and protein fishing mechanism both address the structural background of the enhancement of association rates by structural disorder. The actual molecular mechanism probably has several mechanistic components, such as a relatively nonspecific association enabled by disorder, which increases the lifetime of the encounter complex (Pontius 1993; Pontius and Berg 1991), exposure of PCSs for initial contact (Csizmok et al. 2005), and an increased capture radius of the disordered segment, which provides for an effective spatial search for a partner (i.e., protein fishing) (Evans and Owen 2002). Effective induced folding might be initiated by these initial contacts, as captured theoretically more rigorously in the model of "fly-casting" (Levy, Onuchic, and Wolynes 2007; Shoemaker, Portman, and Wolynes 2000).

Fly-casting treats the process of recognition and subsequent folding as analogous to folding of ordered proteins. The model suggests that an IDP can have an enhanced capture radius for a specific binding site due to which it binds weakly at a relatively large distance and effectively reduces dimensionality of conformational search. Thus, subsequent induced folding is nucleated by the contact between the two molecules, and the occurrence of metastable non-specific bound complexes, which would arise from the ruggedness of the interaction landscape, are avoided.

14.6 ADAPTABILITY AND MOONLIGHTING

The classical structure-function paradigm (see Chapter 1, Section 1.11) holds not only that a protein requires a single well-defined structure to carry out its function, but implicitly also that a protein has only one function. This view turned out to be untenable in light of an ever-increasing number of examples of proteins with more than one, often completely unrelated functions, termed "gene sharing" (Jeffery 2003b), "multitasking" (Jeffery 2004), or "moonlighting" (Jeffery 1999, 2003a). A unique consequence of structural disorder and folding induced upon binding is the capacity of IDPs to adapt to different partners, with potentially different functional outcomes. This possibility was raised in conjunction with $p21^{Cip1}$, which can bind distinct cyclin-Cdk complexes (Dunker et al. 2001; Kriwacki et al. 1996). The possible extent of structural adaptability is demonstrated by the analysis of p53 (Oldfield et al. 2008), a segment of which (residues 374–388) within its regulatory domain can bind four different partners (Cyclin A, sirtuin, CBP, and S100ββ), adopting different structures with all of them (Figure 14.7). Several examples show that such adaptability may result in the capacity of an IDP to have different, sometimes opposing activities with different, or even the same, partner (i.e., to moonlight) (Tompa, Szasz, and Buday 2005).

For example, the IDR regulatory R domain (Baker et al. 2007) of cystic fibrosis transmembrane conductance regulator (CFTR) interacts with the rest of the receptor in two different modes, resulting in either stimulation or inhibition of chloride conductance of the channel (Ma 2000). The random coil C fragment of loop II–III of dihydropyridine receptor (DHPR) interacts with skeletal ryanodine receptor (RyR) in excitation-contraction coupling, in two stochastically alternating modes, with one activating and the other inhibiting the partner (Haarmann et al. 2003). $p21^{Cip1}$ and $p27^{Kip1}$, the classic inhibitors of Cdks, can also activate the kinases (Bagui et al. 2003; Cheng et al. 1999). I2 has been described not only to be able to inhibit but also to activate PP1 (Tung, Wang, and Chan 1995; Yang, Hurley, and Depaoli-Roach 2000). Securin, the inhibitor of separase, is also critical for the proper folding, localization, and activation of the enzyme (Hornig et al. 2002; Jallepalli et al. 2001).

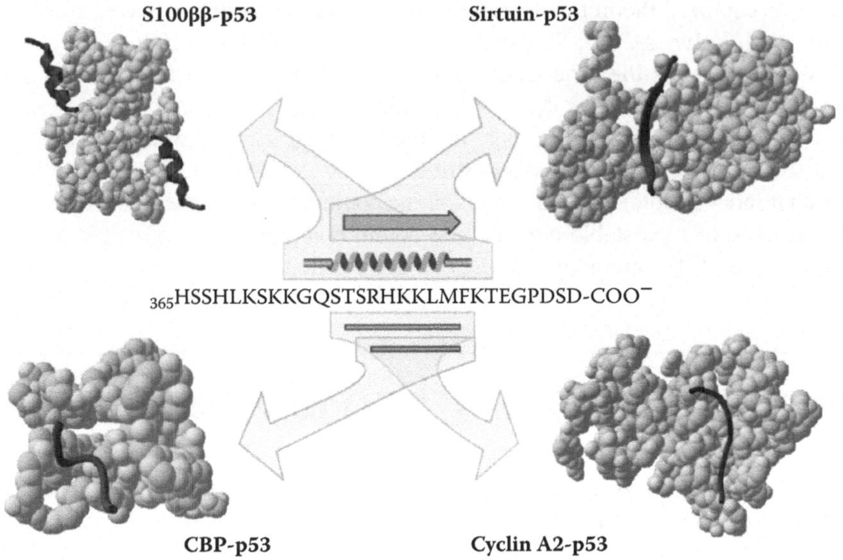

FIGURE 14.7 Structural adaptability of an IDP. Structural analyses of p53 indicate that a segment within its regulatory domain (residues 374–388) can bind four different partners (Cyclin A, sirtuin, CBP, and S100ββ), which show that the same disordered region can adopt different structures. Reproduced with permission from Oldfield et al. (2008), *BMC Genomics* 9, S1, S1. Copyright by the BioMed Central Ltd.

Other disordered moonlighting proteins can bind two different partners with distinct, often opposing activities. For example, the established function of Tβ4 is to sequester G-actin by keeping it in a polymerization-incompetent state (Domanski et al. 2004; Hertzog et al. 2004). The protein can also bind and activate integrin-linked kinase (ILK), which subsequently phosphorylates the survival kinase Akt (Bock-Marquette et al. 2004). EBV-SM not only down-regulates intron-containing mRNA but also up-regulates intron-less mRNA (Ruvolo et al. 1998). PIAS1 not only inhibits activated STAT but can also activate p53 (Liao, Fu, and Shuai 2000; Megidish, Xu, and Xu 2002).

The unifying theme of these and other examples (Tompa 2002; Tompa et al. 2005) is the adaptability of disordered regions for binding distinct partners, or even the very same partner in different modes. Based on the combination of functional, biochemical, and limited structural data, three molecular mechanisms have been proposed for the mechanism of switching between the different functional modes (Figure 14.8) (Tompa et al. 2005). Due to the adaptability of recognition regions, certain moonlighting IDPs can bind different partners via two alternative conformations of the same site, or by two different but overlapping sites (Tβ4). In other cases, the IDP can bind the same partner in two basically different conformations or binding sites, leading to different effects (DHPR C). The third case is when the IDP binds the partner in one mode, but can undergo significant conformational change or reorganization in the bound state, resulting in a distinct functional outcome (I2).

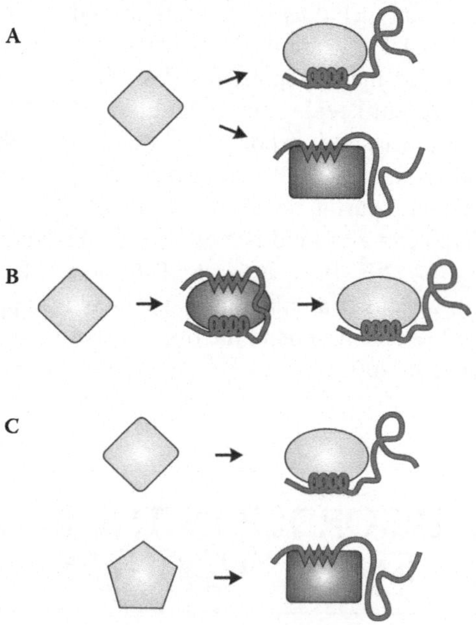

FIGURE 14.8 Mechanisms by which moonlighting IDPs switch between functions. The highly simplified scheme depicts the three basic molecular mechanisms by which IDPs may exert opposing effects. The partner molecule is represented by a light gray diamond, which turns into an oval when it assumes an active conformation and a rectangle when in an inactive conformation. Its light and dark shades indicate activation and inhibition, respectively. Mostly biochemical data suggest that a protein can bind to the same partner in two basically different conformations or binding sites, leading to different effects (A). Another mechanism is when an inhibitor shifts the equilibrium of its partner in favor of its active conformation but blocks its active site. Activation occurs when the inhibitory interaction is partially released due to post-translational modification (B). The protein may also bind two different partners due to structural adaptability of its binding site or by two overlapping (nested) sites (C). Reproduced with permission from Tompa et al. (2005), *Trends Biochem. Sci.* 30, 484–9. Copyright by Elsevier Trends Journals.

14.7 NESTED INTERFACES

IDPs often bind their partners through an elongated interface that makes numerous contacts with the partner. Mutagenesis studies of the resulting complexes suggest that often binding only relies on a few scattered specificity determinants (see also Section 14.2.2), connected by linkers rather free to change. This mode of binding enables binding sites for different partners to be nested or interdigitated.

Well-characterized examples are disordered bone sialoprotein (BSP) and osteopontin (OPN), two members of the SIBLING (Small Integrin-Binding Ligand, N-linked Glycoprotein) family of proteins. Their flexibility enables them to rapidly associate with a number of binding partners including other proteins and the mineral phase of bones and teeth (Fisher et al. 2001). For example, OPN binds to hydroxyapatite (HA) along its entire length, probably via its many Asp groups. This binding is important for mediating cell attachment to HA, yet keeping the RGD motif of OPN available for integrin binding. For example, osteoclasts (i.e., bone cells that remove the bone's mineralized matrix in bone resorption) may use OPN to bridge between the integrins on the cell surface and the mineral phase during resorption (Reinholt et al. 1990). OPN complexed to either integrin or CD44 can also bind Factor H (FH) (Fedarko et al. 2000). Binding of the various partners HA, FH, integrin, and CD44 occurs by partially overlapping binding surfaces, and in various combinations (e.g., HA plus integrin, or FH plus integrin, etc.) that are mutually exclusive due to the overlaps between the respective binding surfaces/motifs (Fisher et al. 2001).

14.8 DISORDER IN THE BOUND STATE: FUZZINESS

A corollary to folding induced upon binding is that IDPs attain a well-defined structure in the bound state (Dyson and Wright 2002a). Although this notion is underscored by structures of many IDPs in complex with their partner, the underlying premise is not generally true. Often, part(s) of the complexes cannot be described by a single conformation because they are structurally ill-defined (i.e., disordered). Because such regions often productively contribute to binding and function (i.e., by all counts, they are integral parts of the complex), the phenomenon of disorder was extended to the bound state and termed "fuzziness" (Tompa and Fuxreiter 2008). Fuzziness is thought to cover two distinct structural phenomena, multiple stable conformations (structural polymorphism), and true disorder when the protein constantly fluctuates between a large number of conformational states (Figure 14.9). Within this simple framework, there are four possible manifestations of fuzziness.

14.8.1 Structural Polymorphism in the Bound State

If the bound IDP adopts a few or multiple alternative well-defined conformations, it is termed "polymorphic," which corresponds to a static type of disorder (Tompa and Fuxreiter 2008). For example, the disordered CBD of Tcf4 binds to β-catenin in an extended conformation, in which its acidic middle segment adopts several distinct conformations (Figure 14.9A) characterized by alternative salt bridges (Graham et al. 2001). Similar behavior was described in the case of the heat-shock protein 90 (Hsp90) carboxy-terminal MEEVD peptide, assuming two alternative conformations in the binding pocket of the tetratricopeptide repeat (TPR) domain of protein phosphatase 5 (PPP5) (Cliff et al. 2006), nuclear localization signals (NLSs), binding to the nuclear import receptor protein importin-α in different conformations

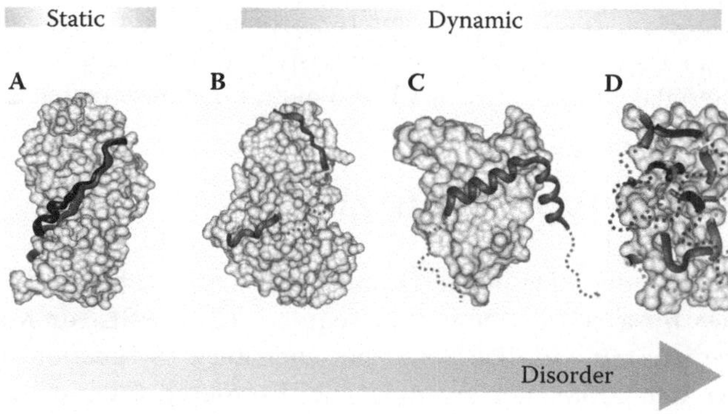

FIGURE 14.9 Fuzzy complexes. Structural view of disorder in protein–protein complexes, denoted as fuzziness. Fuzziness can be either static (A) or dynamic (B–D), shown here in the order of increasing disorder (arrow). Within the complexes, the binding partner is rendered as a gray space-filling model, whereas ribbon(s) represent the IDP, which may be partially disordered (dotted line marks its part that is disordered even in the bound state). In the polymorphic model (A) (pdb 1jdh and 1g3j), there is more than one stable conformation of IDP. In the dynamic cases, such as the clamp (B) (pdb 2f4g), flanking (C) (pdb 1kdx), and random (D) (Sigalov 2004) models, part or the entirety of the IDP remains disordered in the bound state. Reproduced with permission from Tompa and Fuxreiter (2008), *Trends Biochem. Sci.* 33, 2–8. Copyright by Elsevier Trends Journals.

(Fontes, Teh, and Kobe 2000; Fontes et al. 2003) and Pro-rich regions, binding the GYF domain in alternative conformations (Gu et al. 2005).

14.8.2 Clamp-Type of Fuzziness

A different but often encountered mode of disorder in the bound state is when the IDP binds the partner by two recognition elements/domains, but the intervening linker region remains disordered and does not make contact with the partner, such as in a "clamp" (Tompa and Fuxreiter 2008). This mode of binding in fuzziness is directly linked with the concepts of entropic-chain linker functions (Chapter 12, Section 12.1.1), multivalent binding (Section 14.4.3), and probably also with nested interfaces (Section 14.7). For example, the scaffold protein Ste5p (Chapter 12, Section 12.6.2) binds to the MAPK Fus3p via two distinct protein segments (Figure 14.9B), connected by an 8-residue disordered linker (Bhattacharyya et al. 2006). The whole construct binds Fus3p with a K_d of 4 μM, whereas deletion of the linker practically abolishes binding. A similar behavior occurs in the case of the Oct-1 transcription factor, which binds the Ig-κ promoter region in a clamp-like fashion (van Leeuwen et al. 1997). As detailed in Chapter 12, Section 12.1.1, the two DNA binding POU domains of Oct1 recognize a 4–5 base pair sequence, but in the presence of the 23 amino acid-long disordered linker region, they target an octamer DNA sequence with high specificity. Further clamp-type cases of fuzziness in the literature are bipartite NLSs binding to α-importin (Fontes et al. 2000), tripartite binding mode of I2 to PP1 (Hurley et al. 2007), and calpastatin to calpain (Kiss et al.

2008a; Moldoveanu et al. 2008). Clamp-type fuzzy binding with an elongated partner of multiple binding sites may also result in processivity (see Section 14.9), as observed in the case of bacterial cellulase, myosin VI, and matrix metalloproteinase 9 (MMP-9).

14.8.3 Flanking-Type of Fuzziness

Potentially even more disorder in the bound state may be observed in IDPs that bind their partner by a short recognition segment (LM, PSE, or MoRF), when the "flanking" regions remain disordered but contribute to function in some way (Tompa and Fuxreiter 2008). For example, the disordered CTD of measles virus nucleoprotein binds to the XD of viral phosphoprotein by a short recognition element, Box2 (Bourhis et al. 2005). Removal of the apparently non-binding disordered region decreases binding affinity three orders of magnitude. The effect of transcription factor CREB is mediated via interaction of its KID domain with the KIX domain of the co-activator CBP (see Chapter 6, Section 6.3.1, Chapter 10, Section 10.2.3.2, and Section 14.3 of this chapter). KID binds KIX along a segment of 29 amino acids (Radhakrishnan et al. 1997), whereas the rest of CREB remains disordered in the complex (Chapter 6, Figure 6.3, and Figure 14.9C of this chapter). Yet, deletion of these regions lowers the strength of binding fivefold (Zor et al. 2002). Thermodynamic data confirm a similar contribution in the case of the interaction of the splicing factor SF1 with the large subunit of the U2 small nuclear RNA auxiliary factor (U2AF65). The interaction, in which SF1 binds U2AF65 by a short linear motif that becomes structured in the complex (Selenko et al. 2003) with a K_d of 55.6 nM, is important in splice-site selection in pre-mRNA splicing. Contribution of the flanking regions is shown by binding of full-length SF1, which has a significantly lower K_d of 11.8 nM.

14.8.4 Random-Type of Fuzziness

At the extreme of the structural spectrum, proteins in the bound state may appear not to have undergone ordering at all, as suggested by the "random" model (Tompa and Fuxreiter 2008). Three pertinent cases have been reported in the literature (Figure 14.9D). In T-cell signaling, homo-oligomerization of the disordered cytoplasmic domains of T-cell receptor ζ-chains, which leads to the formation of oligomers of the receptor (Sigalov 2004), occurs by fully retaining their disorder, as demonstrated by CD and NMR (Sigalov, Aivazian, and Stern 2004). Although the interactions are rather weak, with K_d values ranging from 10 μM to 1 mM, they do support cellular activation. A similar observation of disorder in the dimeric state is apparent in the case of the product of the umuD gene in *E. coli*, which plays key roles in coordinating the switch from accurate DNA repair to mutagenic translesion DNA synthesis (TLS) during the SOS response to DNA damage (Simon et al. 2008). Elastin, the elastic fiber formed by the self-assembly (co-acervation) of tropoelastin monomers, also has a disordered, complexed state in which a high degree of dynamic disorder is seen by solid-state NMR (Pometun et al. 2004). Recognition in the disordered elastin is probably achieved via a distributed array of short, marginally defined binding motifs, which can even attain binding by alternative patterns, and overall do not lead to a detectable level of ordering of the whole protein.

14.9 PROCESSIVITY OF BINDING

Clamp-type fuzziness may result in processivity of binding if the partner has multiple binding sites, which enable alternating binding events without fully releasing the partner. Three examples have been studied in detail.

As detailed in Chapter 4, Section 4.4.2 and Chapter 12, Section 12.1.1, efficient cellulases have evolved a modular structure of a large catalytic domain linked to a smaller cellulose binding domain by an IDR linker of about 40 amino acids in length (von Ossowski et al. 2005). SAXS studies of a fused double cellulase (von Ossowski et al. 2005) showed that the enzyme acquires a nonrandom continuum of conformations with interdomain separations ranging from 10 Å to 120 Å, biased for compact conformers (see Chapter 4, Figure 4.3). This flexibility of the linker is essential for the processivity of the action of the enzyme, enabled by a combination of tight binding by the cellulose-binding domain and large conformational freedom in search of the catalytic domain for multiple cleavage sites without releasing the substrate.

A similar model has been suggested for MMP-9 (Rosenblum et al. 2007). The proteinase is secreted into the extracellular space, where it is targeted at several substrates, such as collagen. It has a modular structure, with an N-terminal catalytic domain next to three fibronectin type II exosite modules, connected by a 54-residues long Pro/Gly-rich linker (termed OG domain) to a C-terminal hemopexin C domain. By a combination of SAXS and AFM measurements (Rosenblum et al. 2007), the molecule was shown to assume a compact, yet low-density disordered state corresponding to multiple conformations, which may ensure structural adaptation to a large number of substrates with dissimilar structures (see Chapter 5, Section 5.8.1). The enzyme can anchor at the hemoplexin C domain to collagen, extend the linker so that the catalytic domain can search for substrate sites, retract in an inchworm-like manner, and proceed very rapidly along the substrate (Overall and Butler 2007).

Myosin VI is a processive motor that moves along actin filaments (Rock et al. 2005). It has an N-terminal motor domain, a 53-residue unique insert, a single IQ motif, and a putative coiled-coil tail domain. It takes 30–36 nm steps along actin filaments, similar in size to those of myosin V, despite having a shorter lever arm. The large and variable step size suggests that myosin uses diffusive search during stepping. Predictions, EM (Chapter 5, Section 5.7), and single-molecule optical trapping studies of wild-type and mutated constructs show that the proximal tail (an approximately 80-residue segment following the IQ domain) is disordered, which permits the head domains to separate and search diffusively for attachment points in the next step (Rock et al. 2005).

14.10 SEQUENCE INDEPENDENCE IN RECOGNITION

Fuzziness is probably also intimately linked with mutagenesis studies in which resistance of function to randomization of sequences is observed (cf. also Chapter 13, Section 13.4.2). These observations are in sharp contrast with the traditional model of protein–protein recognition.

The classical observation is the sequence-independence of the function of the acidic TAD domain of Gcn4p, which can be replaced with random acidic segments without a major loss of activity (Hope, Mahadevan, and Struhl et al. 1988). A similar behavior was observed in the case of another transcription factor, EWS fusion protein (EFP), which is generated by chromosomal translocation of the Ewing's sarcoma (EWS) oncogene. The TAD of EFP is constituted of multiple (up to 30) copies of imperfect hexapeptide repeat motifs, which can be freely interchanged and their sequences randomized or even reversed, without EFP losing trans-activator function (Ng et al. 2007). Linker histones have sequentially highly diverged CTDs of similar overall amino acid composition, which harbor the binding region of apoptotic nuclease DNA fragmentation factor 40 (DFF40). It was observed that any segment of the CTD of sufficient length can bind and activate the enzyme, regardless of its primary sequence and location in intact CTDs (Hansen et al. 2006).

Amyloid formation by yeast prions Ure2p and Sup35p provide somewhat different examples of the sequence-independence of recognition (Ross et al. 2005). These prions do not harm their host cells, but may impart selective advantages on them, and thus are considered physiological prions (Wickner et al. 1999). They contain Q/N-rich disordered prion domains, the sequences of which can be randomized without the loss of the prion-like property of the protein (Ross et al. 2005).

14.11 ULTRASENSITIVITY OF RECOGNITION

The phenomenon of ultrasensitivity of recognition, another unusual mode of protein–protein interaction enabled by structural disorder, is also in close association with fuzziness. Ultrasensitive binding is observed when numerous suboptimal binding sites are located in close proximity in a disordered segment of a protein, and post-translational modification of several of them is required for productive binding and function, even though there is only a single binding site on the partner. This unusual binding mode results in an ultrasensitive dose-response curve of recognition (Bary et al. 2007), typified by a high Hill coefficient. There are two cases that have been studied in detail: the recognition and ubiquitination of the Cdk inhibitor Sic1 by the SCF ubiquitin ligase subunit of the cell-division cycle protein 4 (Cdc4p), and the binding and inhibition of CFTR by its disordered regulatory (R) domain.

14.11.1 Recognition of Sic1 by Cdc4

The proteasome-mediated degradation of Sicl Cdk inhibitor is critical for cell cycle progression in yeast. Commitment to division, called Start, requires a threshold level of G1 cyclins (Cln1/2/3), which activate Cdc28 (Cdkl) in late G1 phase. As cells progress through Start, B-type cyclins (Clb5/6) activate Cdc28, which is mandatory for the initiation of DNA replication (Schwob et al. 1994). A link between the two events is the phosphorylation of an inhibitor of the Clb-Cdc28 kinases (Sicl), which targets it for degradation (Schneider, Yang, and Futcher 1996; Schwob et al. 1994). Phospho-Sicl is ubiquitinated by the Cdc34-SCF ubiquitin ligase complex, to which it is recruited by an adapter subunit, Cdc4 (Bai et al. 1996; Feldman et al. 1997). The recognition event is mediated by the WD40 domain of Cdc4, which binds with high affinity to the consensus phosphopeptide motif (the Cdc4 phospho-degron, CPD) of Sicl. The unusual behavior comes from Sicl having nine suboptimal CPDs, about six of which needs to be phosphorylated to achieve productive binding that leads to DNA replication. The motifs act in concert to mediate Cdc4 binding; they establish a phosphorylation threshold and prevent premature degradation and DNA replication. Although mutant Sicl with a single "optimal" site is sufficient to mediate strong and specific interaction, ubiquitination, and degradation *in vivo*, it cannot mediate proper Start site function and restrain DNA synthesis due to the improper timing of this reaction (Nash et al. 2001). The requirement for multiple phosphorylation of suboptimal sites probably makes the response highly cooperative and very sensitive to kinase concentration, and the resulting "ultrasensitivity" is key to the proper timing of the initiation of DNA synthesis.

14.11.2 Regulation of CFTR by Its Disordered R Domain

An even more complex case is represented by the CFTR chloride channel, the protein mutated in cystic fibrosis. This channel belongs to the ATP-binding cassette (ABC) superfamily of proteins (Riordan et al. 1989). It has two membrane-spanning domains (MSD1 and MSD2), two nucleotide-binding domains (NBD1 and NBD2), and an intracellular region between the transmembrane segments with a unique cytoplasmic region, the disordered regulatory (R) domain (Ostedgaard et al. 2000). R domain is about 200 residues in length. It lacks a stable structure and has about nine consensus PKA phosphorylation sites, within several short regions that transiently sample helical conformations (Baker et al. 2007). PKA phosphorylation leads to the activation of chloride conductance, but no single site appears to be critical. Phosphorylation of about five sites is required to have an effect, and modifications of individual sites have been found to act synergistically (Chang et al. 1993; Cheng et al. 1991).

NMR studies suggest that several sites in the R region have measurable fractional helical propensity, which is required to mediate their interactions with the NBD domains and keep them apart so that the channel is in an inactive state. Phosphorylation of any of the sites reduces local helicity and decreases affinity of the segment for the NBD domains, but the involvement of several sites is required to have an overall effect of relieving NBD segregation and activating chloride conductance. This observation explains the dependence of CFTR activity on multiple PKA phosphorylation events (Baker et al. 2007).

14.11.3 Electrostatics in Ultrasensitivity

The explanation for the physical basis of these observations of ultrasensitivity probably resides in the effect of an increasing number of charges on the interaction (Borg et al. 2007). A mean-field statistical mechanical approach modeling cumulative electrostatic interactions between a single receptor site and a conformationally disordered polyvalent ligand can adequately describe the threshold in the number of charges required for favorable ligand–receptor contact. This approach also provides a reasonable framework for future more elaborate models, which may also take into account local structural effects, as seen in the case of CFTR R domain (Baker et al. 2007). Such models may eventually also lead to a better understanding of the classical and somewhat enigmatic concept of multisite PTMs in biological function (Cohen 2000; Yang 2005).

14.12 SIGNAL PROPAGATION IN THE STRUCTURAL ENSEMBLE OF IDPS

Most functional models of IDPs incorporate a single binding event, in which part of the IDP binds another protein and modifies its activity and/or localization. The possible structural communication between distinct binding sites within the same IDP molecule is largely neglected, although signal propagation through the structure of IDPs raises the possibility of a higher level of functional integration enabled by disorder. Several observations pertain to this scenario.

14.12.1 The Signaling Conduit $p27^{Kip1}$

$p27^{Kip1}$, the Cdk inhibitor, regulates cell division through complex formation and inhibition of Cyclin A-Cdk2 and Cyclin E-Cdk2 complexes (see Chapter 3, Section 3.7.2; Chapter 10, Section 10.2.3; Section 14.3 of this chapter; and Chapter 15, Section 15.1.3). $p27^{Kip1}$ binds the complex through its amino-terminal KID domain, inserting Tyr^{88} into the ATP-binding pocket of the kinase (Russo et al. 1996). The inhibitor has a CTD of about 100 amino acids in length carrying other functions, such as regulation of degradation by the ubiquitin/proteasome system. Regulation of $p27^{Kip1}$ occurs via Tyr-phosphorylation followed by intramolecular phosphorylation by Cdk2 at Thr^{187}, which eventually leads to ubiquitination and degradation. The enigmatic autophosporylation by an inhibited Cdk2 molecule can be explained by conformational signal propagation within the disordered CTD of the inhibitor. The ternary complex $p27^{Kip1}$-Cyclin A-Cdk2 was characterized by a combination of SAXS and NMR. The C-terminal 100 amino acids (p27-C) of the inhibitor in the complex are locally disordered (i.e., they form neither stable contacts with Cyclin A-Cdk2 nor stable secondary structure on their own). This disorder enables to sample conformational states that allow its phosphorylation by the very same Cdk2 molecule it is bound to, as suggested earlier on biochemical basis (Grimmler et al. 2007). Cyclin A-Cdk2-bound $p27^{Kip1}$ can be phosphorylated at Tyr^{88} by Tyr kinases, such as the Src-family non-receptor Tyr kinase Lyn and the oncogene product BCR-ABL (Grimmler et al. 2007), enabled by the extended and flexible nature of the 100-residue $p27^{Kip1}$ C-terminal domain, even when $p27^{Kip1}$ is bound to Cyclin A-Cdk2 complex (Galea et al. 2008a). Tyr^{88}-phosphorylation of the bound inhibitor results in the ejection of Tyr^{88} from the ATP-binding pocket of the kinase, which yields an active ternary p27-phospho-Tyr88-Cyclin A-Cdk2-complex (Grimmler et al. 2007). Separation of Tyr^{88} from the enzyme enables a large-scale conformational change that brings Thr^{187} at the active site and brings about its phosphorylation. The site is then able to make productive contact with the ubiquitin ligase that ubiquitinates the inhibitor and directs it toward degradation. Thus, the intrinsic flexibility of $p27^{Kip1}$ enables a long-range communication between subsites and the sequential signal transduction by a molecular conduit. It has been suggested that other intrinsically disordered proteins possessing multiple PTM sites may also participate in similar "signaling conduits" (Galea et al. 2008a).

14.12.2 Tailored Auto-Activation of WASP

Long-range propagation of information in the structural ensemble of a disordered protein also enable to engineer its variants for regulatory purposes, as demonstrated by the actin regulatory protein neuronal (N-) WASP (Dueber et al. 2003). WASP displays sophisticated signal integration capacities (Chapter 11, Section 11.6.1 and Figure 14.10). The modular protein contains an output region (VCA domain), which is constitutively active in isolation (i.e., it stimulates actin polymerization by binding and activating the Arp 2/3 complex). In intact WASP, the activity of the VCA domain is repressed by two modular domains: a highly basic (B) motif and a GTPase–binding domain (GBD) through autoinhibitory interactions (Kim et al. 2000a). PIP2 and activated GTPase Cdc42 can bind the B and GBD motifs (Figure 14.3) and disrupt autoinhibition in a highly cooperative manner. The mechanism of release of the autoinhibitory contacts can be completely changed by redesigning the molecule, by replacing the B/GBD module with a PDZ domain (Figure 14.10), which makes the molecule responsive to an engineered cognate peptide. Due to long-range structural communication within the molecule enabled by a long IDR linker region, a series of signaling molecules could be developed, in which the "output" domain was recombined with heterologous autoinhibitory "input" domains (Dueber et al. 2003).

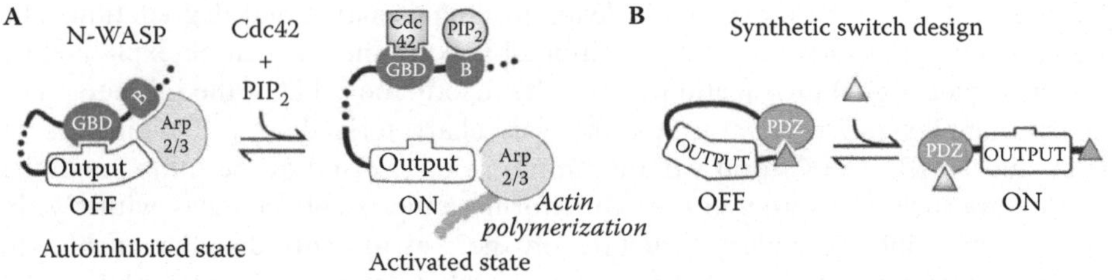

FIGURE 14.10 Redesigning the allosteric N-WASP switch. Disorder of N-WASP enables it to be redesigned as an artificial switch gated by heterologous ligands. (A) N-WASP is a modular allosteric switch, with an output domain signaling to the actin cytoskeleton by stimulating Arp2/3. Its activity is repressed by autoinhibitory interactions involving the endogenous GBD domain and basic motif (marked as B). Ligands can activate N-WASP by disrupting autoinhibitory interactions. (B) A single-input switch could be designed by taking advantage of the allostery enabled by disorder of the long linker region of N-WASP, by placing a PDZ domain-ligand pair flanking the output domain. Reproduced with permission from Dueber et al. (2003), *Science* 301, 1904–8. Copyright by the American Association for the Advancement of Science.

14.12.3 Allostery Mediated by Order–Disorder Transitions

Allosteric signal propagation within an ordered protein can be largely amplified by order-to-disorder transitions in its structure, as shown in the case of the catabolite activator protein (CAP) (Li et al. 2007). CAP is a mostly ordered dimer that consists of two cAMP-binding subunits, each containing a C-terminal DNA-binding domain and an N-terminal ligand-binding domain. Dimeric CAP shows negative cooperativity, because cAMP binding in one molecule impairs cAMP binding in the other. Extensive explicit-solvent MD simulations indicate that the system experiences a switch in motion as a result of cAMP binding, with the DNA-binding module dissociating from the ligand-binding domain, promoted by an order-to-disorder transition within the region connecting ligand-binding and DNA-binding domains. This observed behavior may represent a novel mechanism of allostery in proteins, as also suggested in a theoretical study that showed how transitions between order and disorder coupled with intramolecular recognition can give rise to very effective allosteric switches (Hilser and Thompson 2007).

14.13 DISORDER AND ALTERNATIVE SPLICING

The observation of the correlation between alternative splicing and disorder (Romero et al. 2006) suggests that disorder may have a role in regulated changes in protein functionality. The analysis of a set of 46 differentially spliced genes encoding experimentally characterized human proteins shows that 81% of 75 alternatively spliced fragments are associated with fully (57%) or partially (24%) disordered protein regions, and regions affected by alternative splicing are significantly biased toward disorder-promoting amino acids (Dunker et al. 2001). Disorder predictions are consistent with these experimental data.

14.14 MOLECULAR MIMICRY BY A DISORDERED REGION

IDPs, upon folding, may mimic the structure of other proteins, which results in effective competition phenomena, as demonstrated by colicin binding to its intracellular receptor, TolB (Bonsor et al. 2007; Loftus et al. 2006). The 61-kDa colicin E9 (ColE9) nuclease is a toxin synthesized by *E. coli* to combat competing bacterial cells. The protein is excreted into the extracellular space, where it interacts with outer membrane (OmpF) and periplasmic (TolB) helper proteins of susceptible bacterial cells, translocates into their cytoplasm, and kills them by hydrolyzing their DNA. TolB normally is bound to a globular partner, Pal, and forms a complex that is involved in maintaining the integrity of the outer membrane (OM) in all Gram-negative bacteria that are parasitized by colicins. ColE9 competitively recruits TolB by disrupting its complex with Pal. The N-terminal 83 residues of ColE9 (translocation domain, T), which includes the TolB box, a recognition element for TolB, is intrinsically disordered.

The 16-residue TolB binding epitope folds into a disordered hairpin in complex with TolB (Loftus et al. 2006). Comparison of this structure with that of TolB-Pal (Bonsor et al. 2007) reveals that colicins bind at the very same site where Pal does, with acquiring a local conformation that perfectly mimics Pal residues. Thus, induced folding of this IDP creates a recognition element that is very similar to the folded binding site of an ordered protein, and thus it can competitively displace it in a process termed "competitive recruitment" (Bonsor et al. 2007). Similar cases of "molecular mimicry" by an IDP were also described in 4E-binding protein (4E-BP) binding to eukaryotic translation initiation factor 4E (eIF4E), mimicking the interaction of eIF4G (Chapter 11, Figure 11.4) (Marcotrigiano et al. 1999), and bacterial EspF(U) binding to the autoinhibitory GBD of WASP, mimicking the intramolecular interaction of the VCA region of the latter (Cheng et al. 2008).

14.15 ENTROPY TRANSFER IN CHAPERONE ACTION

A recurring theme in the IDP literature is the reduction of the conformational freedom of an IDP upon binding to the partner in an induced folding process. In this section we examine a model for the chaperone action of disordered proteins, in which the reciprocity of entropy change (i.e., the functional effect of the increase of entropy of the partner concomitant to binding of the disordered protein) is suggested. This mechanism forms the basis of a coherent mechanistic model of the action of disordered chaperones (details on their identity and action in Chapter 12, Section 12.3).

The first element of the model is that disordered proteins/regions provide unique versatility in the recognition process, which may be beneficial in fast, relatively nonspecific and reversible interactions with a range of apparently unrelated partner molecules. The second mechanistic element of the entropy-transfer model is that disordered segments provide a significant effect of solubilization, as demonstrated in many cases of protein aggregation, which are inhibited or sometimes even reversed by disordered chaperones. Because aggregation is usually caused by the association of hydrophobic patches exposed due to inappropriate folding of the substrate, this effect may simply result from shielding by highly hydrophilic disordered segments. Long-range repulsion resulting from entropic exclusion by disordered appendages may add to this effect, because it may physically prevent molecules from approaching each other (entropic bristle/brush mechanism, Chapter 12, Section 12.1.4). The magnitude of this effect has been demonstrated by biophysical measurements in the case of proteins that ensure spacing in the cytoskeleton, such as MAPs (Mukhopadhyay and Hoh 2001) and neurofilament side-arms (Brown and Hoh 1997), and also Nups in NPC gating (Chapter 12, Figure 12.1) (Lim et al. 2006). Similar activity has been suggested for caseins in preventing the aggregation of calcium phosphate nanoclusters (Holt et al. 1996), and its direct involvement in chaperone-related functions was demonstrated in the case of Hsp25, for example (Lindner et al. 2000).

The key mechanistic element of chaperone action by disordered proteins, however, may come from transient ordering of the chaperone upon binding to the substrate. Because kinetically trapped substrates are stuck in a local energy minimum, chaperones assist folding by random disruption of misformed bonds via reciprocal changes in disorder and order. Local loss of flexibility upon substrate binding was seen in the case of GroEL (Gorovits and Horowitz 1995) and α-crystallin (Lindner et al. 1998). A transient increase of flexibility of the misfolded substrate in the presence of the respective chaperone was observed in a group of introns in the presence of StpA (Waldsich, Grossberger, and Schroeder 2002), the TAR element (Azoulay et al. 2003; Bernacchi et al. 2002), and $tRNA^{Lys}$ (Tisne, Roques, and Dardel 2001) in the presence of NCP7, mRNA in the presence of cold-shock proteins (Phadtare, Alsina, and Inouye 1999), and both RuBisCO (Rye et al. 1997) and carbonic anhydrase (Persson et al. 1999) in the presence of GroEL. Because binding by the chaperone keeps different segments/strands of the substrate at a close range, this proximity may also spatially limit subsequent conformational search and speed up the folding process, as directly demonstrated in the case of DNA renaturation facilitated by the disordered CTD of hnRNP A1 (Pontius and Berg 1990). In all, these distinct mechanistic elements of rapid and promiscuous binding, solubilization, local unfolding, and proximal positioning combine into a mechanistic model of the action of disordered chaperones, in which reciprocal changes in order and disorder (i.e., "transfer of entropy") play a key role (Tompa and Csermely 2004).

Structural Disorder and Disease

15

Proteins involved in various diseases, such as cancer and neurodegeneration, have a high frequency of disorder. This general correlation and the possible direct involvement of structural disorder in disease is supported by several bioinformatic analyses and detailed studies on individual proteins. We will discuss the most important diseases and examples, as well as the elevated level of disorder in pathogenic organisms. The chapter will be concluded with how novel structural insight gained from the recognition of structural disorder can be harnessed for the purposes of rational drug design.

15.1 STRUCTURAL DISORDER AND CANCER

15.1.1 Disorder in Cancer-Associated Proteins

Cancer proteins have an elevated level of predicted disorder, as demonstrated by comparing disorder in four protein datasets: human cancer-associated proteins (HCAP), signaling proteins collected by the Alliance for Cellular Signaling (AfCS), eukaryotic proteins from SwissProt, and non-homologous protein segments with well-defined 3-D structures (Iakoucheva et al. 2002). HCAP contain significantly more disorder than proteins in the other datasets, with 79 ± 5%, 66 ± 6%, 47 ± 4%, and 13 ± 4% of the proteins in HCAP, AfCS, SwissProt, and ordered datasets having at least one IDR ≥30 consecutive residues (see Figure 15.1). In addition, only 17.3% of the cancer-associated proteins have sequence alignments with structures in the Protein Data Bank (PDB) covering at least 75% of their lengths, which also points to the prevalence of disorder in the HCAP dataset. The correlation is underscored by some intrinsically disordered proteins (IDPs) involved in cancer, which are among the best characterized proteins for both structural disorder and pathophysiological function. These cases also demonstrate the complexity of the structure-function relationship of modular IDPs.

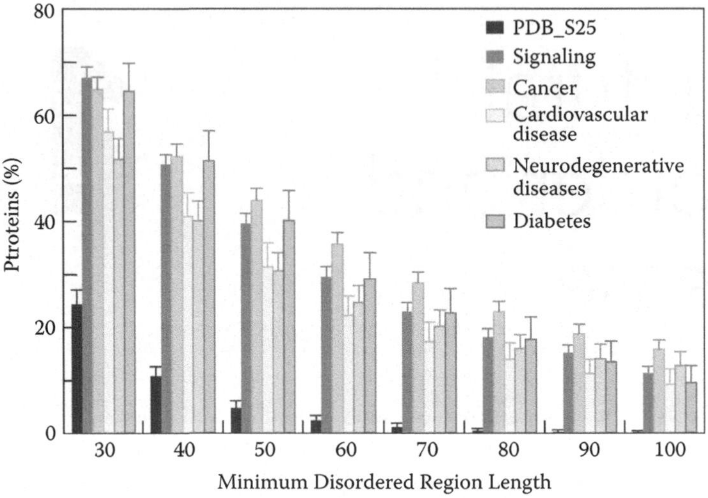

FIGURE 15.1 Predicted disorder of proteins associated with various diseases. The percentage of proteins with at least one IDR of a given minimal length was predicted in six datasets. The datasets are non-homologous protein segments with well-defined 3-D structures from PDB (PDB_S25), human signaling proteins, and proteins implicated in cancer, cardiovascular diseases, neurodegenerative diseases, and diabetes. The error bars represent 95% confidence intervals. Reproduced with permission from Uversky et al. (2008), *Annu. Rev. Biophys.* 37, 215–46. Copyright by Annual Reviews.

15.1.2 p53

p53 is a prime example that without a detailed characterization of structural disorder, even a protein studied in as much detail as p53 cannot be fully understood. p53 plays essential roles in maintaining the integrity of the human genome by controlling apoptosis, cell cycle, deoxyribonucleic acid (DNA) repair, and senescence, and is thus sometimes called the "cellular gatekeeper" (Levine 1997) or "guardian of the genome" (Lane 1992). p53 is directly inactivated in about 50% of human cancers, and in the remainder its activity is lost due to disruption of associated pathways. This protein is a transcription factor that responds to upstream signals generated by stress conditions, such as oncogene activation, DNA damage, and hypoxia, and induces or inhibits about 150 downstream effectors, such as Bax and p53-upregulated modulator of apoptosis (PUMA) (apopotosis), Gadd45, and proliferating cell nuclear antigen (PCNA) (DNA repair), $p21^{Cip1}$ and 14-3-3σ (cell-cycle arrest) and Maspin and brain-specific angiogenesis inhibitor 1 (BAI1) (anti-angiogenesis). At the molecular level, p53 function is regulated by a wide array of post-translational modifications (Joerger and Fersht 2008), and is realized in conjunction with negative (e.g., murine-double minute 2 (MDM2); see Chapter 12, Section 12.6.2.2) and positive (e.g., p300/CBP; see Chapter 11, Section 11.2.2) regulators. MDM2 is an E3 ubiquitin ligase, which directly inhibits its binding functions and promotes its ubiquitin-dependent degradation by the proteasome.

Human p53 is 393 amino acids in length, and it can be divided into four structural and functional regions/domains as detailed in Chapter 4, Section 4.4.3 (see Chapter 9, Figure 9.2 for predicted disorder). Basically, p53 is a homotetramer, with folded tetramerization and core domains that are linked together and flanked by ID domains at the N- and C-termini (Joerger and Fersht 2008). A variety of techniques have shown that the trans-activator domain (TAD) is disordered (Bell et al. 2002; Dawson et al. 2003) but also suggested function-related transient structural organization within regions 15–29 and 39–59, which adopt amphipathic α-helices upon interaction with MDM2 (Kussie et al. 1996) or replication protein A RPA7e (Bochkareva et al. 2005), respectively. The binding region of MDM2 appears as a downward spike on the disorder score (Mohan et al. 2006). Solution studies by multidimensional nuclear magnetic resonance (NMR) have confirmed that unbound full-length p53 TAD populates a helix conformation between residues T^{18}-L^{26} (Lee et al. 2000; Wells et al. 2008).

Paramagnetic resonance enhancement (PRE) experiments confirmed the short-range transient order in p53 TAD, and also have shown its compact dynamic ensemble, in which the regions responsible for MDM and RPA70 binding are separated by an average distance of 10–15 Å, less than the random coil expectation (Vise et al. 2007). Molecular dynamics (MD) simulations restrained by PRE internuclear distances (Lowry et al. 2008b) also suggest a partially collapsed state of TAD, which places the MDM2 and RPA70 binding regions in close proximity, inferring their possible functional interplay. Principal component analysis of the atomic contact maps (Lowry et al. 2008a) suggested that the ensemble is conspicuously nonrandom, with the negative charges uniformly exposed on one face of the clusters. This imbalance of charges may steer other factors in the p53-mediated assembly of complexes.

Structural analysis of other p53 domains also portrays an overall rather complex picture. p53C (DNA-binding region) is well-folded in both DNA-bound and DNA-free forms (Joerger and Fersht 2008), with only marginal stability, however (typical melting temperature is 44–45°C). The tetramerization domain provides yet another interesting structural example, because it is predicted to be fully disordered (Chapter 9, Figure 9.2), yet it is ordered in the tetrameric state. This region is probably an example of two-state complexes, which are disordered in isolation, only to become ordered in the oligomeric state (Gunasekaran, Tsai, and Nussinov 2004). In accord, dimerization of this domain occurs cotranslationally, whereas the tetramers are formed posttranslationally by dimerization of dimers (Nicholls et al. 2002). The C-terminal regulatory domain is fully disordered. This region is highly basic; it can weakly and nonspecifically interact with DNA and modulate binding at specific sites by p53C. This region is subject to extensive regulatory post-translational modifications and shows binding promiscuity (Oldfield et al. 2008), because it can bind several partners, such as S100ββ, CBP, Cyclin A2, and sirtuin, in different local conformations (Chapter 14, Figure 14.7).

The structure of full-length tetrameric p53 (Figure 15.2 and cover picture) has been unveiled by a combination of small-angle X-ray scattering (SAXS), electron microscopy (EM), NMR, and MD simulations (Joerger and Fersht 2008; Wells et al. 2008). In the DNA-free form the protein forms an elongated cross-shaped tetramer, in which core domain dimers are loosely coupled and the N- and C-termini are extended and disordered. In the DNA-bound state, the molecule wraps around the DNA helix, enabled by the flexibility of the linker between p53C and the tetramerization domain. The TADs

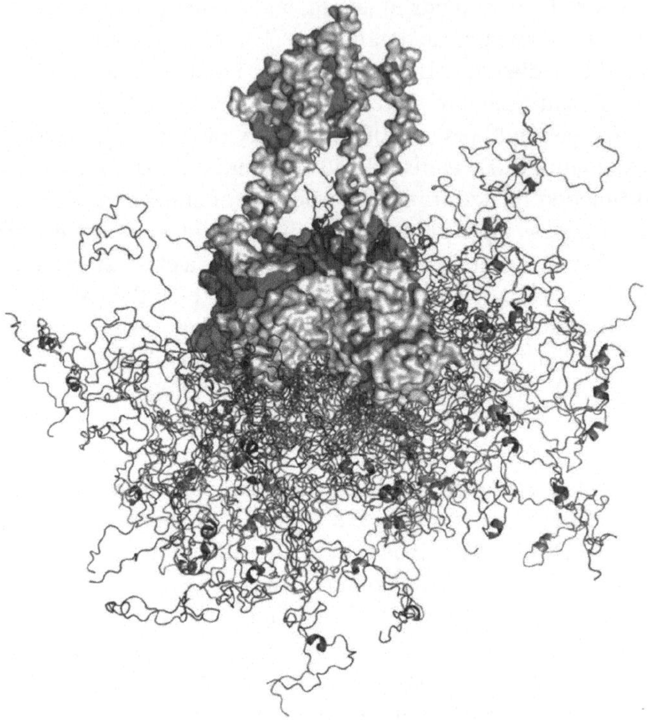

FIGURE 15.2 Disorder in p53. A visual image of tetrameric p53 in complex with DNA was created from the X-ray structure of DNA-bound DBD, the tetramerisation domain (p53CTetD), and the calculated ensemble of the N-terminal domain (see also cover picture). DBD and p53CTetD (light gray) and DNA (dark gray) are shown in space filling model. The flexible CTD is not shown for reasons of clarity. NTDs are modeled by using a conformational sampling model that reproduces NMR residual dipolar coupling (RDC) values; 20 copies for each of the 4 different monomers are shown as thin traces of the polypeptide chain. Reproduced with permission from Wells et al. (2008), *Proc. Natl. Acad. Sci. USA* 105, 5762–7. Copyright by the National Academy of Sciences.

extend away from p53C, probably due to the relative stiffness of their Pro-rich regions, underscoring their role in being the target of a large number of modifications and signaling protein partners (see also Chapter 4, Section 4.4.3).

15.1.3 Cip/Kip Cdk Inhibitors

Progression through the cell-division cycle is regulated by the inhibition of cyclin-dependent kinases (Cdks) by Cip/Kip Cdk inhibitors (CKIs) $p21^{Cip1}$, $p27^{Kip1}$ (Figure 15.3; for details see also Chapter 3, Section 3.7.2, Chapter 10, Section 10.2.3.1, and Chapter 14, Section 14.3 and Section 14.12.1) and $p57^{Kip2}$ (Besson, Dowdy, and Roberts 2008). The inhibitors respond to diverse upstream stimuli: $p21^{Cip1}$ is a transcriptional target of p53 and mediates DNA-damage-induced cell-cycle arrest in G1 and G2; $p27^{Kip1}$ expression

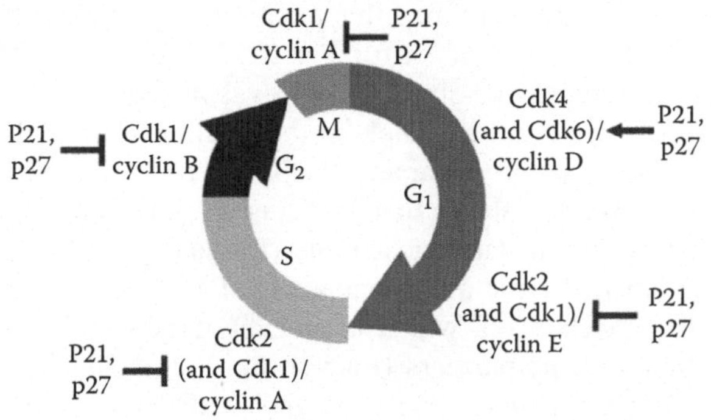

FIGURE 15.3 Regulation of eukaryotic cell division cycle. Scheme of the cell division cycle and the cyclin-Cdk complexes that regulate progression through the different stages. Initiation of cell division in G1 phase requires Cdk4/Cyclin D and Cdk6/Cyclin D activities, whereas progression into S phase requires the operation of Cdk2/Cyclin E and Cdk2/Cyclin A complexes. Cdk1/Cyclin B and Cdk1/Cyclin A are involved in the entry into mitosis (M). Due to their disorder and binding promiscuity, $p21^{Cip1}$ and $p27^{Kip1}$ are involved in inhibiting and activating (the latter indicated by an arrow) various complexes under certain circumstances. Reproduced with permission from Galea et al. (2008), *Biochemistry* 47, 7598–609. Copyright by the American Chemical Society.

is elevated in mitogen-starved cells and is rapidly degraded as cells enter the cell cycle; $p57^{Kip2}$ regulates cell cycle during embryonic development. CKIs are tumor suppressors (Fero et al. 1998; Fero et al. 1996), and they also have other functions in apoptosis, transcriptional regulation, and cytoskeletal dynamics (Besson et al. 2008). However, unlike the classic tumor suppressor genes, oncogenic loss-of-function mutations in CKIs are extremely rare. Instead, their level is down-regulated by distinct mechanisms. Their activity is also regulated by other factors, such as phosphorylation and interaction with protein partners (Galea et al. 2008b; Sherr and Roberts 1999), and they also inhibit cell-cycle progression independent of cyclins and Cdks, via the inhibition of components of the replication machinery (Luo, Hurwitz, and Massague 1995). Both $p21^{Cip1}$ and $p57^{Kip2}$ can bind to PCNA, a DNA polymerase processivity factor, whereas $p27^{Kip1}$ binds minichromosome maintenance deficient 7 (MCM7), which is the subunit of a replication fork helicase (Nallamshetty et al. 2005). In all, CKIs are involved in regulating both cell migration and cell division activated by the same upstream mitogenic signals (Besson et al. 2008) (i.e., they may be involved in the decision between movement and proliferation of cells).

The three CKIs share a conserved, 60-residue N-terminal kinase inhibitory domain (KID, residues 28–90 in $p27^{Kip1}$) but diverge in the remainder of the sequence, which suggests distinct functions and regulation of action. They have nuclear localization signals (NLSs) within their CTDs, $p21^{Cip1}$ and $p57^{Kip2}$ contain a PCNA-binding domain, and $p27^{Kip1}$ and $p57^{Kip2}$ possess a C-terminal QT domain that harbors a Cdk2-dependent phosphorylation site that triggers ubiquitination by Skp, Cullin, I-bat (SCP/Skp) complex (see Chapter 14, Section 14.12.1). NMR studies in the case of $p21^{Cip1}$ (Kriwacki et al. 1996) and CD and fluorescence studies of $p27^{Kip1}$ and $p57^{Kip2}$ (Adkins and Lumb 2002) have shown that the inhibitors are fully random in the solution state.

More detailed studies of $p27^{Kip1}$ refined this picture (Lacy et al. 2004). As described in Section 10.2.3.1, NMR, chemical shift index (CSI), and nuclear Overhauser effect (NOE) values, and restrained MD simulations (Sivakolundu, Bashford, and Kriwacki 2005) suggest that the structural ensemble in the solution state of the protein closely reflects its bound conformation (Chapter 10, Figure 10.3). The structure of $p27^{Kip1}$ KID bound to Cyclin A-Cdk2 shows that Tyr^{88} inserts itself into the catalytic cleft of the kinase, thereby preventing catalytic activity (Russo et al. 1996). The combination of disorder and preformed structural elements (PSEs) in KID enables a sequential "staple"-like binding mechanism, which is probably required for fast, specific, still adaptable interaction with its partners (Lacy et al. 2004). The rest of the molecule remains disordered even in the bound state and mediates other functions, such as PCNA binding (Galea et al. 2008a) and regulation of degradation by the ubiquitin/proteasome machinery (Grimmler et al. 2007). $p21^{Cip1}$ was the first IDP, for which disorder-dependent binding promiscuity was suggested (Kriwacki et al. 1996), which is thought to enable it to regulate distinct cyclin-Cdk complexes that control entry into G1 phase (Cyclin D-Cdk4/6) and progression from G1 to S phase (Cyclin A/E-Cdk2, see Figure 15.3). Promiscuity in binding was implicated in seemingly opposite effects of $p21^{Cip1}$ and $p27^{Kip1}$ on different cyclin-Cdk complexes, promoting the assembly and catalytic activity of some (Cyclin D-Cdk4) and potently inhibiting others (Cyclin A/E-Cdk2 (Cheng et al. 1999; Sherr and Roberts 1999)).

15.1.4 Breast-Cancer 1

Mutations in the breast cancer 1, early onset (BRCA1) gene are implicated in 45% of familial breast cancers and 80% of both familial breast and ovarian cancers (Miki et al. 1994). The product of BRCA1 gene is a multifunctional protein of 1863 amino acids in length (Chapter 12, Section 12.6.2), involved in DNA double-strand break repair, transcription-coupled repair, cell cycle checkpoint control, centrosome duplication, transcription regulation, DNA damage signaling, growth regulation, and the induction of apoptosis (Deng 2006; Venkitaraman 2002). The molecular mechanism of how BRCA1 can carry out these diverse functions is uncertain, but it involves interactions with DNA and a large number of protein partners. Upon UV exposure, BRCA1 localizes to the nucleus in a phosphorylation-dependent manner, which suggests that phosphorylation may play a general role in BRCA1 activation. For example, in response to IR radiation, BRCA1 is phosphorylated by ataxia telangiectasia mutated (ATM) kinase and by ATM-dependent kinase Chk2, whereas upon exposure to UV it is phosphorylated by ATR (ATM and Rad3 related kinase).

BRCA1 has only two small structured domains located at the opposite ends of the protein, both implicated in protein-protein interactions. The N-terminal RING domain (residues 1–103) forms a heterodimer with BRCA1-associated RING domain 1 (BARD1), resulting in an active E3 ubiquitin ligase complex (Brzovic et al. 2003). At the C-terminus, there are two tandem BRCA1 C-terminal domains (BRCT, residues 1646-1863), which might be involved in the DNA damage response signal cascade due to their phosphopeptide binding capacity. The two regions are separated by a long central region of about 1,500 amino acids, which contains no recognizable domains, and is predicted to be largely disordered (Mark et al. 2005). Its 21 overlapping fragments (each about 200 amino acids) were found to be disordered by NMR and CD (Mark et al. 2005). This long IDR harbors binding sites for DNA and more than 50 DNA damage sensors, DNA repair proteins, and signal transduction proteins such as p53, BRCA2, c-Myc, retinoblastoma protein (RB), JunB, Rad50 and Rad51, and the Fanconi anemia group A (FANCA) protein. In all, the observed disorder within the central region of BRCA1 is consistent with an earlier proposal that BRCA1 acts as a scaffold (see Chapter 12, Section 12.6.2) that mediates multiple, weak and possibly transient interactions, thereby integrating multiple signals in the DNA damage response and repair pathway (Foray et al. 2003).

15.1.5 Securin (PTTG)

An IDP critical for the proper execution of cell division cycle is securin (pituitary tumor transforming gene (PTTG) product in humans), which is an oncogene that is overexpressed in several thyroid and colon cancers, and in 90% of pituitary adenomas (Tfelt-Hansen, Kanuparthi, and Chattopadhyay 2006). Securin is an anaphase inhibitor that prevents premature chromosome separation through inhibition of separase, the protease responsible for the physical separation of sister chromatids. The onset of anaphase is signaled by the targeted destruction of securin initiated by ubiquitination at both its D-box and KEN box motifs by the anaphase-promoting complex or cyclosome (APC/C) (Peters 2002). Securin has transforming activity because inappropriate functioning of the separase/securin system causes aneuploidy, a frequent aberration in cancers, and may also up-regulate fibroblast growth factor 2 (FGF-2), a potent mitogenic and angiogenic factor. Securin also acts as a chaperone of separase activity; it targets the enzyme to the nucleus (Hornig et al. 2002; Jallepalli et al. 2001) and has additional functions related to DNA repair and apoptosis (Nagao, Adachi, and Yanagida 2004).

Due to its essential biological role, functional analogues of securin/PTTG have also been described in other species, such as Pds1p in *S. cerevisiae* (Yamamoto, Guacci, and Koshland 1996), Cut2 in *S. pombe* (Hirano et al. 1986), and pimples in *D. melanogaster* (Leismann et al. 2000). These securin analogues, however, are extremely variable in length and sequence, with recognizable sequence similarity restricted to short segments in the N-terminus (Chapter 13, Section 13.4.3). Structural disorder was first shown for the N-terminal half of securin involved in ubiquitination (Cox et al. 2002), and then for full-length securin (Sanchez-Puig, Veprintsev, and Fersht 2005). The role of structural disorder in the structure-function relationship of securin is much less clear than in the case of p53 or CKIs. Full resonance assignment of human securin of 202 amino acids could be achieved by a combination of proton-based and proton-less NMR (see Chapter 6, Section 6.2.4), which confirmed that the protein is entirely disordered, with a transient α-helix between residues D^{150}–F^{159}. This region was confirmed in direct binding assays to be involved in the recognition of separase (Csizmok et al. 2008). Apparently, two recognition elements of securin bind the N-terminal regulatory region and the C-terminal catalytic domain of separase, and prevent their interaction required for activation (Nagao and Yanagida 2006; Waizenegger et al. 2002), in which disorder may be directly involved (Csizmok et al. 2008).

15.1.6 Disorder in Proteins Generated by Chromosomal Translocations

Chromosomal translocations, which represent the major genetic aberration in cancers, such as leukemias, lymphomas, and sarcomas (Futreal et al. 2004; Rabbitts and Stocks 2003), link two distinct chromosomes and either connect a gene with a distinct regulatory region or generate a protein chimera of two unrelated genes. The proteins involved in the translocation event, such as mixed-lineage leukemia (MLL) (Collins and Rabbitts 2002), acute myeloid leukemia (AML)1 (Licht 2001), or CBP (Dyson and Wright 2005), are often long and contain only a few dispersed structural/functional domains (Dyson and Wright 2005; Hess 2004; Licht 2001). IUPred analysis of disorder in 406 human fusion proteins (Hegyi, Buday, and Tompa 2009) show a very high level of predicted disorder (43.3% vs. 20.7% in all human proteins). The actual point of break (known in 255 cases) tends to fall into a locally disordered region (mean IUPred score of the breakpoint is 0.49) and avoids Pfam domains (36.3% vs. 42.5% average Pfam coverage in Swiss-Prot). Apparently, structural disorder within the region that links two distinct proteins enables the fusion protein to evade cellular surveillance mechanisms that usually eliminate misfolded proteins.

Structural disorder also plays a major role in the oncogenic function that emerges upon fusion. Their oncogenic functional elements are connected by long disordered regions in several well-characterized examples, which enable their productive functional combinations that generate an oncogenic signal. There are three basic mechanisms of the underlying structural communication (Table 15.1). In the first type, a substrate sequence gets fused with a Tyr-kinase domain, and disorder between the two motifs enables the protein to effectively phosphorylate itself (e.g., breakpoint cluster region-Abelson leakemia [BCR-ABL], Figure 15.4). In the second type, a dimerisation domain fuses with the Tyr-kinase domain of a growth factor receptor, and disorder enables multiple mutual phosphorylation events between the two subunits within the

TABLE 15.1 Disorder in oncogenic function of fusion proteins*

FUSION PROTEIN (BREAKPOINT 1/BREAKPOINT 2)	*ELEMENTS OF ONCOGENIC FUNCTION IN THE FUSION PROTEINS*	*DISTANCE/ DISORDER BETWEEN ONCOGENIC ELEMENTS*
BCR-ABL (426/26)	Oligomerization domain (BCR, 1–79) Tyr kinase (ABL, 642–893)	562/355
TFG-ALK (240/1057)	Coiled-coil dimerization domain (TFG, 93–124) Tyr-kinase (ALK, 299–566)	175/138
EWS-ATF1 (264/109)	EAD (EWS, 1–86) Leu-zipper (ATF1, 366–425)	280/265

* Functional and structural information on oncogenic fusion proteins generated by chromosomal translocations suggest that their disorder contributes to the emerging oncogenic functions. Three basic types (see text) are shown. Adapted from Hegyi et al. 2009.

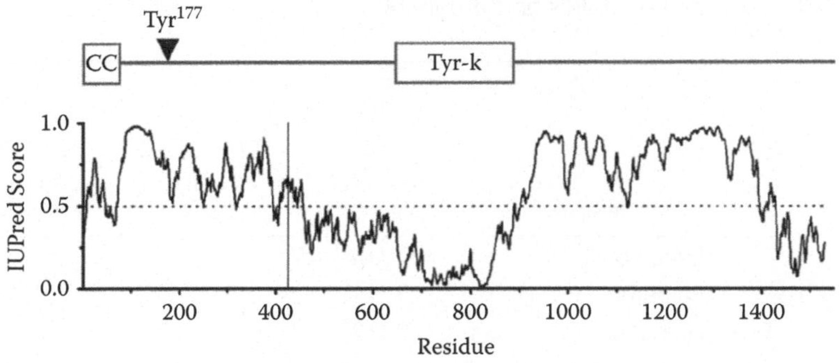

FIGURE 15.4 Predicted disorder and domain structure of BCR-ABL. Structural disorder predicted by the IUPred algorithm and domain structure identified by Pfam of the cancer protein BCR-ABL generated by chromosomal translocation. Disorder score above 0.5 is considered disordered. The position of the break point is marked by a vertical line in the plot, whereas the elements critical for the oncogenic function of the fusion protein are depicted as rectangles (cc, coiled coil; Tyr-k, tyrosine-kinase domain; arrowhead, the position of Tyr^{177} phosphorylation site).

homodimer (e.g., TRK-fused gene-anaplastic lymphoma kinase [TFG-ALK]). In the third type of mechanism, the fusion adds a DNA-binding element to a trans-activator domain, which results in an aberrant transcription factor and misregulation of transcription (e.g., Ewing's sarcoma-activating transcription factor 1 [EWS-ATF1]).

15.2 STRUCTURAL DISORDER IN PROTEINS INVOLVED IN CARDIOVASCULAR DISEASES, DIABETES, AND AUTOIMMUNE DISEASES

Predictions also suggest a high frequency of structural disorder in proteins implicated in cardiovascular disease, CVD (Cheng et al. 2006a). In 487 CVD-related proteins, depletion in the most prominent order-promoting residues (Trp, Phe, Tyr, Ile, and Val) and enrichment in certain disorder-promoting residues (Arg, Gln, Ser, Pro, and Glu) can be observed. CVD-related proteins are enriched in intrinsic disorder (57 ± 4% with at least one IDR ≥30 consecutive residues), reaching levels commensurable with that of signaling proteins (66 ± 6%, see Figure 15.1) and exceeding eukaryotic proteins (47 ± 4%) and structured proteins (13 ± 4%). In general, molecular recognition elements (MoREs) occur in high proportion in proteins involved in CVD, diabetes, autoimmune diseases, neurodegenerative diseases, and cancer (Table 15.2), approaching that of regulatory proteins (Cheng et al. 2006b). Given the noted correlation between short recognition motifs and disorder (see Chapter 14, Section 14.2), this observation underscores the prominence of disorder in disease-associated proteins.

TABLE 15.2 Predicted α-MoREs and druggable targets in genomes, functional classes, and diseases*

	GROUP	*PERCENT OF PROTEINS WITH PREDICTED MoREs*	*NUMBER OF PREDICTED MoREs*	*NUMBER OF PROSPECTIVE DRUGGABLE INTERACTIONS*
Kingdoms	Eukaryotes	21 (±4)		
	Bacteria	3 (±1)		
	Archaea	2 (±2)		
Functional classes	Regulation	48	879	
	Cell division	42	40	
	Differentiation	38	25	
	Cytoskeleton	37	113	
	Membrane	24	88	
	Inhibitor	20	50	
	Transport	18	190	
	Degradation	16	10	
Diseaseses	Cancer	34	1,334	837
	Diabetes	33	176	116
	Autoimmune	32	934	680
	Neurodegenerative disease	24	395	238
	Cardiovascular	21	198	153

* Proteins in various kingdoms, functional categories, and diseases have been predicted for the occurrence of molecular recognition features (α-MoREs). The percent of proteins with α-MoREs, the actual number of predicted α-MoREs, and the number of those examples that possibly represent druggable targets are shown. Adapted from Cheng et al. 2006b.

15.3 STRUCTURAL DISORDER AND NEURODEGENERATIVE DISEASES

Several neurodegenerative diseases are caused by the deposition of insoluble protein aggregates, amyloids (Table 15.3). Because a recurring mechanistic element of these diseases at the molecular level is a large conformational change of some key protein(s), these diseases are also termed "conformational" or "misfolding" diseases. If the deposition occurs in specific tissues (mostly in the brain), the amyloidosis is termed tissue-specific, whereas if aggregates appear in many tissues/organs, it is termed systemic. As appears from discussing the major types of neurodegenerative amyloidoses (Alzheimer's disease, Parkinson's disease, and Gln-repeat (polyGln, polyQ) diseases, such as Huntingon's disease), structural disorder is critically involved in almost all of them.

TABLE 15.3 Amyloid diseases and protein precursors deposited as amyoid*

	PROTEIN INVOLVED	*LENGTH*	*STRUCTURE*
Neurodegenerative			
Alzheimer's	Amyloid β peptide	40–42	IDP
	Tau protein	352–441	IDP
Spongiform encephalopathies	Prion protein	253	IDR (1–120)
Parkinson's	α-synuclein	140	IDP
Huntington's	Huntingtin with polyQ expansion	3144	IDR (polyQ)
Spinocerebellar ataxia	Ataxin with polyQ expansion	816	IDR (polyQ)
Spinal and bulbar muscular atrophy	Androgen receptor with polyQ expansion	919	IDR (polyQ)
Familial British dementia	ABri	23	IDP
Nonneuropathic Systemic Amyloidoses			
AL amyloidosis	Ig light chain	90	all-β, ordered
AA amyloidosis	Serum amyloid A protein	70–140	all-α, ordered
Senile systemic amyloidosisc	Wild-type transthyretin	127	all-β, ordered
Hemodialysis-related amyloidosisc	β2-microglobulin	99	all-β, ordered
ApoAI amyloidosis	N-terminal fragments of apolipoprotein AI	80–93	IDR
Lysozyme amyloidosis	Mutants of lysozyme	130	α + β, ordered
Nonneuropathic Localized Diseases			
Type II diabetes	Amylin (IAPP)	37	IDP
Medullary carcinoma of the thyroid	Calcitonin	32	IDP
Atrial amyloidosis	Atrial natriuretic factor	28	IDP
Cataract	γ-crystallin	variable	all-β, ordered
Pituitary prolactinoma	Prolactin	199	all-α, ordered
Injection-localized amyloidosis	Insulin	21 + 30	all-α, ordered

* Human diseases associated with the formation of extracellular amyloid deposits or intracellular inclusions with amyloid-like characteristics. Within the three major classes (neurodegenerative, nonneuropathic systemic, nonneuropathic localized), the protein (or protein fragment) involved and its length and structural status (disordered or ordered) are shown. The table lists the best characterized diseases and proteins involved. Adapted from Chiti and Dobson 2006.

15.3.1 Alzheimer's Disease

Alzheimer's disease (AD) is a senile cortical neurodegenerative disease and one of the first folding diseases to be recognized (Bertram and Tanzi 2004). AD is the most prevalent, and one of the best known, neurodegenerative amyloidosis, with 2.3 million cases in the U.S. and an estimated 26.6 million worldwide (Brookmeyer, Gray, and Kawas 1998). AD also has an early-onset form, which is diagnosed before the age of 65 (typically in people in their 30s and 40s), accounting for only 5–10% of the cases. About half of early-onset AD cases shows familial pedigree with a discernible genetic predisposition to the disease. Major symptoms of AD are progressive dementia and cognitive decay, confusion, irritability, aggression, mood swings, and eventual and inevitable death in 8–10 years. The major histological lesion is white deposits (plaques) in cortical regions of the brain, composed of β-amyloid peptide (Aβ, 40–42 amino acids in length). Plaques are surrounded by extensions (dystrophic neurites), within which another typical lesion, paired helical filaments (PHFs), can be observed, mostly composed of tau protein (Selkoe 2003). α-synuclein can also be found associated with AD plaques, designated as "non-Aβ component of AD amyloid plaques" (i.e., NACP). This protein is the major component of Lewy-bodies in Parkinson's disease, and will be discussed in Section 15.3.2.1.

15.3.1.1 A β peptide

Mutations in familial AD mostly affect the gene of amyloid precursor protein (APP) and/or presenilin 1 and 2 (PSEN1 and PSEN2). APP is a single-pass transmembrane protein, the metabolism of which under normal conditions is initiated by two proteolytic cleavage events catalyzed by α-secretase (TACE and ADAM10) and γ-secretase (presenilin). This normal product has no propensity for amyloid formation. If cleavage occurs at α- and β-sites due to the action of β-secretase (β-site amyloid precursor protein-cleaving enzyme, BACE), the resulting Aβ peptide (Chiang, Lam, and Luo 2008) displays enhanced amyloidogeneicity and is deposited into plaques, which is the causative event in AD. Mutations in the inherited forms (APP, PSEN1, PSEN2) promote formation of Aβ and lead to the disease. Prior to fibrillation, Aβ monomers exist as predomainantly extended random chains with no α-helical or β-strand structures (Kirkitadze, Condron, and Teplow 2001). A partial refolding to a somewhat more structured state occurs at the earliest stages of fibrillation. Whereas the monomer is not toxic, Aβ becomes neurotoxic to cortical cell cultures when aggregated (Simmons et al. 1994).

15.3.1.2 Tau protein

Also contributing to the etiology of AD is the overactivation of the protein kinase Cdk5 and/or glycogen-synthase kinase (GSK3), which leads to the hyperphosphorylation of tau protein, its dissociation from microtubules and aggregation into PHFs (Iqbal and Grundke-Iqbal 2008). Tau protein belongs to the family of microtubule-associated proteins (MAPs), accessory proteins required for MT polymerization and stability (see Chapter 10, Section 10.2.3.3 and Section 10.4.2, and Chapter 11, Section 11.6.3). Whether the formation of PHFs of soluble tau protein is a cause or a consequence of disease is a matter of debate, because it also appears in other diseases, taupathies, and frontotemporal dementias (e.g., frontotemporal dementia with Parkinsonism linked to chromosome 17, FTDP-17).

Tau protein has several isoforms generated by alternative splicing (Himmler 1989). The longest one is 441 amino acids, and by functional criteria it can be roughly divided into an N-terminal projection domain and a C-terminal repetitive tubulin-binding domain (TBD). Tau protein in isolation is mostly disordered (Schweers et al. 1994), with some short-range structural organization in the MT-binding repeats (Mukrasch et al. 2007a) and also transient long-range tertiary interactions (Jeganathan et al. 2006) (see Figure 10.6).

By a combination of limited proteolysis, generation of overlapping peptides, and aggregation assays, it was shown that a 43-residue fragment within the third repeat of tau is required for self-assembly into filaments (von Bergen et al. 2000). A minimal hexapeptide interaction motif of $V^{306}QIVYK^{311}$ at the beginning of the third internal repeat, which shows the highest predicted β-structure forming potential in tau, is probably the most critical element for aggregation. The importance of local propensity for β-structures in PHF formation was suggested by residual β-structure (see Section 10.2.3.3) for 8–10 residues at the beginning of repeats R2–R4 (Mukrasch et al. 2005). These regions correspond to sequence motifs, which form the core of the cross-β structure of tau PHFs.

15.3.2 Parkinson's Disease

Parkinson's disease (PD) or "shaking palsy" is a chronic degenerative neurological illness that affects 1–2% of the population above 50, with an estimated 1.5 million cases in the U.S. (Thomas and Beal 2007). Clinical features of PD include motor impairments, such as resting tremor, postural instability, slowness in the execution of movement (bradykinesia), gait difficulty, and rigidity. Nonmotoric symptoms involve autonomic, cognitive, and psychiatric problems. Current medications offer symptomatic relief only and cannot stop or revert progression of the disease. PD has a familial etiology in less than 10% of the cases, whereas the majority of the cases are sporadic, and may be related to environmental factors, unknown genetic cause, or both.

PD is characterized by a progressive and massive loss of dopaminergic neurons in the substantia nigra pars compacta. A prevalent neurpathological hallmark of affected neurons is proteinaceous inclusions termed Lewy bodies (Forno 1996) and dystrophic Lewy neurites in surviving neurons. In the familial cases, about a dozen linked genetic loci have been identified (Hardy, Cookson, and Singleton 2003; Thomas and Beal 2007), but only five have an established causative link with PD. The genes are termed PARK1 (α-synuclein), PARK2 (parkin), PARK6, PARK7, and PARK8. α-synuclein is the major component of Lewy bodies, whereas parkin is an E3 ubiquitin ligase, which possibly targets misfolded proteins for proteasomal degradation. Mutations that impair its activity are the major cause of autosomal recessive early-onset PD (Kitada et al. 1998; Shimura et al. 2000).

15.3.2.1 α-synuclein (NACP)

Although PD is not the most prevalent of the neurological disorders discussed, α-synuclein is probably the best characterized IDP involved in them. α-synuclein is also linked to other neurodegenerative disorders collectively termed synucleinopathies (Uversky 2003, 2007), all of which are characterized by Lewy-body depositions. They include dementia with Lewy bodies (DLB), pure autonomic failure (PAF), multiple system atrophy (MSA), and the LB variant of Alzheimer's disease (Bennett 2005; Marti, Tolosa, and Campdelacreu 2003). In AD, a peptide derived from this protein (NAC, from NACP) is also a prominent component of senile plaques (Ueda et al. 1993).

α-synuclein is an acidic protein of 140 amino acids, which occurs primarily in the presynaptic regions of neurons of various brain areas, in close association with synaptic vesicles (Iwai et al. 1995). The protein can be divided into three characteristic regions (Figure 15.5). The amino-terminal 60 residues, dominated by four 11-amino

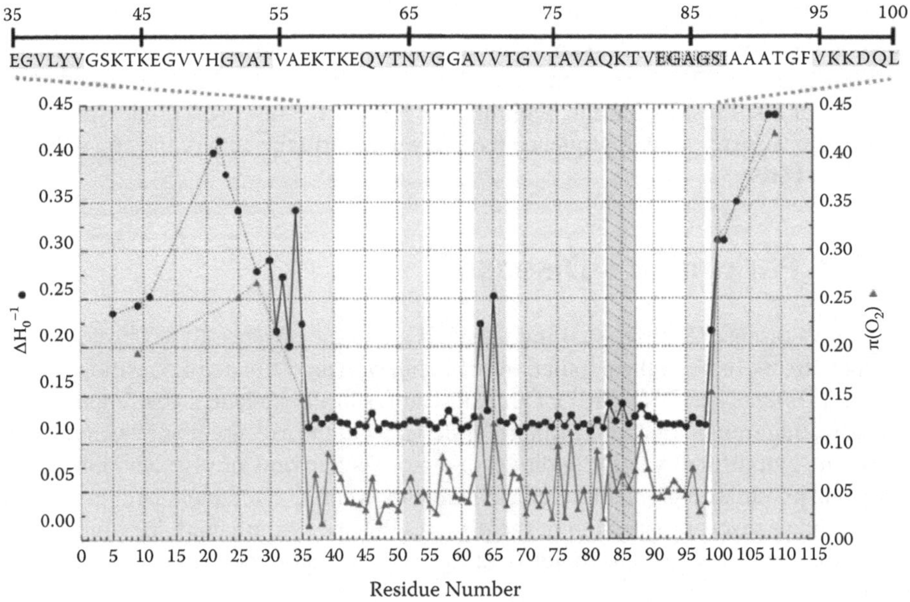

FIGURE 15.5 Residue mobility in the amyloid state of α-synuclein. EPR spectroscopy was used to characterize the mobility and accessibility of residues in α-synuclein fibrils. Data represent a composite of a series of measurements with mutants spin-labeled at positions represented by actual data points. Mobility is represented by the inverse central line widths (ΔH_o^{-1}, black dots), whereas O_2 accessibility is given by π (O_2, gray triangles). Data points are connected by solid lines in the case of consecutive residues; dashed lines are used otherwise. Gray shading indicates areas with increased mobilities and accessibilities. The region 36–98 is rather rigid, forming the core of the fibrils. The linker region 62–67 and flanking regions 5–35 and 99–110 preserve mobility in the amyloid state. Reproduced with permission from Chen et al. (2007), *J. Biol. Chem.* 282, 24970–9. Copyright by the American Society for Biochemistry and Molecular Biology.

acid imperfect repeats, are suggested to constitute a membrane-anchoring region of the protein. This region is similar to the lipid-binding region of apolipoproteins (Clayton and George 1998), it mediates membrane association of the protein (Eliezer et al. 2001; Lee, Choy, and Lee 2002), and it harbors three familial mutations (A30P, E46K, and A53T). The middle segment (the NAC region, residues 61–95) is the most hydrophobic part of the protein and is highly amyloidogenic (el-Agnaf and Irvine 2002). The C-terminal region (96–140) is highly acidic and is enriched in prolines. α-synuclein can interact with a large number of other proteins (Jin et al. 2007) and is a high-affinity inhibitor of phospholipase D2 (Jenco et al. 1998).

α-synuclein was among the first proteins convincingly demonstrated to be disordered by a range of techniques, such as GF, UV, CD, FTIR, and heat stability (Weinreb et al. 1996). α-synuclein is also one of the first IDPs the structural state of which was also studied *in vivo* (McNulty, Young, and Pielak 2006), which suggested that crowding in the *E. coli* periplasm actually stabilizes the disordered state of the protein (see Chapter 8, Section 8.3.2). NMR analyses show that α-synuclein is not fully disordered but contains a transiently ordered short helical region in the membrane-binding region (residues 18–34) (Bussell and Eliezer 2001; Eliezer et al. 2001). Interaction with membranes, which might also belong to the normal physiological function of α-synuclein, also promotes fibril formation (Lee et al. 2002). The N-terminal region of the membrane-bound protein attains stable α-helical conformation, and it has a high aggregation potential (Lee et al. 2002).

Nonrandom tertiary structural organization of α-synuclein was demonstrated in an elegant study by MD simulations restrained by PRE interresidue distances, as detailed in Chapter 10, Section 10.4.2 (Dedmon et al. 2005). Compared to random coil expectations, the resulting structural ensemble is more compact (Chapter 10, Figure 10.7) due to preferential long-range interactions between C-terminal residues 120–140 and the NAC region, which may shield this region from aggregation. In fact, truncation or metal binding of the C-terminal region increase the rate of aggregation (Antony et al. 2003; Crowther et al. 1998; Fernandez et al. 2004; Uversky 2003). Familial point mutations of α-synuclein are located in the first 60 amino acids, and the mutant synucleins do not appear to be structurally different from the wild-type, but they form amyloid-like aggregates more rapidly (Conway, Harpur, and Lansbury 1998; Greenbaum et al. 2005). Transient β-type structural organization possibly critical in promoting aggregation was also ascertained by AFM-induced unfolding experiments (see Chapter 5, Section 5.8.2).

15.3.3 Glutamine-Repeat Diseases

Glutamine-, trinucleotide-, or CAG-repeat diseases, also termed polyQ diseases, are related autosomal dominant neurodegenerative disorders that result from aggregation of a functional protein with a Gln-repeat region that expands to pathological lengths (Table 15.3). The proteins involved are unrelated and result in diseases as diverse as Huntington's disease (huntingtin), spinocerebellar ataxia (ataxin), spinocerebellar ataxia17 (TATA-box binding protein), X-linked spinal and bulbar muscular atrophy, also termed Kennedy's disease (androgen receptor), and hereditary dentatorubral-pallidoluysian atrophy (atropine-1) (Chiti and Dobson 2006; Perutz 1999). Their common Gln-repeat regions of unknown function encompass less than 40 uninterrupted Gln residues under normal conditions. When the region expands to more than 40 residues, it forms insoluble and intractable aggregates.

The diseases show a phenomenon termed anticipation, which refers to their unusual non-Mendelian genetic pattern, in which the symptoms become more severe and show an earlier age of onset with each successive generation (Wells 1996). The cause of anticipation lies in the mechanism of expansion of the coding microsatellite (CAG-repeat) loci, which results from replication slippage (see Chapter 13, Section 13.3.1.1). If DNA polymerase slows down (becomes idle) during replication, the highly repetitive DNA tends to form a hairpin, which makes the polymerase "slip" (i.e., carry on synthesis at a wrong place). The probability of slippage increases with the length of the CAG repeat, which results in a progressive expansion with every passing generation.

The disease is elicited by the transition of a disordered (Crick et al. 2006) polyQ region into an amyloid state, which may be accompanied by either the loss or gain of function. Because homopolymeric Gln regions in proteins might be related to transcription (Gerber et al. 1994), and proteins with multiple long runs of amino acids often function in development and/or transcription regulation (Karlin et al. 2002), the formation of insoluble aggregates may impair transcription function. The loss of function may also extend to other parts of the protein due to decreasing the structural stability of the rest of the protein, as suggested by disorder prediction (Chen 2003) and structural studies of ataxin-3 (Bevivino and Loll 2001). The dominant gain of function may occur by several possible mechanisms, such as disruption of the global balance of protein folding quality control, resulting in the loss of function of diverse metastable proteins (Gidalevitz et al. 2006), or sequestration and inactivation of other Gln-repeat containing transcription factors, such as TATA box binding protein (TBP) and CREB-binding protein (CBP) (Schaffar et al. 2004).

15.3.3.1 Huntington's disease

Huntington's disease (HD) is a neurological disorder characterized by uncoordinated, jerky body movements called chorea, whereas in a few cases very slow movement and stiffness (bradykinesia and dystonia) can occur instead. Cognitive decline and behavioral difficulties also follow, affecting planning, cognitive flexibility, abstract thinking, and initiating actions, whereas at a later stage memory deficits also appear. As the disorder progresses, these symptoms cause complications that reduce life expectancy (Imarisio et al. 2008; Walker 2007). HD is an infrequent disease with less than 1 instance in 10,000. Typically, onset of symptoms is in middle age after affected individuals have had children, but the disorder can manifest at any time between infancy and senescence. When symptoms occur in the 20s, HD is considered juvenile, which usually progresses faster and is more likely to exhibit rigidity and bradykinesia, instead of chorea, and commonly includes seizures. HD is caused by a single causal gene, huntingtin (The Huntington's Disease Collaborative Research Group 1993), which made it one of the first inherited genetic disorders for which an appropriate genetic test was developed.

15.3.3.2 Huntingtin

Huntingtin (HTT) is a large protein of 3,144 amino acids, encoded by a gene consisting of 67 exons. Its normal physiological function is unknown, but knockout mouse models have shown it to be essential for development and survival (Nasir et al. 1995), maybe due to acting as a transcription factor upregulating the expression of brain-derived neurotrophic factor (BDNF), and/or having a role in cytoskeletal anchoring/transport and vesicle trafficking. The protein displays no homology to other proteins and is highly expressed in neurons and in testis (Cattaneo, Zuccato, and Tartari 2005). Its CAG-repeat region encoding for a run of Glns is within exon 1, followed by a Pro-repeat of 11 prolines. When the polyQ region contains fewer than 36 glutamines (usually below 27), it results in a soluble cytoplasmic protein, but when it expands to 36 or more, HTT deposits into aggregates, causing neuronal decay (Katsuno et al. 2008). The number of CAG repeats correlates with age at onset and the rate of progression of symptoms (Kieburtz et al. 1994). With very large repeat counts in about 7% of the cases, HD can even occur under the age of 20.

HTT had a major role in structural studies of the polyQ regions of proteins. CD and NMR data of synthetic peptides or GST fusion constructs showed that polyQ sequences up to the pathogenic length prefer the random coil state (Chen et al. 2001; Masino et al. 2002). The picture was refined in hydrodynamic measurements by fluorescence correlation spectroscopy (FCS) (Crick et al. 2006), which showed that the translational diffusion coefficient of the protein scales with chain length with an exponent 0.32, significantly smaller than the value 0.5 or 0.58 of a random coil chain in an ideal or good solvent (see Chapter 1, Section 1.7). Thus, water is a polymeric poor solvent for polyQ, and the structural ensemble for monomeric polyQ is made up of a heterogeneous collection of collapsed structures.

NMR and CD spectroscopy demonstrated that Gln residues possess a high propensity to adopt the PPII helix conformation (Chellgren et al. 2006) in short tracs up to about 15 residues. Studies of longer stretches are hampered by poor solubility; thus currently there is no evidence that the observed PPII helical structure is a precursor to aggregation, although PPII helix is known to transit easily to other conformational states (see Chapter 1, Section 1.5.1 and Chapter 10, Section 10.2.2) (Blanch et al. 2000).

15.3.4 Prion Diseases

The term *prion disease* (also known as transmissible spongiform encephalopathy, TSE) actually denotes a range of closely related fatal neurodegenerative conditions that afflict mammals. The best-known of these intractable diseases are Creutzfeldt-Jakob disease (CJD), Gerstmann-Sträussler-Scheinker (GSS) disease, and kuru in humans, bovine spongiform encephalopathy (BSE) in cattle, and scrapie in sheep (Legname et al. 2007; Prusiner 1998, 2001). Scrapie has been known for centuries, and it might actually have been noticed thousands of years ago, as suggested by the existence of an ancient Chinese character for "scratching" combined from the characters meaning "sheep" and "itchy" (Wickner 2005). The most unique feature of prion diseases among other neurodegenerative conditions is that they are transmissible, as first demonstrated by Gajdusek (Gajdusek et al. 1968; Gibbs et al. 1968). The disorders cause impairment of brain function, including decline in memory, personality changes, and progressive deterioration of movement. These conditions form a spectrum of diseases with overlapping signs and symptoms. TSEs may be genetic, sporadic, or infectious. Transmission of the pathogen may occur by consumption of infected brains (kuru) or animal products (vCJD), or accidental transmission via corneal transplantation or electrode implantation. Best understood are the hereditary forms of the disease, which are caused by one of about 30 different mutations in the gene of prion protein, PrP.

15.3.4.1 Prion protein

For a long time understanding of the mode of transmission of prion diseases was hampered by the extremely long incubation times and resistance of the pathogen to inactivation by ionizing radiations and UV (Alper et al. 1967). A radical change in concept occurred when Prusiner suggested that the pathogen of these diseases is devoid of a nucleic acid genome and only contains a protein essential for infectivity (Prusiner 1982). This "protein only" hypothesis suggested that TSEs are caused by the transition of the prion protein from its cellular, soluble form (PrP^C) to an altered conformation, the scrapie state (PrP^{SC}). In reflection of the three principal modes of the emergence of the disease, the transition can be either spontaneous (sporadic disease), caused by a mutation of the PrP gene (hereditary or familial), or caused by infection by the scrapie form, which causes the structural conversion of the cellular form (infectious) (Legname et al. 2004; Legname et al. 2007; Prusiner 1998, 2001).

The human prion protein is 254 (230 when processed) amino acids in length and contains an N-terminal endoplasmic reticulum (ER) signal sequence, two Asn-linked glycosylation sites, and a glycosylphosphatidylinositol (GPI) anchor site within its C-terminal half that tethers the mature protein at the extracellular side of neurons. Structural studies of the prion protein from mouse (Riek et al. 1997), bovine (Lopez-Garcia et al. 2000), and humans (Zahn et al. 2000) all agree that the protein is composed of a disordered N-terminal half (residues 1–120) and a globular C-terminal half (residues 121–230). The N-terminal half contains an imperfect polymorphic octapeptide motif (P(H/Q)GGG(G)WGQ) repeated 5 to 13 times (see Chapter 13, Section 13.3.1.4) and an absolutely conserved palindromic sequence (VAGAAAAGAV) within the disordered region. The function of normal PrP has not yet been elucidated, but the octarepeat region constitutes a high-affinity copper ion binding site (Chapter 11, Section 11.8), which raised the idea that PrP is a copper transporter (Brown et al. 1997a) and/or a superoxide dismutase enzyme (Brown et al. 1997b).

There is no difference between either the sequence or posttranslational modification of PrP^C and Prp^{SC} (Stahl et al. 1993), but they basically differ at the secondary structural level. PrP^C contains 40% α-helical structure and negligible β-strand conformation, whereas PrP^{SC} contains about 45% β-sheet (Pan et al. 1993). This difference corroborates that the major causative event in prion diseases is the structural conversion of PrP from the cellular form to the highly β-structured (amyloid) scrapie state. Electron paramagnetic resonance (EPR) spectroscopy of PrP^{SC} suggests that region

160–220, which involves two α-helical regions (helix A and helix B of the C-terminal globular half), is likely involved in the transition to the β-structure.

15.4 SYSTEMIC AMYLOIDOSES

AD, PD, polyQ diseases, and prion diseases may be considered as tissue-specific amyloidoses, because deposition of amyloid fibers affects one tissue, the brain, only. In a range of other amyloid diseases, the fibers are deposited in multiple organs; thus these are termed systemic amyloidoses (Ghiso et al. 2000; Westermark et al. 1999). Different systemic amyloidoses are caused by the misfolding of unrelated proteins and may have different causes (Table 15.3). For example, in reactive systemic (AA) amyloidosis, acute infections or inflammatory diseases may increase serum amyloid A (AA) levels, from which a 76-kDa proteolytic fragment is produced and is deposited primarily in the spleen, kidney, and liver. In hemodialysis-associated amyloidosis, which results from the treatment of acute renal failure, the level of the surface HLA-I invariant light chain (β2-microglobulin) is elevated chronically due to its deficient catabolism by the kidney. A further possible cause may be manifested in senile systemic (ATTR) amyloidosis, a typical disease of the elderly. The amyloid in this case is formed from wild-type transthyretin (TTR), which is deposited in the capillaries and causes a congestive cardiac disorder in the heart. Destabilizing mutations of TTR inherited in an autosomal dominant manner also cause systemic deposition of the protein in familial amyloid polyneuropathy, peripherial neuropathy, cardiomyopathy, and nephropathy in different patients. More than 100 TTR mutations have been identified so far (Ando, Nakamura, and Araki 2005). Mutations of the enzyme lysozyme implicated in autosomal systemic hereditary amyloidosis have also been analyzed in detail. Each mutation (e.g., I56T, F57I, W64R, and D67H) causes the destabilization of the compact and stable enzyme structure (Booth et al. 1997), which leads to lethal deposits in the guts typically causing kidney failure and gastrointestinal bleedings. Although the reasons are not always clear, systemic amyloidoses are primarily caused by the misfolding of ordered proteins.

15.5 COMMON THEMES IN AMYLOID FORMATION

Although conformational (amyloid) diseases are very heterogeneous in terms of practically all aspects of their pathology, they all share three common features:

1. A similar kinetics of progression
2. Prevalence of disorder of the underlying proteins
3. Molecular-structural characteristics of the final aggregated state of the proteins (i.e., the amyloid).

15.5.1 Kinetics of Amyloid Formation

The kinetics of amyloid formation in all the diseases has two common characteristics: (1) the process involves a lag-phase (i.e., the rate-limiting formation of a critical seed), followed by an exponential growth phase, and (2) the lag-phase can be annulated by the addition of preformed seeds. This scheme can be described by two molecular models, "nucleation-polymerization" and "template-assistance" (Come, Fraser, and Lansbury 1993; Jarrett and Lansbury 1993; Rochet and Lansbury 2000), which differ in the thermodynamic nature of the critical step. In nucleation-polymerization, a single transformed molecule is assumed to be less stable than the original form, only to get stabilized in an oligomeric form. The rate-limiting step is the slow assembly of this seed, which can then promote the transformation of further molecules in an oligomerization-assembly process. In the template-assistance model, the transformed state is assumed to be inherently more stable than the cellular state, but it is separated by a high energy barrier and is kinetically not accessible. Transformed molecules, however, can catalyze the conversion by lowering the energy barrier. Rate limiting in this case is the formation of the catalyst. It should be noted that the two models actually mechanistically converge if a monomer seed is assumed in the template-assistance model. In all, the process has many characteristics of a one-dimensional crystal growth, which also explains the disappearance of lag phase in the presence of a preformed seed.

15.5.2 Disorder in Amyloidogenic Proteins

A critical point with respect to the concept of structural disorder is its prevalence in proteins involved in amyloid diseases (Table 15.3). Some of them (Aβ peptide, tau protein, α-synuclein) are fully disordered, and the polyQ regions of huntingtin and androgen receptor are also disordered. Disorder is also critically involved in the prion protein, and in several other proteins involved in amyloidoses, such as apolipoprotein AI (ApoAI amyloidosis) and amylin. Although some proteins of systemic amyloidoses are ordered, there is an apparent enrichment of structural disorder in proteins involved in amyloid diseases (Chiti and Dobson 2006). Because the key structural feature of amyloids is an extended cross-β structure, IDPs might be inherently more prone to form amyloids than globular proteins due to their exposed structural state.

This structural predisposition has led to the suggestion that IDPs have probably undergone evolutionary selection toward amino acid compositions obstructive to amyloid formation (Tompa 2002). Namely, amyloidogeneicity shows a significant positive correlation with hydrophobicity and β-sheet forming potential, and negative correlation with total charge (Chiti et al. 2003). Thus, the characteristic amino acid composition of IDPs (i.e., a low level of hydrophobic amino acids, high charge, and high frequency of Pro), which is known for its β-structure breaking propensity, might each represent elements of the evolutionary strategy of IDPs to escape frequent amyloid formation (Linding et al. 2004; Monsellier and Chiti 2007). Probably for the same reason, of functionally similar amino acids, IDPs tend to use the one with less β-forming propensity, such as Gln instead of Asn and Ser instead of Thr (Monsellier and Chiti 2007).

Estimations with the algorithm developed to assess β-aggregation propensity of polypeptide chains, TANGO (Fernandez-Escamilla et al. 2004), show that globular proteins contain almost three times as much aggregation nucleating regions as IDPs, and the formation of highly structured globular proteins comes at the cost of a higher β-aggregation propensity (Linding et al. 2004). Thus, although IDPs are frequently involved in amyloid diseases, they show signs of evolutionary selection against too often coming down in the form of orderly aggregates.

15.5.3 The Structure of the Amyloid

Under appropriate conditions, practically any protein can form amyloid (i.e., the capacity of amyloid formation is a generic property of polypeptide chains) (Chiti et al. 1999; Dobson 1999, 2002; Fandrich, Fletcher, and Dobson 2001). This is probably also reflected in structural features common to all amyloids. Amyloid fibrils visualized by transmission electron microscopy (TEM) or atomic force microscopy (AFM) usually consist of a number (typically 2–6) of protofilaments, each about 2–5 nm in diameter, twisted together to form rope-like fibrils typically 7–13 nm across (Serpell et al. 2000; Sunde and Blake 1997). The polypeptide chain runs perpendicular to the fiber axis in each protofilament, forming a β-sheet along the fiber. This cross-β structure is highly ordered and is distinguished from general protein aggregates (Rousseau, Schymkowitz, and Serrano 2006). A diagnostic test for amyloids is their binding of specific dyes, such as thioflavin T and Congo red, and emission of characteristic fluorescence.

Structural models of amyloid approaching atomistic resolution could be obtained by different combinations of solid-state NMR, X-ray crystallography, electron microscopy, and EPR spectroscopy (Chiti and Dobson 2006; Rousseau et al. 2006), which all contribute evidence for an extended β-sheet structure composed of parallel, in-register β-strands. The X-ray crystal structure of a model amyloid, the heptapeptide GNNQQNY of the yeast prion Sup35p (Nelson et al. 2005), consists of pairs of parallel β-sheets in which each individual peptide molecule contributes a single β-strand (Figure 15.6). The stacked strands are parallel and are in register in both sheets, whereas the side-chains of the two sheets interdigitate so tightly that water is excluded from the interface, giving rise to a "steric zipper." This model is contrasted with a previous one, "polar zipper" of an extended H-bond network supporting the β-sheet structure of polyQ amyloids (Perutz et al. 1993).

Other amyloid structures have similar features. Solid-state NMR combined with computational energy minimization suggests that the amyloid fibrils formed from Aβ(1–40) have two strands, connected by a 5-residue loop (Petkova et al. 2002). The strands are stacked upon each other; they are parallel and in register, but participate in the formation of two distinct β-sheets within the same protofilament. This model is also supported by EPR spectroscopy, in which spectra from a series of labeled molecules show that the strand regions are highly structured, parallel, and are positioned in register (Torok et al. 2002). EPR was also used to show that fibrils formed from α-synuclein (Chen et al. 2007) and human prion protein (Cobb et al. 2007) are also composed of single-molecule layers that stack on top of one another with parallel, in-register alignment

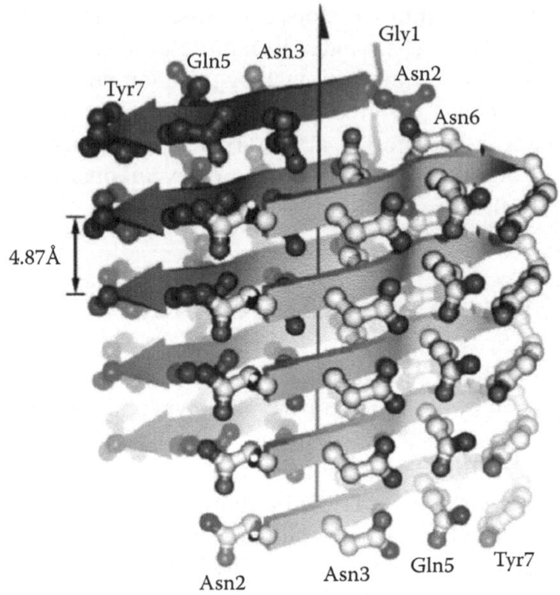

FIGURE 15.6 Crystal structure of a model amyloid. Amyloid-like structure of the heptapeptide GNNQQNY derived from the yeast prion Sup35p. The structure is a sandwich of β-sheets, with each β-strand represented by an arrow. Side-chains protrude from the sheet. Side-chains in the dry interface between the two sheets tightly interdigitate excluding water, whereas side-chains on the wet outside surface form an extensive network of H-bonds. This molecular arrangement is termed a "steric zipper." Reproduced with permission from Nelson et al. (2005), *Nature* 435, 773–8. Copyright by the Nature Publishing group.

of β-strands. A series of mutants and labeling suggest that the tightly packed core region extends from residue 36 to 98 in the case of α-synuclein (see Chapter 5, Section 5.6 and Figure 15.5), whereas in the case of prion protein it extends approximately between residues 160 and 220.

15.5.4 Molecular Mechanism of Transition to the Amyloid State

The remarkable uniformity of amyloid structures suggests mechanistic parallels in the misfolding process that leads to their formation (Uversky and Fink 2004). Because all amyloids possess highly similar cross-β structure, profound conformational rearrangements in the structure of globular precursor proteins have to occur, only possible within a nonnative partially unfolded ensemble. In accord, most mutations associated with accelerated fibrillation destabilize the native structure and increase the steady-state concentration of partially unfolded conformers (Dobson 1999; Uversky and Fink 2004), as shown in the case of lysozyme (Booth et al. 1997), TTR (Lashuel et al. 1999), the p53 tetramerization domain (Lee et al. 2003), and immunoglobulin light chain (Wall et al. 1999). Nonnative conditions destabilizing structure, such as low or high pH, high temperatures, or mild denaturation, also lead to an increased fibrillation, as demonstrated in the case of the SH3 domain of PI3K (Guijarro et al. 1998) and fibronectin type III module of murine fibronectin (Litvinovich et al. 1998), for example. Conversely, amyloidogenicity of a globular protein can be significantly reduced by stabilization of the native structure by ligand binding, for example (Chiti et al. 2001). In the case of IDPs, the primary step of fibrillogenesis is the stabilization of a partially folded conformation, as demonstrated in the case of α-synuclein (Uversky, Li, and Fink 2001b) and IAPP (Kayed et al. 1999). In general, the structural prerequisite of amyloid formation is the transformation of a polypeptide chain into a partially folded disordered conformation, originating from an ordered or disordered initial state.

15.6 DOES STRUCTURAL DISORDER POSE A DANGER?

The abundance of disorder in proteins involved in a wide array of diseases ranging from cancer to neurodegeneration may indicate that structural disorder poses a particular danger to the organism (Dyson and Wright 2005). Of course, the question of a causal link between disorder and disease cannot be answered in general but has different answers in relation to the different proteins and disease conditions. The primary line of division probably lies between diseases that result from upsetting signaling/regulation (e.g., cancer) and diseases that result from protein misfolding (amyloidoses).

In the case of diseases such as cancer, CVD, and diabetes, disorder per se probably poses no particular danger, but its high frequency in proteins of regulatory functions causes a statistical correlation with diseases. Structural disorder is frequent in proteins of signal transduction and regulation of transcription (Iakoucheva et al. 2002; Minezaki et al. 2006; Tompa, Dosztanyi, and Simon 2006b); thus mutations affecting such proteins are likely to cause severe changes in phenotype (i.e., disease, irrespective of their structural status). In p53, for example, missense mutations implicated in cancer cluster in the central ordered DNA-binding domain (Figure 15.7A) and mostly interfere with DNA-binding of the protein (Joerger and Fersht 2008). It is the central function of the protein in DNA repair, cell-cycle regulation, and apoptosis, rather than its disorder, that positions it at a central stage in disease. This argument can be extended by comparing its mutation pattern to that of BRCA1 (Figure 15.7B), because the majority of its oncogenic mutations occur in the central disordered region (Mark et al. 2005). For the lack of details, one might argue that BRCA1 has multiple binding partners, which are recognized by short recognition elements located in local structural disorder (Fuxreiter, Tompa, and Simon 2007). Because these are defined by only a few conserved residues (Neduva and Russell 2005), their function may be affected by disease-causing mutations making it look like disorder is involved in disease. It should be recalled here, however, that IDPs in general are noted for their evolutionary variability (i.e., overall

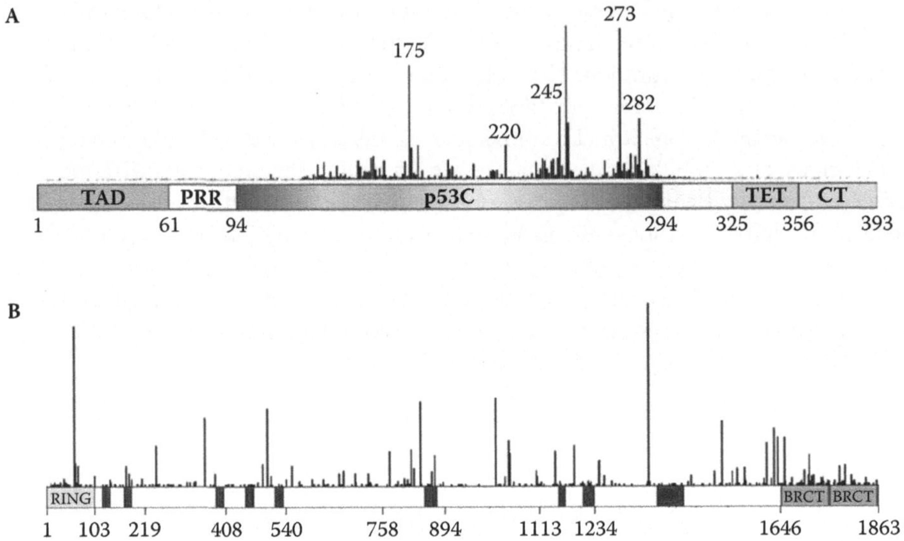

FIGURE 15.7 Oncogenic mutations in p53 and BRCA1. The distribution and relative frequency of oncogenic missense mutations in p53 (A) and BRCA1 (B). In p53, the mutations cluster in the central ordered DNA-binding domain (p536), whereas in BRCA1 they do not dominate in the N- and C-terminal ordered domains, but also appear in large proportions in the central region that is largely disordered. Reproduced with permission from Joerger and Fersht (2008), *Annu. Rev. Biochem.* 77, 557–82, copyright by Annual Reviews, and Mark et al. (2005), *J. Mol. Biol.* 345, 275–87, copyright by Elsevier Inc.

their functions are resistant to mutations). In conclusion, disorder per se might not pose a particular risk in these diseases (Brown et al. 2002; Daughdrill et al. 2007). Of course, disorder appears as a permissive element in the oncogenic function of fusion proteins generated by chromosomal translocations (Section 15.1.6), which establishes a direct link of disorder with cancer.

IDPs/IDRs also appear disproportionately frequently, in about two-thirds of the cases, in amyloid diseases (Table 15.3), and neurodegenerative diseases are caused by disordered precursor proteins without exception. Although the special amino acid composition of IDPs (Tompa 2002) limits their amyloidogeneicity (Linding et al. 2004), their open and exposed structure makes them vulnerable to misfolding/deposition in disease, and in this sense the presence of structural disorder probably does pose a danger to the organism.

15.7 DISORDER IN PATHOGENIC ORGANISMS

Bioinformatic analyses show that structural disorder is prominent in pathogenic organisms, although a direct link of this phenomenon with pathogeneicity has not been generally established. Predictions of the malaria parasites of humans (*P. falciparum* and *P. vivax*), primates (*P. knowlesi*), and rodents (*P. yoelii*, *P. chabaudi*, and *P. berghei*) show that mammalian parasites are more enriched in intrinsic disorder than those of rodents (Feng et al. 2006). In addition, apicomplexan parasites contain the highest level of disorder among distinct phylogenetic groups of species (e.g., fungi, metazoan, archea, and bacteria; see Figure 15.8). Of the different life cycle stages of *P. falciparum*, the sporozoite has the highest disorder content, probably because structural plasticity offers protection against the antibody response of the host and facilitates protein–protein interactions necessary for the attachment to, and invasion of, host cells. An extended study of 19 infectious parasitic protozoa, including the diplomonad *Giardia lamblia* that causes diarrhea, the kinetoplastid *Trypanosoma brucei* that causes sleeping sickness, and the apicomplexan *Plasmodium falciparum*, which causes malaria, corroborated this view (Mohan et al. 2008). Most of these genomes have extensive low-complexity regions, which suggests the prevalence of disorder (Romero et al. 2001; Tompa 2003b). The percentages of proteins with at least one IDR ≥30 consecutive residues is above 50% for 16 out of 19 species (ranging from 87.8% (*T. gondii*) to 24.9% (*V. cholerae*)), above the range of Swiss-Prot (47 ± 4%) and ordered PDB (13 ± 4%) proteins. In addition, *P. falciparum* has an elevated level of predicted α-MoRFs and disorder in its interactome, suggesting that protein–protein interactions may be a prevalent function of its intrinsic disorder.

A possible causative link of disorder and pathogeneicity could be established in the case of human papillomavirus (HPV) and invasive pathogenic bacteria. HPVs cause

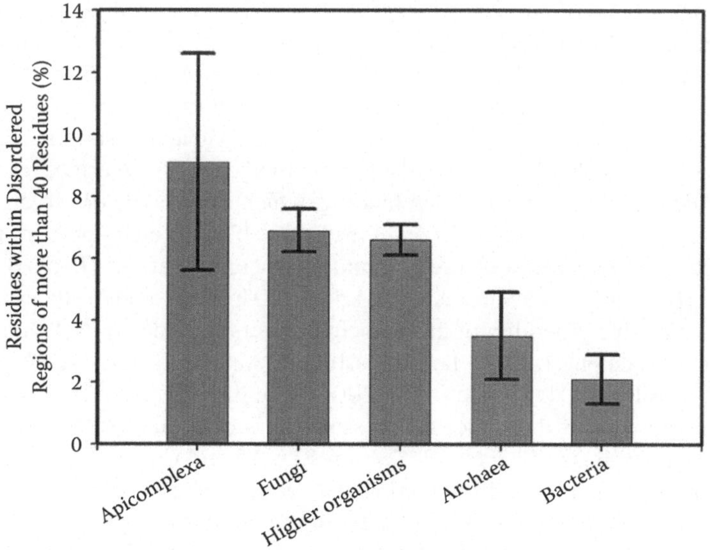

FIGURE 15.8 High predicted disorder in apicomplexan parasites. Comparison of different groups of organisms according to the average percentage of residues within disordered regions more than 40 amino acids in length. Apicomplexan parasites, including *P. falciparum* causing malaria, have an elevated level of disorder. Reproduced with permission from Feng et al. (2006), *Mol. Biochem. Parasitol.* 150, 256–67. Copyright by Elsevier Inc.

benign papillomas and act as cofactors in carcinomas, and they have about 100 different types grouped into low- and high-risk groups according to their association with cancer. Comparative bioinformatics analysis of the two groups, with particular focus on E6 and E7 transforming oncoproteins, showed an increased level of disorder in the high-risk group (Uversky et al. 2006).

The role of disorder in infection is mechanistically established in pathogenic bacteria, which use membrane-anchored extracellular disordered proteins to tether to the extracellular matrix (ECM) of their host, an important step in the mechanism of invasion (Patti et al. 1994). These proteins termed MSCRAMMs (see Chapter 10, Section 10.2.3.4 for details) are single transmembrane-spanning helix proteins, with highly repetitive disordered extracellular tail regions that undergo disorder-to-order transition upon binding to the cognate ECM component (House-Pompeo et al. 1996; Schwarz-Linek et al. 2004; Schwarz-Linek et al. 2003).

15.8 RATIONAL DRUG DESIGN BASED ON PROTEIN DISORDER

Whereas the interfaces of protein–protein complexes are attractive targets for drug development, they are usually not readily amenable to interference by small-molecule chemical compounds. Thus, current drugs on the market preferentially target the active site of enzymes (Sebolt-Leopold and English 2006) or ligand binding sites of receptors (Lagerstrom and Schioth 2008), and most attempts to develop drug molecules that block protein–protein interactions have so far failed (Drews 2000), with only eight such drugs known (Arkin 2005; Arkin and Wells 2004). As detailed in Chapter 14, Section 14.2.5, IDPs are often engaged in special types of interactions, when their short recognition element binds in a hydrophobic groove of the partner molecule. The interface resembles that of receptor-ligand or enzyme-substrate binding, which can usually be targeted by small molecules. In accord, four out of the eight drugs that block protein–protein interactions involve a complex between a disordered and a structured partner (BAK/Bcl-xL, p53/MDM2, Tcf/β-catenin, and Smac/XIAP (Arkin 2005; Arkin and Wells 2004)), which points to the possible generality of this principle (Uversky, Oldfield, and Dunker 2008).

This point is emphasized by the successful targeting of the p53-MDM2 interaction (Klein and Vassilev 2004; Vassilev et al. 2004). The importance of this relation is that the level and activity of the tumor suppressor p53 is controlled by MDM2 through a feedback mechanism (see Chapter 12, Section 12.6.2.2 and Chapter 15, Section 15.1.2). Thus, inhibiting their physical interaction may reactivate p53 in tumors and provide a novel therapeutic strategy against cancer. In the complex, MDM2 presents a well-defined hydrophobic pocket filled by three side-chains of the binding segment of p53 TAD, against which potent and selective small-molecule antagonists, the Nutlins, could be raised (Klein and Vassilev 2004; Vassilev et al. 2004). Nutlins bind MDM2 in the p53-binding pocket and activate the p53 pathway in cancer cells, leading to cell cycle arrest and apoptosis *in vitro*, and growth inhibition of human tumor xenografts in nude mice *in vivo*.

The number of such targets estimated by combining disorder and α-MoRE predictions is in the range of thousands (Cheng et al. 2006b). These "druggable" targets in the human proteome are predicted to form interfaces with structured partners that can be likely mimicked by small molecules (Table 15.2), because they engage in a weak interaction of an isolated helical segment, which is likely to fit into a groove or pocket. The amphipathic nature of the helix entails a complementary concave hydrophobic binding surface on the partner that can be targeted by a small molecule. About 2,000 such elements are found in proteins involved in cancer, diabetes, and autoimmune, neurodegenerative, and cardiovascular diseases.

References

Abdul-Manan, N., B. Aghazadeh, G. A. Liu, et al. 1999. Structure of Cdc42 in complex with the GTPase-binding domain of the "Wiskott–Aldrich syndrome" protein. *Nature* 399: 379–83.

Abel, K., M. D. Yoder, R. Hilgenfeld, and F. Jurnak. 1996. An alpha to beta conformational switch in EF-Tu. *Structure* 4: 1153–9.

Adams, V. H., S. J. Mcbryant, P. A. Wade, C. L. Woodcock, and J. C. Hansen. 2007. Intrinsic disorder and autonomous domain function in the multifunctional nuclear protein, MeCP2. *J. Biol. Chem.* 282: 15057–64.

Adkins, J. N. and K. J. Lumb. 2002. Intrinsic structural disorder and sequence features of the cell cycle inhibitor p57Kip2. *Proteins* 46: 1–7.

Adzhubei, A. A. and M. J. Sternberg. 1993. Left-handed polyproline II helices commonly occur in globular proteins. *J. Mol. Biol.* 229: 472–93.

Agarwal, R. and O. Cohen-Fix. 2002. Phosphorylation of the mitotic regulator Pds1/securin by Cdc28 is required for efficient nuclear localization of Esp1/separase. *Genes Dev.* 16: 1371–82.

Ahn, M., S. Kim, M. Kang, Y. Ryu, and T. D. Kim. 2006. Chaperone-like activities of alpha-synuclein: alpha-synuclein assists enzyme activities of esterases. *Biochem. Biophys. Res. Commun.* 346: 1142–9.

Alarcon-Vargas, D. and Z. Ronai. 2002. p53-Mdm2—the affair that never ends. *Carcinogenesis* 23: 541–7.

Alattia, J. R., H. Kurokawa, and M. Ikura. 1999. Structural view of cadherin-mediated cell–cell adhesion. *Cell. Mol. Life Sci.* 55: 359–67.

Alber, F., S. Dokudovskaya, L. M. Veenhoff, et al. 2007. The molecular architecture of the nuclear pore complex. *Nature* 450: 695–701.

Alber, T., W. A. Gilbert, D. R. Ponzi, and G. A. Petsko. 1983. The role of mobility in the substrate binding and catalytic machinery of enzymes. *Ciba Found. Symp.* 93: 4–24.

Ali, M. M., S. M. Roe, C. K. Vaughan, et al. 2006. Crystal structure of an Hsp90-nucleotide-p23/Sba1 closed chaperone complex. *Nature* 440: 1013–7.

Alioth, S., S. Meyer, R. Dutzler, and K. Pervushin. 2007. The cytoplasmic domain of the chloride channel ClC-0: structural and dynamic characterization of flexible regions. *J. Mol. Biol.* 369: 1163–9.

Aloy, P. and R. B. Russell. 2004. Ten thousand interactions for the molecular biologist. *Nat Biotechnol.* 22: 1317–21.

Alper, T., W. A. Cramp, D. A. Haig, and M. C. Clarke. 1967. Does the agent of scrapie replicate without nucleic acid? *Nature* 214: 764–6.

Alsheikh, M. K., B. J. Heyen, and S. K. Randall. 2003. Ion binding properties of the dehydrin ERD14 are dependent upon phosphorylation. *J. Biol. Chem.* 278: 40882–9.

Ando, Y., M. Nakamura, and S. Araki. 2005. Transthyretin-related familial amyloidotic polyneuropathy. *Arch. Neurol.* 62: 1057–62.

Andrews, A. L., D. Atkinson, M. T. A. Evans, et al. 1979. The conformation and aggregation of bovine beta-casein A. I. Molecular aspects of thermal aggregation. *Biopolymers* 18: 1105–21.

Anfinsen, C. B. 1973. Principles that govern the folding of protein chains. *Science* 181: 223–30.

Antony, T., W. Hoyer, D. Cherny, G. Heim, T. M. Jovin, and V. Subramaniam. 2003. Cellular polyamines promote the aggregation of alpha-synuclein. *J. Biol. Chem.* 278: 3235–40.

Aranda-Espinoza, H., P. Carl, J. F. Leterrier, P. Janmey, and D. E. Discher. 2002. Domain unfolding in neurofilament sidearms: effects of phosphorylation and ATP. *FEBS Lett.* 531: 397–401.

Arifuzzaman, M., M. Maeda, A. Itoh, et al. 2006. Large-scale identification of protein–protein interaction of *Escherichia coli* K-12. *Genome Res.* 16: 686–91.

Arkin, M. 2005. Protein–protein interactions and cancer: small molecules going in for the kill. *Curr. Opin. Chem. Biol.* 9: 317–24.

Arkin, M. R. and J. A. Wells. 2004. Small-molecule inhibitors of protein–protein interactions: progressing towards the dream. *Nat. Rev. Drug Discov.* 3: 301–17.

Asher, G., N. Reuven, and Y. Shaul. 2006. 20S proteasomes and protein degradation "by default." *BioEssays* 28: 844–9.

Asher, G., P. Tsvetkov, C. Kahana, and Y. Shaul. 2005. A mechanism of ubiquitin-vindependent proteasomal degradation of the tumor suppressors p53 and p73. *Genes Dev.* 19: 316–21.

Asturias, F. J., Y. W. Jiang, L. C. Myers, C. M. Gustafsson, and R. D. Kornberg. 1999. Conserved structures of mediator and RNA polymerase II holoenzyme. *Science* 283: 985–87.

Averna, M., R. De Tullio, M. Passalacqua, F. Salamino, S. Pontremoli, and E. Melloni. 2001. Changes in intracellular calpastatin localization are mediated by reversible phosphorylation. *Biochem J.* 354: 25–30.

Avila, J. 1989. *Microtubule Proteins.* Boca Raton, FL: CRC Press.

Azen, E. A. 1993. Genetics of salivary protein polymorphisms. *Crit. Rev. Oral Biol. Med.* 4: 479–85.

Azoulay, J., J. P. Clamme, J. L. Darlix, B. P. Roques, and Y. Mely. 2003. Destabilization of the HIV-1 complementary sequence of TAR by the nucleocapsid protein through activation of conformational fluctuations. *J. Mol. Biol.* 326: 691–700.

Babu, Y. S., C. E. Bugg, and W. J. Cook. 1988. Structure of calmodulin refined at 2.2 A resolution. *J. Mol. Biol.* 204: 191–204.

Bagui, T. K., S. Mohapatra, E. Haura, and W. J. Pledger. 2003. P27Kip1 and p21Cip1 are not required for the formation of active D cyclin-cdk4 complexes. *Mol. Cell Biol.* 23: 7285–90.

Bai, C., P. Sen, K. Hofmann, et al. 1996. SKP1 connects cell cycle regulators to the ubiquitin proteolysis machinery through a novel motif, the F-box. *Cell* 86: 263–74.

Bai, Y., J. S. Milne, L. Mayne, and S. W. Englander. 1993. Primary structure effects on peptide group hydrogen exchange. *Proteins* 17: 75–86.

Baker, J. M., R. P. Hudson, V. Kanelis, et al. 2007. CFTR regulatory region interacts with NBD1 predominantly via multiple transient helices. *Nat. Struct. Mol. Biol.* 14: 738–45.

Balázs, A., V. Csizmok, et al. 2009. High levels of structural disorder in scaffold proteins as exemplified by a novel neuronal protein, Caskin1. *FEBS J*, in press:

Baldwin, R. L. 2007. Energetics of protein folding. *J. Mol. Biol.* 371: 283–301.

Ban, N., P. Nissen, J. Hansen, P. B. Moore, and T. A. Steitz. 2000. The complete atomic structure of the large ribosomal subunit at 2.4 A resolution. *Science* 289: 905–20.

Barabasi, A. L. and Z. N. Oltvai. 2004. Network biology: understanding the cell's functional organization. *Nat. Rev. Genet.* 5: 101–13.

Barron, L. D., E. W. Blanch, and L. Hecht. 2002. Unfolded proteins studied by Raman optical activity. *Adv. Protein. Chem.* 62: 51–90.

Barron, L. D., L. Hecht, E. W. Blanch, and A. F. Bell. 2000. Solution structure and dynamics of biomolecules from Raman optical activity. *Prog. Biophys. Mol. Biol.* 73: 1–49.

Barth, A. 2007. Infrared spectroscopy of proteins. *Biochim. Biophys. Acta.* 1767: 1073–101.

Barzegar, A., R. Yousefi, A. Sharifzadeh, et al. 2008. Chaperone activities of bovine and camel beta-caseins: importance of their surface hydrophobicity in protection against alcohol dehydrogenase aggregation. *Int. J. Biol. Macromol.* 42: 392–9.

Baskakov, I. and D. W. Bolen. 1998. Forcing thermodynamically unfolded proteins to fold. *J. Biol. Chem.* 273: 4831–4.

Bateman, A., E. Birney, L. Cerruti, et al. 2002. The Pfam protein families database. *Nucleic. Acids Res.* 30: 276–80.

Bates, I. R., J. B. Feix, J. M. Boggs, and G. Harauz. 2004. An immunodominant epitope of myelin basic protein is an amphipathic alpha-helix. *J. Biol. Chem.* 279: 5757–64.

Batra-Safferling, R., K. Abarca-Heidemann, H. G. Korschen, et al. 2006. Glutamic acid-rich proteins of rod photoreceptors are natively unfolded. *J. Biol. Chem.* 281: 1449–60.

Baxa, U., P. D. Ross, R. B. Wickner, and A. C. Steven. 2004. The N-terminal prion domain of Ure2p converts from an unfolded to a thermally resistant conformation upon filament formation. *J. Mol. Biol.* 339: 259–64.

Baxter, N. J., T. H. Lilley, E. Haslam, and M. P. Williamson. 1997. Multiple interactions between polyphenols and a salivary proline-rich protein repeat result in complexation and precipitation. *Biochemistry* 36: 5566–77.

Bell, S., C. Klein, L. Muller, S. Hansen, and J. Buchner. 2002. p53 Contains large unstructured regions in its native state. *J. Mol. Biol.* 322: 917–27.

Belle, A., A. Tanay, L. Bitincka, R. Shamir, and E. K. O'Shea. 2006. Quantification of protein half-lives in the budding yeast proteome. *Proc. Natl. Acad. Sci. USA* 103: 13004–9.

Belmont, L. D. and T. J. Mitchison. 1996. Identification of a protein that interacts with tubulin dimers and increases the catastrophe rate of microtubules. *Cell* 84: 623–31.

Bennett, M. C. 2005. The role of alpha-synuclein in neurodegenerative diseases. *Pharmacol. Ther.* 105: 311–31.

Bentrop, D., M. Beyermann, R. Wissmann, and B. Fakler. 2001. NMR structure of the "ball-and-chain" domain of KCNMB2, the beta 2-subunit of large conductance Ca2+- and voltage-activated potassium channels. *J. Biol. Chem.* 276: 42116–21.

Bernacchi, S., S. Stoylov, E. Piemont, et al. 2002. HIV-1 nucleocapsid protein activates transient melting of least stable parts of the secondary structure of TAR and its complementary sequence. *J. Mol. Biol.* 317: 385–99.

Bernado, P., L. Blanchard, P. Timmins, D. Marion, R. W. Ruigrok, and M. Blackledge. 2005. A structural model for unfolded proteins from residual dipolar couplings and small-angle x-ray scattering. *Proc. Natl. Acad. Sci. USA* 102: 17002–7.

Bernado, P., E. Mylonas, M. V. Petoukhov, M. Blackledge, and D. I. Svergun. 2007. Structural characterization of flexible proteins using small-angle X-ray scattering. *J. Am. Chem. Soc.* 129: 5656–64.

Bertoncini, C. W., Y. S. Jung, C. O. Fernandez, et al. 2005. Release of long-range tertiary interactions potentiates aggregation of natively unstructured alpha-synuclein. *Proc. Natl. Acad. Sci. USA* 102: 1430–5.

Bertoncini, C. W., R. M. Rasia, G. R. Lamberto, et al. 2007. Structural characterization of the intrinsically unfolded protein beta-synuclein, a natural negative regulator of alpha-synuclein aggregation. *J. Mol. Biol.* 372: 708–22.

Bertram, L. and R. E. Tanzi. 2004. Alzheimer's disease: one disorder, too many genes? *Hum. Mol. Genet.* 13 Spec. No. 1: R135–41.

Besson, A., S. F. Dowdy, and J. M. Roberts. 2008. CDK inhibitors: cell cycle regulators and beyond. *Dev. Cell* 14: 159–69.

Betts, R., S. Weinsheimer, G. E. Blouse, and J. Anagli. 2003. Structural determinants of the calpain inhibitory activity of calpastatin peptide B27-WT. *J. Biol. Chem.* 278: 7800–9.

Bevivino, A. E. and P. J. Loll. 2001. An expanded glutamine repeat destabilizes native ataxin-3 structure and mediates formation of parallel beta-fibrils. *Proc. Natl. Acad. Sci. USA* 98: 11955–60.

Bhattacharyya, J. and K. P. Das. 1999. Molecular chaperone-like properties of an unfolded protein, alpha(s)-casein. *J. Biol. Chem.* 274: 15505–9.

Bhattacharyya, R. P., A. Remenyi, M. C. Good, C. J. Bashor, A. M. Falick, and W. A. Lim. 2006. The Ste5 scaffold allosterically modulates signaling output of the yeast mating pathway. *Science* 311: 822–6.

Bhaumik, S. R., E. Smith, and A. Shilatifard. 2007. Covalent modifications of histones during development and disease pathogenesis. *Nat. Struct. Mol. Biol.* 14: 1008–16.

Bienkiewicz, E. A., J. N. Adkins, and K. J. Lumb. 2002. Functional consequences of preorganized helical structure in the intrinsically disordered cell-cycle inhibitor p27(Kip1). *Biochemistry* 41: 752–9.

Bienkiewicz, E. A., A. M. Woody, and R. W. Woody. 2000. Conformation of the RNA polymerase II C-terminal domain: circular dichroism of long and short fragments. *J. Mol. Biol.* 297: 119–33.

Bjorklund, A. K., D. Ekman, and A. Elofsson. 2006. Expansion of protein domain repeats. *PLoS Comput. Biol.* 2: e114.

Black, J. C., J. E. Choi, S. R. Lombardo, and M. Carey. 2006. A mechanism for coordinating chromatin modification and preinitiation complex assembly. *Mol. Cell* 23: 809–18.

Blake, C. C., D. F. Koenig, G. A. Mair, A. C. North, D. C. Phillips, and V. R. Sarma. 1965. Structure of hen egg-white lysozyme. A three-dimensional Fourier synthesis at 2 Angstrom resolution. *Nature* 206: 757–61.

Blanch, E. W., D. D. Kasarda, L. Hecht, K. Nielsen, and L. D. Barron. 2003. New insight into the solution structures of wheat gluten proteins from Raman optical activity. *Biochemistry* 42: 5665–73.

Blanch, E. W., L. A. Morozova-Roche, D. A. Cochran, A. J. Doig, L. Hecht, and L. D. Barron. 2000. Is polyproline II helix the killer conformation? A Raman optical activity study of the amyloidogenic prefibrillar intermediate of human lysozyme. *J. Mol. Biol.* 301: 553–63.

Blander, G. and L. Guarente. 2004. The Sir2 family of protein deacetylases. *Annu. Rev. Biochem.* 73: 417–35.

Blencowe, B. J. 2006. Alternative splicing: new insights from global analyses. *Cell* 126: 37–47.

Blom, N., S. Gammeltoft, and S. Brunak. 1999. Sequence and structure-based prediction of eukaryotic protein phosphorylation sites. *J. Mol. Biol.* 294: 1351–62.

Bloomfield, V. A. and T. K. Lim. 1978. Quasi-elastic laser light scattering. *Methods Enzymol.* 48: 415–94.

Bochicchio, B. and A. M. Tamburro. 2002. Polyproline II structure in proteins: identification by chiroptical spectroscopies, stability, and functions. *Chirality* 14: 782–92.

Bochkareva, E., L. Kaustov, A. Ayed, et al. 2005. Single-stranded DNA mimicry in the p53 transactivation domain interaction with replication protein A. *Proc. Natl. Acad. Sci. USA* 102: 15412–7.

Bochtler, M., L. Ditzel, M. Groll, C. Hartmann, and R. Huber. 1999. The proteasome. *Annu. Rev. Biophys. Biomol. Struct.* 28: 295–317.

Bock-Marquette, I., A. Saxena, M. D. White, J. Michael Dimaio, and D. Srivastava. 2004. Thymosin beta4 activates integrin-linked kinase and promotes cardiac cell migration, survival and cardiac repair. *Nature* 432: 466–72.

Bodart, J. F., J. M. Wieruszeski, L. Amniai, et al. 2008. NMR observation of Tau in Xenopus oocytes. *J. Magn. Reson.* 192: 252–7.

Bode, W., P. Schwager, and R. Huber. 1978. The transition of bovine trypsinogen to a trypsin-like state upon strong ligand binding. The refined crystal structures of the bovine trypsinogen-pancreatic trypsin inhibitor complex and of its ternary complex with Ile-Val at 1.9 A resolution. *J. Mol. Biol.* 118: 99–112.

Bois, P. and A. J. Jeffreys. 1999. Minisatellite instability and germline mutation. *Cell. Mol. Life Sci.* 55: 1636–48.

Bois, P. R. 2003. Hypermutable minisatellites, a human affair? *Genomics* 81: 349–55.

Bokor, M., V. Csizmok, D. Kovacs, et al. 2005. NMR relaxation studies on the hydrate layer of intrinsically unstructured proteins. *Biophys J.* 88: 2030–7.

Bonin, I., R. Muhlberger, G. P. Bourenkov, et al. 2004. Structural basis for the interaction of Escherichia coli NusA with protein N of phage lambda. *Proc. Natl. Acad. Sci. USA* 101: 13762–7.

Bonsor, D. A., I. Grishkovskaya, E. J. Dodson, and C. Kleanthous. 2007. Molecular mimicry enables competitive recruitment by a natively disordered protein. *J. Am. Chem. Soc.* 129: 4800–7.

Booth, D. R., M. Sunde, V. Bellotti, et al. 1997. Instability, unfolding and aggregation of human lysozyme variants underlying amyloid fibrillogenesis. *Nature* 385: 787–93.

Bordoli, L., F. Kiefer, and T. Schwedel. 2007. Assessment of disorder predictions in CASP7. *Proteins* 69 (Suppl 8): 129–36.

Borg, M., T. Mittag, T. Pawson, M. Tyers, J. D. Forman-Kay, and H. S. Chan. 2007. Polyelectrostatic interactions of disordered ligands suggest a physical basis for ultrasensitivity. *Proc. Natl. Acad. Sci. USA* 104: 9650–5.

Bourhis, J. M., B. Canard, and S. Longhi. 2006. Structural disorder within the replicative complex of measles virus: functional implications. *Virology* 344: 94–110.

Bourhis, J. M., V. Receveur-Brechot, M. Oglesbee, et al. 2005. The intrinsically disordered C-terminal domain of the measles virus nucleoprotein interacts with the C-terminal domain of the phosphoprotein via two distinct sites and remains predominantly unfolded. *Protein Sci.* 14: 1975–92.

Bracken, C., L. M. Iakoucheva, P. R. Romero, and A. K. Dunker. 2004. Combining prediction, computation and experiment for the characterization of protein disorder. *Curr. Opin. Struct. Biol.* 14: 570–6.

Braig, K., Z. Otwinowski, R. Hegde, et al. 1994. The crystal structure of the bacterial chaperonin GroEL at 2.8 A. *Nature* 371: 578–86.

Bray, E. A. 1993. Molecular responses to water deficit. *Plant Physiol.* 103: 1035–40.

Breidenbach, M. A. and A. T. Brunger. 2004. Substrate recognition strategy for botulinum neurotoxin serotype A. *Nature* 432: 925–9.

Brenner, S. E. 2000. Target selection for structural genomics. *Nat. Struct. Biol.* 7 Suppl: 967–9.

Bretscher, A. 1984. Smooth muscle caldesmon. Rapid purification and F-actin cross-linking properties. *J. Biol. Chem.* 259: 12873–80.

Bright, J. N., T. B. Woolf, and J. H. Hoh. 2001. Predicting properties of intrinsically unstructured proteins. *Prog. Biophys. Mol. Biol.* 76: 131–73.

Brookmeyer, R., S. Gray, and C. Kawas. 1998. Projections of Alzheimer's disease in the United States and the public health impact of delaying disease onset. *Am. J. Public. Health.* 88: 1337–42.

Brooks, C. L. and W. Gu. 2006. p53 ubiquitination: Mdm2 and beyond. *Mol. Cell* 21: 307–15.

Brown, C. J., S. Takayama, A. M. Campen, et al. 2002. Evolutionary rate heterogeneity in proteins with long disordered regions. *J. Mol. Evol.* 55: 104–10.

Brown, D. R., K. Qin, J. W. Herms, et al. 1997a. The cellular prion protein binds copper in vivo. *Nature* 390: 684–7.

Brown, D. R., W. J. Schulz-Schaeffer, B. Schmidt, and H. A. Kretzschmar. 1997b. Prion protein-deficient cells show altered response to oxidative stress due to decreased SOD-1 activity. *Exp. Neurol.* 146: 104–12.

Brown, H. G. and J. H. Hoh. 1997. Entropic exclusion by neurofilament sidearms: a mechanism for maintaining interfilament spacing. *Biochemistry* 36: 15035–40.

Brunger, A. T., M. A. Breidenbach, R. Jin, A. Fischer, J. S. Santos, and M. Montal. 2007. Botulinum neurotoxin heavy chain belt as an intramolecular chaperone for the light chain. *PLoS Pathog.* 3: 1191–4.

Brzovic, P. S., J. R. Keeffe, H. Nishikawa, et al. 2003. Binding and recognition in the assembly of an active BRCA1/BARD1 ubiquitin-ligase complex. *Proc. Natl. Acad. Sci. USA* 100: 5646–51.

Buard, J. and A. J. Jeffreys. 1997. Big, bad minisatellites. *Nat. Genet.* 15: 327–8.

Bubunenko, M. G., S. V. Chuikov, and A. T. Gudkov. 1992. The length of the interdomain region of the L7/L12 protein is important for its function. *FEBS Lett.* 313: 232–4.

Buday, L. 1999. Membrane-targeting of signalling molecules by SH2/SH3 domain-containing adaptor proteins. *Biochim. Biophys. Acta.* 1422: 187–204.

Buee, L., T. Bussiere, V. Buee-Scherrer, A. Delacourte, and P. R. Hof. 2000. Tau protein isoforms, phosphorylation and role in neurodegenerative disorders. *Brain Res. Brain Res. Rev.* 33: 95–130.

Bushnell, D. A., K. D. Westover, R. E. Davis, and R. D. Kornberg. 2004. Structural basis of transcription: an RNA polymerase II-TFIIB cocrystal at 4.5 angstroms. *Science* 303: 983–8.

Bussell, R. Jr. and D. Eliezer. 2001. Residual structure and dynamics in Parkinson's disease-associated mutants of alpha-synuclein. *J. Biol. Chem.* 276: 45996–6003.

Bustos, D. M. and A. A. Iglesias. 2006. Intrinsic disorder is a key characteristic in partners that bind 14-3-3 proteins. *Proteins* 63: 35–42.

Callaghan, A. J., J. P. Aurikko, L. L. Ilag, et al. 2004. Studies of the RNA degradosome-organizing domain of the *Escherichia coli* ribonuclease RNase E. *J. Mol. Biol.* 340: 965–79.

Campbell, K. M., A. R. Terrell, P. J. Laybourn, and K. J. Lumb. 2000. Intrinsic structural disorder of the C-terminal activation domain from the bZIP transcription factor Fos. *Biochemistry* 39: 2708–13.

Carafoli, E., L. Santella, D. Branca, and M. Brini. 2001. Generation, control, and processing of cellular calcium signals. *Crit. Rev. Biochem. Mol. Biol.* 36: 107–260.

Carlson, D. M. 1993. Salivary proline-rich proteins: biochemistry, molecular biology, and regulation of expression. *Crit. Rev. Oral. Biol. Med.* 4: 495–502.

Caron, E. 2002. Regulation of Wiskott–Aldrich syndrome protein and related molecules. *Curr. Opin. Cell Biol.* 14: 82–7.

Carr, C. M. and P. S. Kim. 1993. A spring-loaded mechanism for the conformational change of influenza hemagglutinin. *Cell* 73: 823–32.

Carrard, G., A. Koivula, H. Soderlund, and P. Beguin. 2000. Cellulose-binding domains promote hydrolysis of different sites on crystalline cellulose. *Proc. Natl. Acad. Sci. USA* 97: 10342–7.

Casares, S., M. Sadqi, O. Lopez-Mayorga, F. Conejero-Lara, and N. A. Van Nuland. 2004. Detection and characterization of partially unfolded oligomers of the SH3 domain of alpha-spectrin. *Biophys J.* 86: 2403–13.

Cattaneo, E., C. Zuccato, and M. Tartari. 2005. Normal huntingtin function: an alternative approach to Huntington's disease. *Nat. Rev. Neurosci.* 6: 919–30.

Chakrabortee, S., C. Boschetti, L. J. Walton, S. Sarkar, D. C. Rubinsztein, and A. Tunnacliffe. 2007. Hydrophilic protein associated with desiccation tolerance exhibits broad protein stabilization function. *Proc. Natl. Acad. Sci. USA* 104: 18073–78.

Chang, J. F., K. Phillips, T. Lundback, M. Gstaiger, J. E. Ladbury, and B. Luisi. 1999. Oct-1 POU and octamer DNA co-operate to recognize the Bob-1 transcription co-activator via induced folding. *J. Mol. Biol.* 288: 941–52.

Chang, X. B., J. A. Tabcharani, Y. X. Hou, et al. 1993. Protein kinase A (PKA) still activates CFTR chloride channel after mutagenesis of all 10 PKA consensus phosphorylation sites. *J. Biol. Chem.* 268: 11304–11.

Charlton, A. J., N. J. Baxter, T. H. Lilley, E. Haslam, C. J. Mcdonald, and M. P. Williamson. 1996. Tannin interactions with a full-length human salivary proline-rich protein display a stronger affinity than with single proline-rich repeats. *FEBS Lett.* 382: 289–92.

Chatterjee, A., A. Kumar, J. Chugh, S. Srivastava, N. S. Bhavesh, and R. V. Hosur. 2005. NMR of unfolded proteins. *J. Chem. Sci.* 117: 3–21.

Chattopadhyay, K., E. L. Elson, and C. Frieden. 2005. The kinetics of conformational fluctuations in an unfolded protein measured by fluorescence methods. *Proc. Natl. Acad. Sci. USA* 102: 2385–9.

Chehab, N. H., A. Malikzay, E. S. Stavridi, and T. D. Halazonetis. 1999. Phosphorylation of Ser-20 mediates stabilization of human p53 in response to DNA damage. *Proc. Natl. Acad. Sci. USA* 96: 13777–82.

Chellgren, B. W., A. F. Miller, and T. P. Creamer. 2006. Evidence for Polyproline ii helical structure in short polyglutamine tracts. *J. Mol. Biol.* 361: 362–71.

Chen, B. S. and K. W. Roche. 2007. Regulation of NMDA receptors by phosphorylation. *Neuropharmacology* 53: 362–8.

Chen, D., M. Li, J. Luo, and W. Gu. 2003. Direct interactions between HIF-1a and Mdm2 modulate p53 function. *J. Biol. Chem.* 278: 13595–98.

Chen, H. T. and S. Hahn. 2004. Mapping the location of TFIIB within the RNA polymerase II transcription preinitiation complex: a model for the structure of the PIC. *Cell* 119: 169–80.

Chen, J. W., P. Romero, V. N. Uversky, and A. K. Dunker. 2006a. Conservation of intrinsic disorder in protein domains and families: I. A database of conserved predicted disordered regions. *J. Proteome. Res.* 5: 879–87.

Chen, J. W., P. Romero, V. N. Uversky, and A. K. Dunker. 2006b. Conservation of intrinsic disorder in protein domains and families: II. Functions of conserved disorder. *J. Proteome. Res.* 5: 888–98.

Chen, K., Z. Liu and N. R. Kallenbach. 2004. The polyproline II conformation in short alanine peptides is noncooperative. *Proc. Natl. Acad. Sci. USA* 101: 15352–7.

Chen, M., M. Margittai, J. Chen, and R. Langen. 2007. Investigation of alpha-synuclein fibril structure by site-directed spin labeling. *J. Biol. Chem.* 282: 24970–9.

Chen, S., V. Berthelier, W. Yang, and R. Wetzel. 2001. Polyglutamine aggregation behavior in vitro supports a recruitment mechanism of cytotoxicity. *J. Mol. Biol.* 311: 173–82.

Chen, Y. W. 2003. Local protein unfolding and pathogenesis of polyglutamine-expansion diseases. *Proteins* 51: 68–73.

Cheng, H. C., B. M. Skehan, K. G. Campellone, J. M. Leong, and M. K. Rosen. 2008. Structural mechanism of WASP activation by the enterohaemorrhagic *E. coli* effector EspF(U). *Nature* 454: 1009–13.

Cheng, M., P. Olivier, J. A. Diehl, et al. 1999. The p21(Cip1) and p27(Kip1) CDK "inhibitors" are essential activators of cyclin D-dependent kinases in murine fibroblasts. *EMBO J.* 18: 1571–83.

Cheng, S. H., D. P. Rich, J. Marshall, R. J. Gregory, M. J. Welsh, and A. E. Smith. 1991. Phosphorylation of the R domain by cAMP-dependent protein kinase regulates the CFTR chloride channel. *Cell* 66: 1027–36.

Cheng, Y., T. Legall, C. J. Oldfield, A. K. Dunker, and V. N. Uversky. 2006a. Abundance of intrinsic disorder in protein associated with cardiovascular disease. *Biochemistry* 45: 10448–60.

Cheng, Y., T. Legall, C. J. Oldfield, et al. 2006b. Rational drug design via intrinsically disordered protein. *Trends Biotechnol.* 24: 435–42.

Cheng, Y., C. J. Oldfield, J. Meng, P. Romero, V. N. Uversky, and A. K. Dunker. 2007. Mining alpha-helix-forming molecular recognition features with cross species sequence alignments. *Biochemistry* 46: 13468–77.

Chereau, D., F. Kerff, P. Graceffa, Z. Grabarek, K. Langsetmo, and R. Dominguez. 2005. Actin-bound structures of Wiskott–Aldrich syndrome protein (WASP)-homology domain 2 and the implications for filament assembly. *Proc. Natl. Acad. Sci. USA* 102: 16644–9.

Chiang, P. K., M. A. Lam, and Y. Luo. 2008. The many faces of amyloid beta in Alzheimer's disease. *Curr. Mol. Med.* 8: 580–4.

Chien, P., J. S. Weissman, and A. H. Depace. 2004. Emerging principles of conformation-based prion inheritance. *Annu. Rev. Biochem.* 73: 617–56.

Chirita, C. N., E. E. Congdon, H. Yin, and J. Kuret. 2005. Triggers of full-length tau aggregation: a role for partially folded intermediates. *Biochemistry* 44: 5862–72.

Chiti, F. and C. M. Dobson. 2006. Protein misfolding, functional amyloid, and human disease. *Annu. Rev. Biochem.* 75: 333–66.

Chiti, F., M. Stefani, N. Taddei, G. Ramponi, and C. M. Dobson. 2003. Rationalization of the effects of mutations on peptide and protein aggregation rates. *Nature* 424: 805–8.

Chiti, F., N. Taddei, M. Stefani, C. M. Dobson, and G. Ramponi. 2001. Reduction of the amyloidogenicity of a protein by specific binding of ligands to the native conformation. *Protein Sci.* 10: 879–86.

Chiti, F., P. Webster, N. Taddei, et al. 1999. Designing conditions for in vitro formation of amyloid protofilaments and fibrils. *Proc. Natl. Acad. Sci. USA* 96: 3590–4.

Chong, P. A., B. Ozdamar, J. L. Wrana, and J. D. Forman-Kay. 2004. Disorder in a target for the Smad2 mad homology 2 domain and its implications for binding and specificity. *J. Biol. Chem.* 279: 40707–14.

Chothia, C., J. Gough, C. Vogel, and S. A. Teichmann. 2003. Evolution of the protein repertoire. *Science* 300: 1701–3.

Chou, P. Y. and G. D. Fasman. 1978. Prediction of the secondary structure of proteins from their amino acid sequence. *Adv. Enzymol. Relat. Areas Mol. Biol.* 47: 45–148.

Chung, W. H., J. L. Craighead, W. H. Chang, et al. 2003. RNA polymerase II/TFIIF structure and conserved organization of the initiation complex. *Mol. Cell* 12: 1003–13.

Clapier, C. R., G. Langst, D. F. Corona, P. B. Becker, and K. P. Nightingale. 2001. Critical role for the histone H4 N terminus in nucleosome remodeling by ISWI. *Mol. Cell Biol.* 21: 875–83.

Clayton, D. F. and J. M. George. 1998. The synucleins: a family of proteins involved in synaptic function, plasticity, neurodegeneration and disease. *Trends Neurosci.* 21: 249–54.

Cliff, M. J., R. Harris, D. Barford, J. E. Ladbury, and M. A. Williams. 2006. Conformational diversity in the TPR domain-mediated interaction of protein phosphatase 5 with Hsp90. *Structure* 14: 415–26.

Cobb, N. J., F. D. Sonnichsen, H. Mchaourab, and W. K. Surewicz. 2007. Molecular architecture of human prion protein amyloid: a parallel, in-register beta-structure. *Proc. Natl. Acad. Sci. USA* 104: 18946–51.

Coeytaux, K. and A. Poupon. 2005. Prediction of unfolded segments in a protein sequence based on amino acid composition. *Bioinformatics* 21: 1891–900.

Cohen, P. 2000. The regulation of protein function by multisite phosphorylation—a 25-year update. *Trends Biochem. Sci.* 25: 596–601.

Cohen, P. T. 1997. Novel protein serine/threonine phosphatases: variety is the spice of life. *Trends Biochem. Sci.* 22: 245–51.

Cohen, P. T., M. X. Chen and C. G. Armstrong. 1996. Novel protein phosphatases that may participate in cell signaling. *Adv. Pharmacol.* 36: 67–89.

Collins, E. C. and T. H. Rabbitts. 2002. The promiscuous MLL gene links chromosomal translocations to cellular differentiation and tumour tropism. *Trends Mol. Med.* 8: 436–42.

Come, J. H., P. E. Fraser, and P. T. Lansbury Jr. 1993. A kinetic model for amyloid formation in the prion diseases: importance of seeding. *Proc. Natl. Acad. Sci. USA* 90: 5959–63.

Conrad, B. and S. E. Antonarakis. 2007. Gene duplication: a drive for phenotypic diversity and cause of human disease. *Annu. Rev. Genomics. Hum. Genet.* 8: 17–35.

Consortium. 2004. Finishing the euchromatic sequence of the human genome. *Nature* 431: 931–45.

Conway, K. A., J. D. Harper, and P. T. Lansbury. 1998. Accelerated in vitro fibril formation by a mutant alpha-synuclein linked to early-onset Parkinson disease. *Nat. Med.* 4: 1318–20.

Conway, K. A., J. D. Harper, and P. T. Lansbury Jr. 2000. Fibrils formed in vitro from alpha-synuclein and two mutant forms linked to Parkinson's disease are typical amyloid. *Biochemistry* 39: 2552–63.

Copley, R. R., T. Doerks, I. Letunic, and P. Bork. 2002. Protein domain analysis in the era of complete genomes. *FEBS Lett.* 513: 129–34.

Copley, R. R., L. Goodstadt, and C. Ponting. 2003. Eukaryotic domain evolution inferred from genome comparisons. *Curr. Opin. Genet. Dev.* 13: 623–8.

Cordero, O. J., C. S. Sarandeses, J. L. Lopez, and M. Nogueira. 1992. On the anomalous behaviour on gel-filtration and SDS-electrophoresis of prothymosin-alpha. *Biochem. Int.* 28: 1117–24.

Corradi, G. R. and H. P. Adamo. 2007. Intramolecular fluorescence resonance energy transfer between fused autofluorescent proteins reveals rearrangements of the N- and C-terminal segments of the plasma membrane Ca2+ pump involved in the activation. *J. Biol. Chem.* 282: 35440–8.

Cortese, M. S., J. P. Baird, V. N. Uversky, and A. K. Dunker. 2005. Uncovering the unfoldome: enriching cell extracts for unstructured proteins by acid treatment. *J. Proteome. Res.* 4: 1610–8.

Cox, C. J., K. Dutta, E. T. Petri, et al. 2002. The regions of securin and cyclin B proteins recognized by the ubiquitination machinery are natively unfolded. *FEBS Lett.* 527: 303–8.

Cramer, P., D. A. Bushnell, and R. D. Kornberg. 2001. Structural basis of transcription: RNA polymerase II at 2.8 angstrom resolution. *Science* 292: 1863–76.

Creamer, L. K., T. Richardson, and D. A. Parry. 1981. Secondary structure of bovine alpha s1- and beta-casein in solution. *Arch. Biochem. Biophys.* 211: 689–96.

Crevel, G. and S. Cotterill. 1995. DF 31, a sperm decondensation factor from *Drosophila melanogaster*: purification and characterization. *EMBO J.* 14: 1711–7.

Crevel, G., H. Huikeshoven, and S. Cotterill. 2001. Df31 is a novel nuclear protein involved in chromatin structure in *Drosophila melanogaster. J. Cell Sci.* 114: 37–47.

Crick, S. L., M. Jayaraman, C. Frieden, R. Wetzel, and R. V. Pappu. 2006. Fluorescence correlation spectroscopy shows that monomeric polyglutamine molecules form collapsed structures in aqueous solutions. *Proc. Natl. Acad. Sci. USA* 103: 16764–9.

Cristofari, G. and J. L. Darlix. 2002. The ubiquitous nature of RNA chaperone proteins. *Prog. Nucleic Acid Res. Mol. Biol.* 72: 223–68.

Cristofari, G., C. Gabus, D. Ficheux, M. Bona, S. F. Le Grice, and J. L. Darlix. 1999. Characterization of active reverse transcriptase and nucleoprotein complexes of the yeast retrotransposon Ty3 in vitro. *J. Biol. Chem.* 274: 36643–8.

Crowther, R. A., R. Jakes, M. G. Spillantini, and M. Goedert. 1998. Synthetic filaments assembled from C-terminally truncated alpha-synuclein. *FEBS Lett.* 436: 309–12.

Csermely, P. 1997. Proteins, RNAs and chaperones in enzyme evolution: a folding perspective. *Trends Biochem. Sci.* 22: 147–9.

Csermely, P. 1999. Chaperone-percolator model: a possible molecular mechanism of Anfinsen-cage-type chaperones. *Bioessays* 21: 959–65.

Csermely, P., T. Schnaider, C. Soti, Z. Prohaszka, and G. Nardai. 1998. The 90-kDa molecular chaperone family: structure, function, and clinical applications. A comprehensive review. *Pharmacol. Ther.* 79: 129–68.

Csizmok, V., M. Bokor, P. Banki, et al. 2005. Primary contact sites in intrinsically unstructured proteins: the case of calpastatin and microtubule-associated protein 2. *Biochemistry* 44: 3955–64.

Csizmok, V., I. Felli, P. Tompa, L. Banci, and I. Bertini. 2008. Structural and dynamic characterization of intrinsically disordered human securin by NMR spectroscopy. *J. Am. Chem. Soc.* 130: 16873–9.

Csizmok, V., E. Szollosi, P. Friedrich, and P. Tompa. 2006. A novel two-dimensional electrophoresis technique for the identification of intrinsically unstructured proteins. *Mol. Cell. Proteomics.* 5: 265–73.

Curran, J. and D. Kolakofsky. 1999. Replication of paramyxoviruses. *Adv. Virus Res.* 54: 403–22.

Dafforn, T. R. and C. J. Smith. 2004. Natively unfolded domains in endocytosis: hooks, lines and linkers. *EMBO Rep.* 5: 1046–52.

Daggett, V. and A. R. Fersht. 2003. Is there a unifying mechanism for protein folding? *Trends Biochem. Sci.* 28: 18–25.

Dai, M. S. and H. Lu. 2004. Inhibition of MDM2-mediated p53 ubiquitination and degradation by ribosomal protein L5. *J. Biol. Chem.* 279: 44475–82.

Dai, P., H. Akimaru, Y. Tanaka, et al. 1996. CBP as a transcriptional coactivator of c-Myb. *Genes Dev.* 10: 528–40.

Dames, S. A., M. Martinez-Yamout, R. N. De Guzman, H. J. Dyson, and P. E. Wright. 2002. Structural basis for Hif-1 alpha/CBP recognition in the cellular hypoxic response. *Proc. Natl. Acad. Sci. USA* 99: 5271–6.

Daniels, D. L., K. Eklof-Spink, and W. I. Weis. 2001. Beta-catenin: molecular plasticity and drug design. *Trends Biochem. Sci.* 26: 672–8.

Daughdrill, G. W., M. S. Chadsey, J. E. Karlinsey, K. T. Hughes, and F. W. Dahlquist. 1997. The C-terminal half of the anti-sigma factor, FlgM, becomes structured when bound to its target, sigma 28. *Nat. Struct. Biol.* 4: 285–91.

Daughdrill, G. W., L. J. Hanely, and F. W. Dahlquist. 1998. The C-terminal half of the anti-sigma factor FlgM contains a dynamic equilibrium solution structure favoring helical conformations. *Biochemistry* 37: 1076–82.

Daughdrill, G. W., P. Narayanaswami, S. H. Gilmore, A. Belczyk, and C. J. Brown. 2007. Dynamic behavior of an intrinsically unstructured linker domain is conserved in the face of negligible amino acid sequence conservation. *J. Mol. Evol.* 65: 277–88.

Davey, N. E., D. C. Shields, and R. J. Edwards. 2006. SLiMDisc: short, linear motif discovery, correcting for common evolutionary descent. *Nucleic. Acids Res.* 34: 3546–54.

David, D. C., R. Layfield, L. Serpell, Y. Narain, M. Goedert, and M. G. Spillantini. 2002. Proteasomal degradation of tau protein. *J. Neurochem.* 83: 176–85.

Davies, K. J. 2001. Degradation of oxidized proteins by the 20S proteasome. *Biochimie* 83: 301–10.

Dawson, R., L. Muller, A. Dehner, C. Klein, H. Kessler, and J. Buchner. 2003. The N-terminal domain of p53 is natively unfolded. *J. Mol. Biol.* 332: 1131–41.

De Guzman, R. N., M. Martinez-Yamout, H. J. Dyson, and P. E. Wright. 2004. Interaction of the TAZ1 domain of CREBbinding protein with the activation domain of CITED2: regulation by competition between intrinsically unstructured ligands for non-identical binding sites. *J. Biol. Chem.* 279: 3042–49.

Dedmon, M. M., K. Lindorff-Larsen, J. Christodoulou, M. Vendruscolo, and C. M. Dobson. 2005. Mapping long-range interactions in alpha-synuclein using spin-label NMR and ensemble molecular dynamics simulations. *J. Am. Chem. Soc.* 127: 476–7.

Dedmon, M. M., C. N. Patel, G. B. Young, and G. J. Pielak. 2002. FlgM gains structure in living cells. *Proc. Natl. Acad. Sci. USA* 99: 12681–4.

Demarest, S. J., S. Deechongkit, H. J. Dyson, R. M. Evans, and P. E. Wright. 2004. Packing, specificity, and mutability at the binding interface between the p160 coactivator and CREB-binding protein. *Protein Sci.* 13: 203–10.

Demarest, S. J., M. Martinez-Yamout, J. Chung, et al. 2002. Mutual synergistic folding in recruitment of CBP/p300 by p160 nuclear receptor coactivators. *Nature* 415: 549–53.

Demchenko, A. P. 2001. Recognition between flexible protein molecules: induced and assisted folding. *J. Mol. Recognit.* 14: 42–61.

Deng, C. X. 2006. BRCA1: cell cycle checkpoint, genetic instability, DNA damage response and cancer evolution. *Nucleic. Acids Res.* 34: 1416–26.

Denning, D. P., S. S. Patel, V. Uversky, A. L. Fink, and M. Rexach. 2003. Disorder in the nuclear pore complex: the FG repeat regions of nucleoporins are natively unfolded. *Proc. Natl. Acad. Sci. USA* 100: 2450–5.

Denning, D. P. and M. F. Rexach. 2007. Rapid evolution exposes the boundaries of domain structure and function in natively unfolded FG nucleoporins. *Mol. Cell. Proteomics.* 6: 272–82.

Denning, D. P., V. Uversky, S. S. Patel, A. L. Fink, and M. Rexach. 2002. The *Saccharomyces cerevisiae* nucleoporin Nup2p is a natively unfolded protein. *J. Biol. Chem.* 277: 33447–55.

Dill, K. A. and H. S. Chan. 1997. From Levinthal to pathways to funnels. *Nat. Struct. Biol.* 4: 10–9.

Dill, K. A. and D. Shortle. 1991. Denatured states of proteins. *Annu. Rev. Biochem.* 60: 795–825.

Dingwall, C., S. M. Dilworth, S. J. Black, S. E. Kearsey, L. S. Cox, and R. A. Laskey. 1987. Nucleoplasmin cDNA sequence reveals polyglutamic acid tracts and a cluster of sequences homologous to putative nuclear localization signals. *EMBO J.* 6: 69–74.

Dinitto, J. P. and P. W. Huber. 2003. Mutual induced fit binding of *Xenopus* ribosomal protein L5 to 5S rRNA. *J. Mol. Biol.* 330: 979–92.

Dobson, C. M. 1993. Flexible friends. *Current Biology* 3: 530–32.

Dobson, C. M. 1999. Protein misfolding, evolution and disease. *Trends Biochem. Sci.* 24: 329–32.

Dobson, C. M. 2002. Getting out of shape. *Nature* 418: 729–30.

Domanski, M., M. Hertzog, J. Coutant, et al. 2004. Coupling of folding and binding of thymosin beta4 upon interaction with monomeric actin monitored by nuclear magnetic resonance. *J. Biol. Chem.* 279: 23637–45.

Donaldson, L. and J. P. Capone. 1992. Purification and characterization of the carboxyl-terminal transactivation domain of Vmw65 from herpes simplex virus type 1. *J. Biol. Chem.* 267: 1411–4.

Donne, D. G., J. H. Viles, D. Groth, et al. 1997. Structure of the recombinant full-length hamster prion protein PrP(29-231): the N terminus is highly flexible. *Proc. Natl. Acad. Sci. USA* 94: 13452–7.

Dosztanyi, Z., J. Chen, A. K. Dunker, I. Simon, and P. Tompa. 2006. Disorder and sequence repeats in hub proteins and their implications for network evolution. *J. Proteome. Res.* 5: 2985–95.

Dosztanyi, Z., V. Csizmok, P. Tompa, and I. Simon. 2005a. IUPred: web server for the prediction of intrinsically unstructured regions of proteins based on estimated energy content. *Bioinformatics* 21: 3433–4.

Dosztanyi, Z., V. Csizmok, P. Tompa, and I. Simon. 2005b. The pair-wise energy content estimated from amino acid composition discriminates between folded and instrinsically unstructured proteins. *J. Mol. Biol.* 347: 827–39.

Dosztanyi, Z., M. Sandor, P. Tompa, and I. Simon. 2007. Prediction of protein disorder at the domain level. *Curr. Protein Pept. Sci.* 8: 161–71.

Drenth, J. 2006. *Principles of Protein X-Ray Crystallography*. New York: Springer.

Drews, J. 2000. Drug discovery: a historical perspective. *Science* 287: 1960–4.

Dueber, J. E., B. J. Yeh, K. Chak, and W. A. Lim. 2003. Reprogramming control of an allosteric signaling switch through modular recombination. *Science* 301: 1904–8.

Dunker, A. K. 2007. Another window into disordered protein function. *Structure* 15: 1026–8.

Dunker, A. K., C. J. Brown, J. D. Lawson, L. M. Iakoucheva, and Z. Obradovic. 2002. Intrinsic disorder and protein function. *Biochemistry* 41: 6573–82.

Dunker, A. K., M. S. Cortese, P. Romero, L. M. Iakoucheva, and V. N. Uversky. 2005. Flexible nets. The roles of intrinsic disorder in protein interaction networks. *FEBS J.* 272: 5129–48.

Dunker, A. K., E. Garner, S. Guillot, et al. 1998. Protein disorder and the evolution of molecular recognition: theory, predictions and observations. *Pac Symp Biocomputing* 3: 473–84.

Dunker, A. K., J. D. Lawson, C. J. Brown, et al. 2001. Intrinsically disordered protein. *J. Mol. Graphics. Modelling* 19: 26–59.

Dunker, A. K., Z. Obradovic, P. Romero, E. C. Garner, and C. J. Brown. 2000. Intrinsic protein disorder in complete genomes. *Genome Inform. Ser. Workshop Genome Inform.* 11: 161–71.

Dunker, A. K., C. J. Oldfield, J. Meng, et al. 2008. The unfoldomics decade: an update on intrinsically disordered proteins. *BMC Genomics* 9 Suppl 2: S1.

Dyson, H. J. and P. E. Wright. 2002a. Coupling of folding and binding for unstructured proteins. *Curr. Opin. Struct. Biol.* 12: 54–60.

Dyson, H. J. and P. E. Wright. 2002b. Insights into the structure and dynamics of unfolded proteins from nuclear magnetic resonance. *Adv. Protein Chem.* 62: 311–40.

Dyson, H. J. and P. E. Wright. 2004. Unfolded proteins and protein folding studied by NMR. *Chem. Rev.* 104: 3607–22.

Dyson, H. J. and P. E. Wright. 2005. Intrinsically unstructured proteins and their functions. *Nat. Rev. Mol. Cell Biol.* 6: 197–208.

Ebert, M. O., S. H. Bae, H. J. Dyson, and P. E. Wright. 2008. NMR relaxation study of the complex formed between CBP and the activation domain of the nuclear hormone receptor coactivator ACTR. *Biochemistry* 47: 1299–308.

Ekman, D., S. Light, A. K. Bjorklund, and A. Elofsson. 2006. What properties characterize the hub proteins of the protein–protein interaction network of *Saccharomyces cerevisiae*? *Genome Biol.* 7: R45.

El-Agnaf, O. M. and G. B. Irvine. 2002. Aggregation and neurotoxicity of alpha-synuclein and related peptides. *Biochem. Soc. Trans.* 30: 559–65.

Eliezer, D., P. Barre, M. Kobaslija, D. Chan, X. Li, and L. Heend. 2005. Residual structure in the repeat domain of tau: echoes of microtubule binding and paired helical filament formation. *Biochemistry* 44: 1026–36.

Eliezer, D., E. Kutluay, R. Bussell Jr., and G. Browne. 2001. Conformational properties of alpha-synuclein in its free and lipid-associated states. *J. Mol. Biol.* 307: 1061–73.

Elion, E. A. 2001. The Ste5p scaffold. *J. Cell Sci.* 114: 3967–78.

Elkins, J. M., K. S. Hewitson, L. A. Mcneill, et al. 2003. Structure of factor-inhibiting hypoxia-inducible factor (HIF) reveals mechanism of oxidative modification of HIF-1alpha. *J. Biol. Chem.* 278: 1802–06.

Ellis, R. J. 2001. Macromolecular crowding: obvious but underappreciated. *Trends Biochem. Sci.* 26: 597–604.

Ellis, R. J. 2006. Molecular chaperones: assisting assembly in addition to folding. *Trends Biochem. Sci.* 31: 395–401.

Etoh, Y., M. Simon, and H. Green. 1986. Involucrin acts as a transglutaminase substrate at multiple sites. *Biochem. Biophys. Res. Commun.* 136: 51–6.

Evans, P. R. and D. J. Owen. 2002. Endocytosis and vesicle trafficking. *Curr. Opin. Struct. Biol.* 12: 814–21.

Fabrega, C., V. Shen, S. Shuman, and C. D. Lima. 2003. Structure of an mRNA capping enzyme bound to the phosphorylated carboxy-terminal domain of RNA polymerase II. *Mol. Cell* 11: 1549–61.

Fandrich, M., M. A. Fletcher, and C. M. Dobson. 2001. Amyloid fibrils from muscle myoglobin. *Nature* 410: 165–6.

Fazzio, T. G., M. E. Gelbart, and T. Tsukiyama. 2005. Two distinct mechanisms of chromatin interaction by the Isw2 chromatin remodeling complex in vivo. *Mol. Cell Biol.* 25: 9165–74.

Fedarko, N. S., B. Fohr, P. G. Robey, M. F. Young, and L. W. Fisher. 2000. Factor H binding to bone sialoprotein and osteopontin enables tumor cell evasion of complement-mediated attack. *J. Biol. Chem.* 275: 16666–72.

Feldman, R. M., C. C. Correll, K. B. Kaplan, and R. J. Deshaies. 1997. A complex of Cdc4p, Skp1p, and Cdc53p/cullin catalyzes ubiquitination of the phosphorylated CDK inhibitor Sic1p. *Cell* 91: 221–30.

Felsenstein, J. 1997. An alternating least squares approach to inferring phylogenies from pairwise distances. *Syst. Biol.* 46: 101–11.

Feng, Z. P., X. Zhang, P. Han, N. Arora, R. F. Anders, and R. S. Norton. 2006. Abundance of intrinsically unstructured proteins in *P. falciparum* and other apicomplexan parasite proteomes. *Mol. Biochem. Parasitol.* 150: 256–67.

Fernandez-Escamilla, A. M., F. Rousseau, J. Schymkowitz, and L. Serrano. 2004. Prediction of sequence-dependent and mutational effects on the aggregation of peptides and proteins. *Nat. Biotechnol.* 22: 1302–6.

Fernandez, C. O., W. Hoyer, M. Zweckstetter, et al. 2004. NMR of alpha-synuclein-polyamine complexes elucidates the mechanism and kinetics of induced aggregation. *EMBO J.* 23: 2039–46.

Fero, M. L., E. Randel, K. E. Gurley, J. M. Roberts, and C. J. Kemp. 1998. The murine gene p27Kip1 is haplo-insufficient for tumour suppression. *Nature* 396: 177–80.

Fero, M. L., M. Rivkin, M. Tasch, et al. 1996. A syndrome of multiorgan hyperplasia with features of gigantism, tumorigenesis, and female sterility in p27(Kip1)-deficient mice. *Cell* 85: 733–44.

Ferraro, D. M., N. D. Lazo, and A. D. Robertson. 2004. EX1 hydrogen exchange and protein folding. *Biochemistry* 43: 587–94.

Ferreon, J. C. and V. J. Hilser. 2004. Thermodynamics of binding to SH3 domains: the energetic impact of polyproline II (P(II)) helix formation. *Biochemistry* 43: 7787–97.

Ferron, F., S. Longhi, B. Canard, and D. Karlin. 2006. A practical overview of protein disorder prediction methods. *Proteins* 65: 1–14.

Fersht, A. 1985. *Enzyme Structure and Mechanism*. New York: W.H. Freeman and Co.

Fields, S. 2005. High-throughput two-hybrid analysis. The promise and the peril. *FEBS J.* 272: 5391–9.

Fink, A. L. 2005. Natively unfolded proteins. *Curr. Opin. Struct. Biol.* 15: 35–41.

Fisher, E. 1894. Einfuss der Configuration auf die Wirkung der Enzyme. *Ber. Dt. Chem. Ges.* 27: 2985–93.

Fisher, L. W., D. A. Torchia, B. Fohr, M. F. Young, and N. S. Fedarko. 2001. Flexible structures of SIBLING proteins, bone sialoprotein, and osteopontin. *Biochem. Biophys. Res. Commun.* 280: 460–5.

Fitzkee, N. C. and G. D. Rose. 2004. Reassessing random-coil statistics in unfolded proteins. *Proc. Natl. Acad. Sci. USA* 101: 12497–502.

Flaugh, S. L. and K. J. Lumb. 2001. Effects of macromolecular crowding on the intrinsically disordered proteins c-Fos and p27(Kip1). *Biomacromolecules* 2: 538–40.

Fletcher, C. M. and G. Wagner. 1998. The interaction of eIF4E with 4E-BP1 is an induced fit to a completely disordered protein. *Protein Sci.* 7: 1639–42.

Flory, P. J. 1969. *Statistical Mechanics of Chain Molecules*. New York: Wiley.

Fontana, A., G. Fassina, C. Vita, D. Dalzoppo, M. Zamai, and M. Zambonin. 1986. Correlation between sites of limited proteolysis and segmental mobility in thermolysin. *Biochemistry* 25: 1847–51.

Fontana, A., P. Polverino De Laureto, V. De Filippis, E. Scaramella, and M. Zambonin. 1997a. Probing the partly folded states of proteins by limited proteolysis. *Fold. Des.* 2: R17–26.

Fontana, A., M. Zambonin, P. Polverino De Laureto, V. De Filippis, A. Clementi, and E. Scaramella. 1997b. Probing the conformational state of apomyoglobin by limited proteolysis. *J. Mol. Biol.* 266: 223–30.

Fontes, M. R., T. Teh, and B. Kobe. 2000. Structural basis of recognition of monopartite and bipartite nuclear localization sequences by mammalian importin-alpha. *J. Mol. Biol.* 297: 1183–94.

Fontes, M. R., T. Teh, G. Toth, et al. 2003. Role of flanking sequences and phosphorylation in the recognition of the simian-virus-40 large T-antigen nuclear localization sequences by importin-alpha. *Biochem J.* 375: 339–49.

Foray, N., D. Marot, A. Gabriel, et al. 2003. A subset of ATM- and ATR-dependent phosphorylation events requires the BRCA1 protein. *EMBO J.* 22: 2860–71.

Forno, L. S. 1996. Neuropathology of Parkinson's disease. *J. Neuropathol. Exp. Neurol.* 55: 259–72.

Forster, T. 1948. Intermolecular energy migration and fluorescence. *Ann. Phys. (Leipzig)* 2: 55–75.

Fowler, D. M., A. V. Koulov, W. E. Balch, and J. W. Kelly. 2007. Functional amyloid-from bacteria to humans. *Trends Biochem. Sci.* 32: 217–24.

Frebel, K. and S. Wiese. 2006. Signalling molecules essential for neuronal survival and differentiation. *Biochem. Soc. Trans.* 34: 1287–90.

Frey, S., R. P. Richter, and D. Gorlich. 2006. FG-rich repeats of nuclear pore proteins form a three-dimensional meshwork with hydrogel-like properties. *Science* 314: 815–7.

Frieden, C., K. Chattopadhyay, and E. L. Elson. 2002. What fluorescence correlation spectroscopy can tell us about unfolded proteins. *Adv. Protein Chem.* 62: 91–109.

Futreal, P. A., L. Coin, M. Marshall, et al. 2004. A census of human cancer genes. *Nat. Rev. Cancer.* 4: 177–83.

Fuxreiter, M., I. Simon, P. Friedrich, and P. Tompa. 2004. Preformed structural elements feature in partner recognition by intrinsically unstructured proteins. *J. Mol. Biol.* 338: 1015–26.

Fuxreiter, M., P. Tompa, and I. Simon. 2007. Structural disorder imparts plasticity on linear motifs. *Bioinformatics* 23: 950–6.

Gabus, C., E. Derrington, P. Leblanc, et al. 2001. The prion protein has RNA binding and chaperoning properties characteristic of nucleocapsid protein NCP7 of HIV-1. *J. Biol. Chem.* 276: 19301–9.

Gabus, C., R. Mazroui, S. Tremblay, E. W. Khandjian, and J. L. Darlix. 2004. The fragile X mental retardation protein has nucleic acid chaperone properties. *Nucleic. Acids Res.* 32: 2129–37.

Gajdusek, D. C., C. J. Gibbs Jr., D. M. Asher, and E. David. 1968. Transmission of experimental kuru to the spider monkey (*Ateles geoffreyi*). *Science* 162: 693–4.

Galea, C. A., A. Nourse, Y. Wang, S. G. Sivakolundu, W. T. Heller, and R. W. Kriwacki. 2008a. Role of intrinsic flexibility in signal transduction mediated by the cell cycle regulator, p27(Kip1). *J. Mol. Biol.* 376: 827–38.

Galea, C. A., V. R. Pagala, J. C. Obenauer, C. G. Park, C. A. Slaughter, and R. W. Kriwacki. 2006. Proteomic studies of the intrinsically unstructured mammalian proteome. *J. Proteome. Res.* 5: 2839–48.

Galea, C. A., Y. Wang, S. G. Sivakolundu, and R. W. Kriwacki. 2008b. Regulation of cell division by intrinsically unstructured proteins: intrinsic flexibility, modularity, and signaling conduits. *Biochemistry* 47: 7598–609.

Galzitskaya, O. V., S. O. Garbuzynskiy, and M. Y. Lobanov. 2006. FoldUnfold: web server for the prediction of disordered regions in protein chain. *Bioinformatics* 22: 2948–9.

Ganesh, O. K., T. B. Green, A. S. Edison, and S. J. Hagen. 2006. Characterizing the residue level folding of the intrinsically unstructured IA3. *Biochemistry* 45: 13585–96.

Garay-Arroyo, A., J. M. Colmenero-Flores, A. Garciarrubio, and A. A. Covarrubias. 2000. Highly hydrophilic proteins in prokaryotes and eukaryotes are common during conditions of water deficit. *J. Biol. Chem.* 275: 5668–74.

Garbuzynskiy, S. O., M. Y. Lobanov, and O. V. Galzitskaya. 2004. To be folded or to be unfolded? *Protein Sci.* 13: 2871–7.

Garner, E., P. Cannon, P. Romero, Z. Obradovic, and A. K. Dunker. 1998. Predicting Disordered Regions from Amino Acid Sequence: Common Themes Despite Differing Structural Characterization. *Genome Inform. Ser. Workshop Genome Inform.* 9: 201–13.

Garner, E., P. Romero, A. K. Dunker, C. Brown, and Z. Obradovic. 1999. Predicting Binding Regions within Disordered Proteins. *Genome Inform. Ser. Workshop Genome Inform.* 10: 41–50.

Garrett, R. H. and C. M. Grisham. 2007. *Biochemistry*. Belmont, CA: Thomson Brooks/Cole.

Gast, K., H. Damaschun, K. Eckert, et al. 1995. Prothymosin alpha: a biologically active protein with random coil conformation. *Biochemistry* 34: 13211–8.

Gavin, A. C., P. Aloy, P. Grandi, et al. 2006. Proteome survey reveals modularity of the yeast cell machinery. *Nature* 440: 631–6.

Gavin, A. C., M. Bosche, R. Krause, et al. 2002. Functional organization of the yeast proteome by systematic analysis of protein complexes. *Nature* 415: 141–7.

George, R. A. and J. Heringa. 2002. An analysis of protein domain linkers: their classification and role in protein folding. *Protein Eng.* 15: 871–9.

Gerber, H. P., K. Seipel, O. Georgiev, et al. 1994. Transcriptional activation modulated by homopolymeric glutamine and proline stretches. *Science* 263: 808–11.

Ghiso, J., R. Vidal, A. Rostagno, et al. 2000. A newly formed amyloidogenic fragment due to a stop codon mutation causes familial British dementia. *Ann. NY Acad. Sci.* 903: 129–37.

Giaever, G., A. M. Chu, L. Ni, et al. 2002. Functional profiling of the Saccharomyces cerevisiae genome. *Nature* 418: 387–91.

Gianni, S., N. R. Guydosh, F. Khan, et al. 2003. Unifying features in protein-folding mechanisms. *Proc. Natl. Acad. Sci. USA* 100: 13286–91.

Gibbs, C. J. Jr., D. C. Gajdusek, D. M. Asher, et al. 1968. Creutzfeldt–Jakob disease (spongiform encephalopathy): transmission to the chimpanzee. *Science* 161: 388–9.

Gidalevitz, T., A. Ben-Zvi, K. H. Ho, H. R. Brignull, and R. I. Morimoto. 2006. Progressive disruption of cellular protein folding in models of polyglutamine diseases. *Science* 311: 1471–4.

Giesecke, H., J. C. Barale, G. Langsley, and A. W. Cornelissen. 1991. The C-terminal domain of RNA polymerase II of the malaria parasite *Plasmodium berghei. Biochem. Biophys. Res. Commun.* 180: 1350–5.

Gigant, B., P. A. Curmi, C. Martin-Barbey, et al. 2000. The 4 A X-ray structure of a tubulin:stathmin-like domain complex. *Cell* 102: 809–16.

Gill, G. and M. Ptashne. 1987. Mutants of GAL4 protein altered in an activation function. *Cell* 51: 121–6.

Gillespie, J. R. and D. Shortle. 1997. Characterization of long-range structure in the denatured state of staphylococcal nuclease. II. Distance restraints from paramagnetic relaxation and calculation of an ensemble of structures. *J. Mol. Biol.* 268: 170–84.

Girdwood, D. and E. A. T. B. Specified. 2003. p300 transcriptional repression is mediated by SUMO modification. *Cell* 11: 1043–54.

Goldberg, M. E., G. V. Semisotnov, B. Friguet, K. Kuwajima, O. B. Ptitsyn, and S. Sugai. 1990. An early immunoreactive folding intermediate of the tryptophan synthease beta 2 subunit is a "molten globule." *FEBS Lett.* 263: 51–6.

Goldfarb, L. G., P. Brown, W. R. Mccombie, et al. 1991. Transmissible familial Creutzfeldt–Jakob disease associated with five, seven, and eight extra octapeptide coding repeats in the PRNP gene. *Proc. Natl. Acad. Sci. USA* 88: 10926–30.

Goldgur, Y., S. Rom, R. Ghirlando, et al. 2007. Desiccation and zinc binding induce transition of tomato abscisic acid stress ripening 1, a water stress- and salt stress-regulated plant-specific protein, from unfolded to folded state. *Plant Physiol.* 143: 617–28.

Gooding, J. M., K. L. Yap, and M. Ikura. 2004. The cadherin–catenin complex as a focal point of cell adhesion and signalling: new insights from three-dimensional structures. *Bioessays* 26: 497–511.

Goodman, R. H. and S. Smolik. 2000. CBP/p300 in cell growth, transformation, and development. *Genes Dev.* 14: 1553–77.

Gorovits, B. M. and P. M. Horowitz. 1995. The molecular chaperonin cpn60 displays local flexibility that is reduced after binding with an unfolded protein. *J. Biol. Chem.* 270: 13057–62.

Goyal, K., L. Tisi, A. Basran, et al. 2003. Transition from Natively unfolded to folded state induced by desiccation in an anhydrobiotic nematode protein. *J. Biol. Chem.* 278: 12977–84.

Goyal, K., L. J. Walton, and A. Tunnacliffe. 2005. LEA proteins prevent protein aggregation due to water stress. *Biochem J.* 388: 151–7.

Graciet, E., P. Gans, N. Wedel, S. Lebreton, J. M. Camadro, and B. Gontero. 2003. The small protein CP12: a protein linker for supramolecular complex assembly. *Biochemistry* 42: 8163–70.

Graham, T. A., D. M. Ferkey, F. Mao, D. Kimelman, and W. Xu. 2001. Tcf4 can specifically recognize beta-catenin using alternative conformations. *Nat. Struct. Biol.* 8: 1048–52.

Graham, T. A., C. Weaver, F. Mao, D. Kimelman, and W. Xu. 2000. Crystal structure of a beta-catenin/Tcf complex. *Cell* 103: 885–96.

Granzier, H. and S. Labeit. 2002. Cardiac titin: an adjustable multi-functional spring. *J. Physiol.* 541: 335–42.

Greaser, M. 2001. Identification of new repeating motifs in titin. *Proteins* 43: 145–9.

Green, T. B., O. Ganesh, K. Perry, et al. 2004. IA3, an aspartic proteinase inhibitor from *Saccharomyces cerevisiae*, is intrinsically unstructured in solution. *Biochemistry* 43: 4071–81.

Greenbaum, E. A., C. L. Graves, A. J. Mishizen-Eberz, et al. 2005. The E46K mutation in alpha-synuclein increases amyloid fibril formation. *J. Biol. Chem.* 280: 7800–7.

Greenblatt, J. and J. Li. 1982. Properties of the N gene transcription antitermination protein of bacteriophage lambda. *J. Biol. Chem.* 257: 362–5.

Greene, L. H., R. Wijesinha-Bettoni, and C. Redfield. 2006. Characterization of the molten globule of human serum retinol-binding protein using NMR spectroscopy. *Biochemistry* 45: 9475–84.

Grimmler, M., Y. Wang, T. Mund, et al. 2007. Cdk-inhibitory activity and stability of p27Kip1 are directly regulated by oncogenic tyrosine kinases. *Cell* 128: 269–80.

Grosschedl, R., K. Giese, and J. Pagel. 1994. HMG domain proteins: architectural elements in the assembly of nucleoprotein structures. *Trends Genet.* 10: 94–100.

Grossman, S. R., M. Perez, A. L. Kung, et al. 1998. p300/MDM2 complexes participate in MDM2-mediated p53 degradation. *Mol. Cell* 2: 405–15.

Group, The Huntington's Disease Collaborative Research. 1993. A novel gene containing a trinucleotide repeat that is expanded and unstable on Huntington's disease chromosomes. *Cell* 72: 971–83.

Gu, W., M. Kofler, I. Antes, C. Freund, and V. Helms. 2005. Alternative binding modes of proline-rich peptides binding to the GYF domain. *Biochemistry* 44: 6404–15.

Guijarro, J. I., M. Sunde, J. A. Jones, I. D. Campbell, and C. M. Dobson. 1998. Amyloid fibril formation by an SH3 domain. *Proc. Natl. Acad. Sci. USA* 95: 4224–8.

Gunasekaran, K., C. J. Tsai, S. Kumar, D. Zanuy, and R. Nussinov. 2003. Extended disordered proteins: targeting function with less scaffold. *Trends Biochem. Sci.* 28: 81–5.

Gunasekaran, K., C. J. Tsai, and R. Nussinov. 2004. Analysis of ordered and disordered protein complexes reveals structural features discriminating between stable and unstable monomers. *J. Mol. Biol.* 341: 1327–41.

Guo, J. T., J. W. Jaromczyk, and Y. Xu. 2007. Analysis of chameleon sequences and their implications in biological processes. *Proteins* 67: 548–58.

Gusev, N. B., J. Hajdu, and P. Friedrich. 1979. Motility of the N-terminal tail of phosphorylase b as revealed by cross-linking. *Biochem. Biophys. Res. Commun.* 90: 70–7.

Gutierrez-Cruz, G., A. H. Van Heerden, and K. Wang. 2001. Modular motif, structural folds and affinity profiles of the PEVK segment of human fetal skeletal muscle titin. *J. Biol. Chem.* 276: 7442–9.

Haarmann, C. S., D. Green, M. G. Casarotto, D. R. Laver, and A. F. Dulhunty. 2003. The random-coil "C" fragment of the dihydropyridine receptor II–III loop can activate or inhibit native skeletal ryanodine receptors. *Biochem J.* 372: 305–16.

Hackel, M., H. J. Hinz, and G. R. Hedwig. 1999. A new set of peptide-based group heat capacities for use in protein stability calculations. *J. Mol. Biol.* 291: 197–213.

Hackel, M., T. Konno, and H. Hinz. 2000. A new alternative method to quantify residual structure in "unfolded" proteins. *Biochim. Biophys. Acta.* 1479: 155–65.

Hagerman, A. E. and L. G. Butler. 1981. The specificity of proanthocyanidin–protein interactions. *J. Biol. Chem.* 256: 4494–7.

Hagestedt, T., B. Lichtenberg, H. Wille, E. M. Mandelkow, and E. Mandelkow. 1989. Tau protein becomes long and stiff upon phosphorylation: correlation between paracrystalline structure and degree of phosphorylation. *J. Cell Biol.* 109: 1643–51.

Hai, C. M. and Z. Gu. 2006. Caldesmon phosphorylation in actin cytoskeletal remodeling. *Eur. J. Cell Biol.* 85: 305–9.

Hajdu, J., V. Dombradi, G. Bot, and P. Friedrich. 1979. Structural changes in glycogen phosphorlase as revealed by cross-linking with bifunctional diimidates: phosphorylase b. *Biochemistry* 18: 4037–41.

Han, J. D., N. Bertin, T. Hao, et al. 2004. Evidence for dynamically organized modularity in the yeast protein–protein interaction network. *Nature* 430: 88–93.

Han, J. H., S. Batey, A. A. Nickson, S. A. Teichmann, and J. Clarke. 2007. The folding and evolution of multidomain proteins. *Nat. Rev. Mol. Cell Biol.* 8: 319–30.

Hanna, R. A., B. E. Garcia-Diaz, and P. L. Davies. 2007. Calpastatin simultaneously binds four calpains with different kinetic constants. *FEBS Lett.* 581: 2894–8.

Hansen, J. C. 2002. Conformational dynamics of the chromatin fiber in solution: determinants, mechanisms and functions. *Annu. Rev. Biophys. Biomol. Struct.* 31: 361–92.

Hansen, J. C., X. Lu, E. D. Ross, and R. W. Woody. 2006. Intrinsic protein disorder, amino Acid composition, and histone terminal domains. *J. Biol. Chem.* 281: 1853–6.

Hansen, J. C., C. Tse, and A. P. Wolffe. 1998. Structure and function of the core histone N-termini: more than meets the eye. *Biochemistry* 37: 17637–41.

Harauz, G., N. Ishiyama, C. M. Hill, I. R. Bates, D. S. Libich, and C. Fares. 2004. Myelin basic protein-diverse conformational states of an intrinsically unstructured protein and its roles in myelin assembly and multiple sclerosis. *Micron* 35: 503–42.

Hardy, J., M. R. Cookson, and A. Singleton. 2003. Genes and parkinsonism. *Lancet Neurol* 2: 221–8.

Haritos, A. A., P. P. Yialouris, E. P. Heimer, A. M. Felix, E. Hannappel, and M. A. Rosemeyer. 1989. Evidence for the monomeric nature of thymosins. *FEBS Lett.* 244: 287–90.

Hartlepp, K. F., C. Fernandez-Tornero, A. Eberharter, T. Grune, C. W. Muller, and P. B. Becker. 2005. The histone fold subunits of *Drosophila* CHRAC facilitate nucleosome sliding through dynamic DNA interactions. *Mol. Cell Biol.* 25: 9886–96.

Hashimoto, M., T. Ichimura, H. Mizoguchi, et al. 2005. Cell size and nucleoid organization of engineered *Escherichia coli* cells with a reduced genome. *Mol. Microbiol.* 55: 137–49.

Hauer, J. A., P. Barthe, S. S. Taylor, J. Parello, and A. Padilla. 1999a. Two well-defined motifs in the cAMP-dependent protein kinase inhibitor (PKI alpha) correlate with inhibitory and nuclear export function. *Protein Sci.* 8: 545–53.

Hauer, J. A., S. S. Taylor, and D. A. Johnson. 1999b. Binding-dependent disorder–order transition in PKI alpha: a fluorescence anisotropy study. *Biochemistry* 38: 6774–80.

Hayashi, K., K. Kanda, F. Kimizuka, I. Kato, and K. Sobue. 1989. Primary structure and functional expression of h-caldesmon complementary DNA. *Biochem. Biophys. Res. Commun.* 164: 503–11.

Haynes, C. and L. M. Iakoucheva. 2006. Serine/arginine-rich splicing factors belong to a class of intrinsically disordered proteins. *Nucleic. Acids Res.* 34: 305–12.

Haynes, C., C. J. Oldfield, F. Ji, et al. 2006. Intrinsic disorder is a common feature of hub proteins from four eukaryotic interactomes. *PLoS Comput. Biol.* 2: e100.

He, Z., A. K. Dunker, C. R. Wesson, and W. R. Trumble. 1993. Ca(2+)-induced folding and aggregation of skeletal muscle sarcoplasmic reticulum calsequestrin. The involvement of the trifluoperazine-binding site. *J. Biol. Chem.* 268: 24635–41.

Heery, D. M., S. Hoare, S. Hussain, M. G. Parker, and H. Sheppard. 2001. Core LXXLL motif sequences in CREB-binding protein, SRC1, and RIP140 define affinity and selectivity for steroid and retinoid receptors. *J. Biol. Chem.* 276: 6695–702.

Hegyi, H., L. Buday, and P. Tompa. 2009. Intrinsic structural disorder confers cellular viability on oncogenic fusion proteins. *PLoS Comput. Biol.*, in press

Hegyi, H., E. Schad, and P. Tompa. 2007. Structural disorder promotes assembly of protein complexes. *BMC Struct. Biol.* 7: 65.

Hemmings, H. C. Jr., A. C. Nairn, D. W. Aswad, and P. Greengard. 1984. DARPP-32, a dopamine- and adenosine 3′:5′-monophosphate-regulated phosphoprotein enriched in dopamine-innervated brain regions. II. Purification and characterization of the phosphoprotein from bovine caudate nucleus. *J. Neurosci.* 4: 99–110.

Hemmings, H. C. Jr., A. C. Nairn, J. I. Elliott, and P. Greengard. 1990. Synthetic peptide analogs of DARPP-32 (Mr 32,000 dopamine- and cAMP-regulated phosphoprotein), an inhibitor of protein phosphatase-1. Phosphorylation, dephosphorylation, and inhibitory activity. *J. Biol. Chem.* 265: 20369–76.

Heringa, J. 1998. Detection of internal repeats: how common are they? *Curr. Opin. Struct. Biol.* 8: 338–45.

Hernandez, M. A., J. Avila, and J. M. Andreu. 1986. Physicochemical characterization of the heat-stable microtubule-associated protein MAP2. *Eur. J. Biochem.* 154: 41–8.

Herschlag, D. 1995. RNA chaperones and the RNA folding problem. *J. Biol. Chem.* 270: 20871–4.

Hershey, P. E., S. M. Mcwhirter, J. D. Gross, G. Wagner, T. Alber, and A. B. Sachs. 1999. The Cap-binding protein eIF4E promotes folding of a functional domain of yeast translation initiation factor eIF4G1. *J. Biol. Chem.* 274: 21297–304.

Hershko, A. and A. Ciechanover. 1998. The ubiquitin system. *Annu. Rev. Biochem.* 67: 425–79.

Hertzog, M., C. Van Heijenoort, D. Didry, et al. 2004. The beta-thymosin/WH2 domain; structural basis for the switch from inhibition to promotion of actin assembly. *Cell* 117: 611–23.

Hess, J. L. 2004. MLL: a histone methyltransferase disrupted in leukemia. *Trends Mol. Med.* 10: 500–7.

Hess, S. T., S. Huang, A. A. Heikal, and W. W. Webb. 2002. Biological and chemical applications of fluorescence correlation spectroscopy: a review. *Biochemistry* 41: 697–705.

Heyen, B. J., M. K. Alsheikh, E. A. Smith, C. F. Torvik, D. F. Seals, and S. K. Randall. 2002. The calcium-binding activity of a vacuole-associated, dehydrin-like protein is regulated by phosphorylation. *Plant Physiol.* 130: 675–87.

Hill, C. M., I. R. Bates, G. F. White, F. R. Hallett, and G. Harauz. 2002. Effects of the osmolyte trimethylamine-N-oxide on conformation, self-association, and two-dimensional crystallization of myelin basic protein. *J. Struct. Biol.* 139: 13–26.

Hilser, V. J. and E. B. Thompson. 2007. Intrinsic disorder as a mechanism to optimize allosteric coupling in proteins. *Proc. Natl. Acad. Sci. USA* 104: 8311–5.

Himmler, A. 1989. Structure of the bovine tau gene: alternatively spliced transcripts generate a protein family. *Mol. Cell Biol.* 9: 1389–96.

Hirano, T., S. I. Funahashi, T. Uemura, and M. Yanagida. 1986. Isolation and characterization of Schizosaccharomyces pombe cutmutants that block nuclear division but not cytokinesis. *EMBO J.* 5: 2973–79.

Hiroaki, H., T. Ago, T. Ito, H. Sumimoto, and D. Kohda. 2001. Solution structure of the PX domain, a target of the SH3 domain. *Nat. Struct. Biol.* 8: 526–30.

Hollenbeck, J. J., D. L. McClain, and M. G. Oakley. 2002. The role of helix stabilizing residues in GCN4 basic region folding and DNA binding. *Protein Sci.* 11: 2740–7.

Holt, C. and L. Sawyer. 1993. Caseins as rheomorphic proteins: interpretation of primary and secondary structures of the alpha(s1)-, beta- and kappa-caseins. *J. Chem. Soc. Faraday Trans.* 89: 2683–92.

Holt, C., N. M. Wahlgren, and T. Drakenberg. 1996. Ability of a beta-casein phosphopeptide to modulate the precipitation of calcium phosphate by forming amorphous dicalcium phosphate nanoclusters. *Biochem J.* 314: 1035–9.

Honnappa, S., W. Jahnke, J. Seelig, and M. O. Steinmetz. 2006. Control of intrinsically disordered stathmin by multisite phosphorylation. *J. Biol. Chem.* 281: 16078–83.

Hope, I. A., S. Mahadevan, and K. Struhl. 1988. Structural and functional characterization of the short acidic transcriptional activation region of yeast GCN4 protein. *Nature* 333: 635–40.

Hornig, N. C., P. P. Knowles, N. Q. Mcdonald, and F. Uhlmann. 2002. The dual mechanism of separase regulation by securin. *Curr. Biol.* 12: 973–82.

Hoshi, T., W. N. Zagotta, and R. W. Aldrich. 1990. Biophysical and molecular mechanisms of Shaker potassium channel inactivation. *Science* 250: 533–8.

Hotopp, J. C., M. E. Clark, D. C. Oliveira, et al. 2007. Widespread lateral gene transfer from intracellular bacteria to multicellular eukaryotes. *Science* 317: 1753–6.

House-Pompeo, K., Y. Xu, D. Joh, P. Speziale, and M. Hook. 1996. Conformational changes in the fibronectin binding MSCRAMMs are induced by ligand binding. *J. Biol. Chem.* 271: 1379–84.

Howard, M. B., N. A. Ekborg, L. E. Taylor, S. W. Hutcheson, and R. M. Weiner. 2004. Identification and analysis of polyserine linker domains in prokaryotic proteins with emphasis on the marine bacterium *Microbulbifer degradans*. *Protein Sci.* 13: 1422–5.

Hoyt, M. A., J. Zich, J. Takeuchi, M. Zhang, C. Govaerts, and P. Coffino. 2006. Glycine-alanine repeats impair proper substrate unfolding by the proteasome. *EMBO J.* 25: 1720–9.

Hua, Q. X., W. H. Jia, B. P. Bullock, J. F. Habener, and M. A. Weiss. 1998. Transcriptional activator–coactivator recognition: nascent folding of a kinase-inducible transactivation domain predicts its structure on coactivator binding. *Biochemistry* 37: 5858–66.

Huang, H. D., J. T. Horng, F. M. Lin, Y. C. Chang, and C. C. Huang. 2005. SpliceInfo: an information repository for mRNA alternative splicing in human genome. *Nucleic. Acids Res.* 33: D80–5.

Hubbard, S. J., R. J. Beynon, and J. M. Thornton. 1998. Assessment of conformational parameters as predictors of limited proteolytic sites in native protein structures. *Protein Eng.* 11: 349–59.

Hubbard, S. J., F. Eisenmenger, and J. M. Thornton. 1994. Modeling studies of the change in conformation required for cleavage of limited proteolytic sites. *Protein Sci.* 3: 757–68.

Hubbell, W. L., C. Altenbach, C. M. Hubbell, and H. G. Khorana. 2003. Rhodopsin structure, dynamics, and activation: a perspective from crystallography, site-directed spin labeling, sulfhydryl reactivity, and disulfide cross-linking. *Adv. Protein Chem.* 63: 243–90.

Huber, A. H., D. B. Stewart, D. V. Laurents, W. J. Nelson, and W. I. Weis. 2001. The cadherin cytoplasmic domain is unstructured in the absence of beta-catenin. A possible mechanism for regulating cadherin turnover. *J. Biol. Chem.* 276: 12301–9.

Huber, A. H. and W. I. Weis. 2001. The structure of the beta-catenin/E-cadherin complex and the molecular basis of diverse ligand recognition by beta-catenin. *Cell* 105: 391–402.

Hunter, T. 1987. A thousand and one protein kinases. *Cell* 50: 823–9.

Hurley, T. D., J. Yang, L. Zhang, et al. 2007. Structural basis for regulation of protein phosphatase 1 by inhibitor-2. *J. Biol. Chem.* 282: 28874–83.

Hurst, L. D. 2002. The Ka/Ks ratio: diagnosing the form of sequence evolution. *Trends Genet.* 18: 486.

Huth, J. R., C. A. Bewley, M. S. Nissen, et al. 1997. The solution structure of an HMG-I(Y)-DNA complex defines a new architectural minor groove binding motif. *Nat. Struct. Biol.* 4: 657–65.

Iakoucheva, L., C. Brown, J. Lawson, Z. Obradovic, and A. Dunker. 2002. Intrinsic disorder in cell-signaling and cancer-associated proteins. *J. Mol. Biol.* 323: 573–84.

Iakoucheva, L. M., A. L. Kimzey, C. D. Masselon, et al. 2001a. Identification of intrinsic order and disorder in the DNA repair protein XPA. *Protein Sci.* 10: 560–71.

Iakoucheva, L. M., A. L. Kimzey, C. D. Masselon, R. D. Smith, A. K. Dunker, and E. J. Ackerman. 2001b. Aberrant mobility phenomena of the DNA repair protein XPA. *Protein Sci.* 10: 1353–62.

Iakoucheva, L. M., P. Radivojac, C. J. Brown, et al. 2004. The importance of intrinsic disorder for protein phosphorylation. *Nucleic. Acids Res.* 32: 1037–49.

Ikemoto, N., B. Antoniu, J. J. Kang, L. G. Meszaros, and M. Ronjat. 1991. Intravesicular calcium transient during calcium release from sarcoplasmic reticulum. *Biochemistry* 30: 5230–7.

Ikura, M. and J. B. Ames. 2006. Genetic polymorphism and protein conformational plasticity in the calmodulin superfamily: two ways to promote multifunctionality. *Proc. Natl. Acad. Sci. USA* 103: 1159–64.

Imarisio, S., J. Carmichael, V. Korolchuk, et al. 2008. Huntington's disease: from pathology and genetics to potential therapies. *Biochem J.* 412: 191–209.

Iqbal, K. and I. Grundke-Iqbal. 2008. Alzheimer neurofibrillary degeneration: significance, etiopathogenesis, therapeutics and prevention. *J. Cell. Mol. Med.* 12: 38–55.

Irar, S., E. Oliveira, M. Pages, and A. Goday. 2006. Towards the identification of late-embryogenic-abundant phosphoproteome in *Arabidopsis* by 2-DE and MS. *Proteomics* 6 Suppl 1: S175–85.

Irobi, E., A. H. Aguda, M. Larsson, et al. 2004. Structural basis of actin sequestration by thymosin-beta4: implications for WH2 proteins. *EMBO J.* 23: 3599–608.

Ishida, T. and K. Kinoshita. 2007. PrDOS: prediction of disordered protein regions from amino acid sequence. *Nucleic. Acids Res.* 35: W460–4.

Ishida, T. and K. Kinoshita. 2008. Prediction of disordered regions in proteins based on the meta approach. *Bioinformatics* 24: 1344–8.

Ivanyi-Nagy, R., L. Davidovic, E. W. Khandjian, and J. L. Darlix. 2005. Disordered RNA chaperone proteins: from functions to disease. *Cell. Mol. Life Sci.* 62: 1409–17.

Ivanyi-Nagy, R., J. P. Lavergne, C. Gabus, D. Ficheux, and J. L. Darlix. 2007. RNA chaperoning and intrinsic disorder in the core proteins of Flaviviridae. *Nucleic. Acids Res.* 36: 712–25.

Iwai, A., E. Masliah, M. Yoshimoto, et al. 1995. The precursor protein of non-A beta component of Alzheimer's disease amyloid is a presynaptic protein of the central nervous system. *Neuron* 14: 467–75.

Iwakuma, T. and G. Lozano. 2003. MDM2, an introduction. *Mol. Cancer Res.* 1: 993–1000.

Jackson, G. S., I. Murray, L. L. Hosszu, et al. 2001. Location and properties of metal-binding sites on the human prion protein. *Proc. Natl. Acad. Sci. USA* 98: 8531–5.

Jackson, S. E. and A. R. Fersht. 1991. Folding of chymotrypsin inhibitor 2. 1. Evidence for a two-state transition. *Biochemistry* 30: 10428–35.

Jallepalli, P. V., I. C. Waizenegger, F. Bunz, et al. 2001. Securin is required for chromosomal stability in human cells. *Cell* 105: 445–57.

James, L. C., P. Roversi, and D. S. Tawfik. 2003. Antibody multispecificity mediated by conformational diversity. *Science* 299: 1362–7.

James, L. C. and D. S. Tawfik. 2003. Conformational diversity and protein evolution—a 60-year-old hypothesis revisited. *Trends Biochem. Sci.* 28: 361–8.

Jarrett, J. T. and P. T. Lansbury Jr. 1993. Seeding "one-dimensional crystallization" of amyloid: a pathogenic mechanism in Alzheimer's disease and scrapie? *Cell* 73: 1055–8.

Jeffery, C. J. 1999. Moonlighting proteins. *Trends Biochem. Sci.* 24: 8–11.

Jeffery, C. J. 2003a. Moonlighting proteins: old proteins learning new tricks. *Trends Genet.* 19: 415–7.

Jeffery, C. J. 2003b. Multifunctional proteins: examples of gene sharing. *Ann. Med.* 35: 28–35.

Jeffery, C. J. 2004. Molecular mechanisms for multitasking: recent crystal structures of moonlighting proteins. *Curr. Opin. Struct. Biol.* 14: 663–8.

Jeganathan, S., M. Von Bergen, H. Brutlach, H. J. Steinhoff, and E. Mandelkow. 2006. Global hairpin folding of tau in solution. *Biochemistry* 45: 2283–93.

Jencks, W. P. 1981. On the attribution and additivity of binding energies. *Proc. Natl. Acad. Sci. USA* 78: 4046–50.

Jenco, J. M., A. Rawlingson, B. Daniels, and A. J. Morris. 1998. Regulation of phospholipase D2: selective inhibition of mammalian phospholipase D isoenzymes by alpha- and beta-synucleins. *Biochemistry* 37: 4901–9.

Jenuwein, T. and C. D. Allis. 2001. Translating the histone code. *Science* 293: 1074–80.

Jeong, H., S. P. Mason, A. L. Barabasi, and Z. N. Oltvai. 2001. Lethality and centrality in protein networks. *Nature* 411: 41–2.

Jin, J., G. J. Li, J. Davis, et al. 2007. Identification of novel proteins associated with both alpha-synuclein and DJ-1. *Mol. Cell. Proteomics.* 6: 845–59.

Jin, Y. and R. L. Dunbrack Jr. 2005. Assessment of disorder predictions in CASP6. *Proteins* 61 Suppl 7: 167–75.

Jin, Y., H. Lee, S. X. Zeng, M. S. Dai, and H. Lu. 2003. MDM2 promotes p21waf1/cip1 proteasomal turnover independently of ubiquitination. *EMBO J.* 22: 6365–77.

Joerger, A. C. and A. R. Fersht. 2008. Structural biology of the tumor suppressor p53. *Annu. Rev. Biochem.* 77: 557–82.

Johnson, S. A. and T. Hunter. 2005. Kinomics: methods for deciphering the kinome. *Nat Methods.* 2: 17–25.

Jones, D. T. and J. J. Ward. 2003. Prediction of disordered regions in proteins from position specific score matrices. *Proteins* 53 Suppl 6: 573–8.

Jones, J. A., D. K. Wilkins, L. J. Smith, and C. M. Dobson. 1997. Characterization of protein unfolding by NMR diffusion measurements. *J. Biomolec. NMR* 10: 199–203.

Jones, S. and J. M. Thornton. 1996. Principles of protein–protein interactions. *Proc. Natl. Acad. Sci. USA* 93: 13–20.

Kabsch, W. and C. Sander. 1983. Dictionary of protein secondary structure: pattern recognition of hydrogen-bonded and geometrical features. *Biopolymers* 22: 2577–637.

Kabsch, W. and C. Sander. 1984. On the use of sequence homologies to predict protein structure: identical pentapeptides can have completely different conformations. *Proc. Natl. Acad. Sci. USA* 81: 1075–8.

Kalthoff, C. 2003. A novel strategy for the purification of recombinantly expressed unstructured protein domains. *J. Chromatogr. B* 786: 247–54.

Kalthoff, C., J. Alves, C. Urbanke, R. Knorr, and E. J. Ungewickell. 2002. Unusual structural organization of the endocytic proteins AP180 and epsin 1. *J. Biol. Chem.* 277: 8209–16.

Karlin, D., F. Ferron, B. Canard, and S. Longhi. 2003. Structural disorder and modular organization in *Paramyxovirinae* N and P. *J. Gen. Virol.* 84: 3239–52.

Karlin, S., L. Brocchieri, A. Bergman, J. Mrazek, and A. J. Gentles. 2002. Amino acid runs in eukaryotic proteomes and disease associations. *Proc. Natl. Acad. Sci. USA* 99: 333–8.

Karlin, S. and C. Burge. 1996. Trinucleotide repeats and long homopeptides in genes and proteins associated with nervous system disease and development. *Proc. Natl. Acad. Sci. USA* 93: 1560–5.

Karush, F. 1950. Heterogeneity of the binding sites of bovine serum albumin. *J. Am. Chem. Soc.* 72: 2705–13.

Katsuno, M., H. Banno, K. Suzuki, et al. 2008. Molecular genetics and biomarkers of polyglutamine diseases. *Curr. Mol. Med.* 8: 221–34.

Kayed, R., J. Bernhagen, N. Greenfield, et al. 1999. Conformational transitions of islet amyloid polypeptide (IAPP) in amyloid formation in vitro. *J. Mol. Biol.* 287: 781–96.

Kellermayer, M. S., S. B. Smith, H. L. Granzier, and C. Bustamante. 1997. Folding–unfolding transitions in single titin molecules characterized with laser tweezers. *Science* 276: 1112–6.

Kelly, J. W. 1998. The alternative conformations of amyloidogenic proteins and their multi-step assembly pathways. *Curr. Opin. Struct. Biol.* 8: 101–6.

Kendrew, J. C., G. Bodo, H. M. Dintzis, R. G. Parrish, H. Wyckoff, and D. C. Phillips. 1958. A three-dimensional model of the myoglobin molecule obtained by X-ray analysis. *Nature* 181: 662–6.

Keskin, O., B. Ma, and R. Nussinov. 2005. Hot regions in protein—protein interactions: the organization and contribution of structurally conserved hot spot residues. *J. Mol. Biol.* 345: 1281–94.

Khan, A. N. and P. N. Lewis. 2005. Unstructured conformations are a substrate requirement for the Sir2 family of NAD-dependent protein deacetylases. *J. Biol. Chem.* 280: 36073–8.

Khaymina, S. S., J. M. Kenney, M. M. Schroeter, and J. M. Chalovich. 2007. Fesselin is a natively unfolded protein. *J. Proteome. Res.* 6: 3648–54.

Kieburtz, K., M. Macdonald, C. Shih, et al. 1994. Trinucleotide repeat length and progression of illness in Huntington's disease. *J. Med. Genet.* 31: 872–4.

Kim, A. S., L. T. Kakalis, N. Abdul-Manan, G. A. Liu, and M. K. Rosen. 2000a. Autoinhibition and activation mechanisms of the Wiskott–Aldrich syndrome protein. *Nature* 404: 151–8.

Kim, E., A. Magen, and G. Ast. 2007. Different levels of alternative splicing among eukaryotes. *Nucleic. Acids Res.* 35: 125–31.

Kim, P. S. and R. L. Baldwin. 1990. Intermediates in the folding reactions of small proteins. *Annu. Rev. Biochem.* 59: 631–60.

Kim, T. D., H. J. Ryu, H. I. Cho, C. H. Yang, and J. Kim. 2000b. Thermal behavior of proteins: heat-resistant proteins and their heat-induced secondary structural changes. *Biochemistry* 39: 14839–46.

King, R. W., M. Glotzer, and M. W. Kirschner. 1996. Mutagenic analysis of the destruction signal of mitotic cyclins and structural characterization of ubiquitinated intermediates. *Mol. Biol. Cell* 7: 1343–57.

Kirkitadze, M. D., M. M. Condron, and D. B. Teplow. 2001. Identification and characterization of key kinetic intermediates in amyloid beta-protein fibrillogenesis. *J. Mol. Biol.* 312: 1103–19.

Kiss, R., Z. Bozoky, D. Kovacs, et al. 2008a. Calcium-induced tripartite binding of intrinsically disordered calpastatin to its cognate enzyme, calpain. *FEBS Lett.* 582: 2149–54.

Kiss, R., D. Kovacs, P. Tompa, and A. Perczel. 2008b. Local structural preferences of calpastatin, the intrinsically unstructured protein inhibitor of calpain. *Biochemistry* 47: 6936–45.

Kissinger, C. R., H. E. Parge, D. R. Knighton, et al. 1995. Crystal structures of human calcineurin and the human FKBP12-FK506-calcineurin complex. *Nature* 378: 641–4.

Kitada, T., S. Asakawa, N. Hattori, et al. 1998. Mutations in the Parkin gene cause autosomal recessive juvenile parkinsonism. *Nature* 392: 605–8.

Klein, C. and L. T. Vassilev. 2004. Targeting the p53–MDM2 interaction to treat cancer. *Br. J. Cancer* 91: 1415–9.

Kleinschmidt, J. A., C. Dingwall, G. Maier, and W. W. Franke. 1986. Molecular characterization of a karyophilic, histone-binding protein: cDNA cloning, amino acid sequence and expression of nuclear protein N1/N2 of *Xenopus laevis*. *EMBO J.* 5: 3547–52.

Koag, M. C., R. D. Fenton, S. Wilkens, and T. J. Close. 2003. The binding of maize DHN1 to lipid vesicles. Gain of structure and lipid specificity. *Plant Physiol.* 131: 309–16.

Kohn, J. E., I. S. Millett, J. Jacob, et al. 2004. Random-coil behavior and the dimensions of chemically unfolded proteins. *Proc. Natl. Acad. Sci. USA* 101: 12491–6.

Konno, T., N. Tanaka, M. Kataoka, E. Takano, and M. Maki. 1997. A circular dichroism study of preferential hydration and alcohol effects on a denatured protein, pig calpastatin domain I. *Biochim. Biophys. Acta.* 1342: 73–82.

Kornberg, R. D. 2005. Mediator and the mechanism of transcriptional activation. *Trends Biochem. Sci.* 30: 235–9.

Koshland, D. E. 1958. Application of a theory of enzyme specificity to protein synthesis. *Proc. Natl. Acad. Sci. USA* 44: 98–104.

Kovacs, D., E. Kalmar, Z. Torok, and P. Tompa. 2008. Chaperone activity of ERD10 and ERD14, two disordered stress-related plant proteins. *Plant Physiol.* 147: 381–90.

Kovacs, D., M. Rakacs, B. Agoston, et al. 2009. Janus chaperones: assistance of both RNA- and protein-folding by ribosomal proteins. *FEBS Lett.* 583: 88–92.

Kovacs, G. G., L. Laszlo, J. Kovacs, et al. 2004. Natively unfolded tubulin polymerization promoting protein TPPP/p25 is a common marker of alpha-synucleinopathies. *Neurobiol. D.* 17: 155–62.

Kreil, D. P. and G. Kreil. 2000. Asparagine repeats are rare in mammalian proteins. *Trends Biochem. Sci.* 25: 270–1.

Kreimer, D. I., R. Szosenfogel, D. Goldfarb, I. Silman, and L. Weiner. 1994. Two-state transition between molten globule and unfolded states of acetylcholinesterase as monitored by electron paramagnetic resonance spectroscopy. *Proc. Natl. Acad. Sci. USA* 91: 12145–9.

Kretsinger, R. H. and C. E. Nockolds. 1973. Carp muscle calcium-binding protein. II. Structure determination and general description. *J. Biol. Chem.* 248: 3313–26.

Krimm, S. and J. Bandekar. 1986. Vibrational spectroscopy and conformation of peptides, polypeptides, and proteins. *Adv. Protein Chem.* 38: 181–364.

Kriwacki, R. W., L. Hengst, L. Tennant, S. I. Reed, and P. E. Wright. 1996. Structural studies of p21Waf1/Cip1/Sdi1 in the free and Cdk2-bound state: conformational disorder mediates binding diversity. *Proc. Natl. Acad. Sci. USA* 93: 11504–9.

Kriwacki, R. W., J. Wu, L. Tennant, P. E. Wright, and G. Siuzdak. 1997. Probing protein structure using biochemical and biophysical methods. Proteolysis, matrix-assisted laser desorption/ionization mass spectrometry, high-performance liquid chromatography and size-exclusion chromatography of p21Waf1/Cip1/Sdi1. *J. Chromatogr. A* 777: 23–30.

Kumar, N., S. Shukla, S. Kumar, et al. 2008. Intrinsically disordered protein from a pathogenic mesophile *Mycobacterium* tuberculosis adopts structured conformation at high temperature. *Proteins 71: 1123–33*

Kumar, R., R. Betney, J. Li, E. B. Thompson, and I. J. Mcewan. 2004. Induced alpha-helix structure in AF1 of the androgen receptor upon binding transcription factor TFIIF. *Biochemistry* 43: 3008–13.

Kumar, R., S. R. Pavithra, and U. Tatu. 2007. Three-dimensional structure of heat shock protein 90 from *Plasmodium falciparum*: molecular modelling approach to rational drug design against malaria. *J. Biosci.* 32: 531–6.

Kurdistani, S. K. and M. Grunstein. 2003. Histone acetylation and deacetylation in yeast. *Nat Rev Mol. Cell Biol.* 4: 276–84.

Kussie, P. H., S. Gorina, V. Marechal, et al. 1996. Structure of the MDM2 oncoprotein bound to the p53 tumor suppressor transactivation domain. *Science* 274: 948–53.

Kyte, J. and R. F. Doolittle. 1982. A simple method for displaying the hydropathic character of a protein. *J. Mol. Biol.* 157: 105–32.

Labeit, S. and B. Kolmerer. 1995. Titins: giant proteins in charge of muscle ultrastructure and elasticity. *Science* 270: 293–6.

Lacy, E. R., I. Filippov, W. S. Lewis, et al. 2004. p27 binds cyclin-CDK complexes through a sequential mechanism involving binding-induced protein folding. *Nat. Struct. Mol. Biol.* 11: 358–64.

Laemmli, U. K. 1970. Cleavage of structural proteins during the assembly of the head of bacteriophage T4. *Nature* 227: 680–5.

Lagerstrom, M. C. and H. B. Schioth. 2008. Structural diversity of G protein-coupled receptors and significance for drug discovery. *Nat. Rev. Drug Discov.* 7: 339–57.

Lakowicz, J. R. 2006. *Principles of Fluorescence Spectroscopy*, 3rd ed. New York: Springer.

Lane, D. P. 1992. Cancer. p53, guardian of the genome. *Nature* 358: 15–6.

Langowski, J., W. Kremer, and U. Kapp. 1992. Dynamic light scattering for study of solution conformation and dynamics of superhelical DNA. *Methods Enzymol.* 211: 430–48.

Lariviere, L., S. Geiger, S. Hoeppner, S. Rother, K. Strasser, and P. Cramer. 2006. Structure and TBP binding of the mediator head subcomplex Med8-Med18-Med20. *Nat. Struct. Mol. Biol.* 13: 895–901.

Lashuel, H. A., C. Wurth, L. Woo, and J. W. Kelly. 1999. The most pathogenic transthyretin variant, L55P, forms amyloid fibrils under acidic conditions and protofilaments under physiological conditions. *Biochemistry* 38: 13560–73.

Le Gall, T., P. R. Romero, M. S. Cortese, V. N. Uversky, and A. K. Dunker. 2007. Intrinsic disorder in the protein data bank. *J. Biomol. Struct. Dyn.* 24: 325–42.

Lebowitz, J., M. S. Lewis, and P. Schuck. 2002. Modern analytical ultracentrifugation in protein science: a tutorial review. *Protein Sci.* 11: 2067–79.

Lee, A. S., C. Galea, E. L. Digiammarino, et al. 2003. Reversible amyloid formation by the p53 tetramerization domain and a cancer-associated mutant. *J. Mol. Biol.* 327: 699–709.

Lee, H., K. H. Mok, R. Muhandiram, et al. 2000. Local structural elements in the mostly unstructured transcriptional activation domain of human p53. *J. Biol. Chem.* 275: 29426–32.

Lee, H. J., C. Choi, and S. J. Lee. 2002. Membrane-bound alpha-synuclein has a high aggregation propensity and the ability to seed the aggregation of the cytosolic form. *J. Biol. Chem.* 277: 671–8.

Lee, L., E. Stollar, J. Chang, et al. 2001. Expression of the Oct-1 transcription factor and characterization of its interactions with the Bob1 coactivator. *Biochemistry* 40: 6580–8.

Lee, M. K. and D. W. Cleveland. 1996. Neuronal intermediate filaments. *Annu. Rev. Neurosci.* 19: 187–217.

Legault, P., J. Li, J. Mogridge, L. E. Kay, and J. Greenblatt. 1998. NMR structure of the bacteriophage lambda N peptide/boxB RNA complex: recognition of a GNRA fold by an arginine-rich motif. *Cell* 93: 289–99.

Legname, G., I. V. Baskakov, H. O. Nguyen, et al. 2004. Synthetic mammalian prions. *Science* 305: 673–6.

Legname, G., S. J. Dearmond, F. Cohen, and S. B. Prusiner. 2007. Pathogenesis of prion diseases. In *Protein Misfolding, Aggregation and Conformational Diseases*. New York: Springer.

Lehn, D. A., T. S. Elton, K. R. Johnson, and R. Reeves. 1988. A conformational study of the sequence specific binding of HMG-I (Y) with the bovine interleukin-2 cDNA. *Biochem Int.* 16: 963–71.

Leismann, O., A. Herzig, S. Heidmann, and C. F. Lehner. 2000. Degradation of *Drosophila* PIM regulates sister chromatid separation during mitosis. *Genes Dev.* 14: 2192–205.

Letunic, I., L. Goodstadt, N. J. Dickens, et al. 2002. Recent improvements to the SMART domain-based sequence annotation resource. *Nucleic. Acids Res.* 30: 242–4.

Levine, A. J. 1997. p53, the cellular gatekeeper for growth and division. *Cell* 88: 323–31.

Levinthal, C. 1969. How to Fold Graciously. In *Mossbauer Spectroscopy in Biological Systems* (eds DeBrunner, J. T. P and E. Munck) pp. 22–24.

Levy, Y., J. N. Onuchic, and P. G. Wolynes. 2007. Fly-casting in protein-DNA binding: frustration between protein folding and electrostatics facilitates target recognition. *J. Am. Chem. Soc.* 129: 738–9.

Li, H., A. F. Oberhauser, S. D. Redick, M. Carrion-Vazquez, H. P. Erickson, and J. M. Fernandez. 2001. Multiple conformations of PEVK proteins detected by single-molecule techniques. *Proc. Natl. Acad. Sci. USA* 98: 10682–6.

Li, L. and S. Lindquist. 2000. Creating a protein-based element of inheritance. *Science* 287: 661–4.

Li, L., V. N. Uversky, A. K. Dunker, and S. O. Meroueh. 2007. A computational investigation of allostery in the catabolite activator protein. *J. Am. Chem. Soc.* 129: 15668–76.

Li, M. and J. Song. 2007. The N- and C-termini of the human Nogo molecules are intrinsically unstructured: bioinformatics, CD, NMR characterization, and functional implications. *Proteins* 68: 100–8.

Li, X., Z. Obradovic, C. J. Brown, E. Garner, and A. K. Dunker. 2000. Comparing predictors of disordered protein. *Genome Inform. Ser. Workshop Genome Inform.* 11: 172–84.

Li, X., P. Romero, M. Rani, A. K. Dunker, and Z. Obradovic. 1999. Predicting protein disorder for N-, C-, and internal regions. *Genome Inform. Ser. Workshop Genome Inform.* 10: 30–40.

Liao, J., Y. Fu and K. Shuai. 2000. Distinct roles of the NH2- and COOH-terminal domains of the protein inhibitor of activated signal transducer and activator of transcription (STAT) 1 (PIAS1) in cytokine-induced PIAS1-Stat1 interaction. *Proc. Natl. Acad. Sci. USA* 97: 5267–72.

Libich, D. S. and G. Harauz. 2008. Backbone dynamics of the 18.5 kDa isoform of myelin basic protein reveals transient alpha-helices and a calmodulin-binding site. *Biophys J.* 94: 4847–66.

Licht, J. D. 2001. AML1 and the AML1-ETO fusion protein in the pathogenesis of t(8;21) AML. *Oncogene* 20: 5660–79.

Liebovitch, L. S., L. Y. Selector, and R. P. Kline. 1992. Statistical properties predicted by the ball and chain model of channel inactivation. *Biophys J.* 63: 1579–85.

Lieutaud, P., B. Canard, and S. Longhi. 2008. MeDor: a metaserver for predicting protein disorder. *BMC Genomics* 9 Suppl 2: S25.

Lim, R. Y., N. P. Huang, J. Koser, et al. 2006. Flexible phenylalanine-glycine nucleoporins as entropic barriers to nucleocytoplasmic transport. *Proc. Natl. Acad. Sci. USA* 103: 9512–7.

Linding, R., L. J. Jensen, F. Diella, P. Bork, T. J. Gibson, and R. B. Russell. 2003a. Protein disorder prediction: implications for structural proteomics. *Structure* 11: 1453–9.

Linding, R., R. B. Russell, V. Neduva, and T. J. Gibson. 2003b. GlobPlot: exploring protein sequences for globularity and disorder. *Nucleic. Acids Res.* 31: 3701–8.

Linding, R., J. Schymkowitz, F. Rousseau, F. Diella, and L. Serrano. 2004. A comparative study of the relationship between protein structure and beta-aggregation in globular and intrinsically disordered proteins. *J. Mol. Biol.* 342: 345–53.

Lindner, R. A., J. A. Carver, M. Ehrnsperger, et al. 2000. Mouse Hsp25, a small shock protein. The role of its C-terminal extension in oligomerization and chaperone action. *Eur. J. Biochem.* 267: 1923–32.

Lindner, R. A., A. Kapur, M. Mariani, S. J. Titmuss, and J. A. Carver. 1998. Structural alterations of alpha-crystallin during its chaperone action. *Eur. J. Biochem.* 258: 170–83.

Lindquist, S. 1997. Mad cows meet psi-chotic yeast: the expansion of the prion hypothesis. *Cell* 89: 495–8.

Lipari, G. and A. Szabo. 1982. Model-free approach to the interpretation of nuclear magnetic resonance relaxation in macromolecules 1. Theory and range of validity. *J. Am. Chem. Soc.* 104: 4546–59.

Lippens, G., A. Sillen, C. Smet, et al. 2006. Studying the natively unfolded neuronal tau protein by solution NMR spectroscopy. *Protein Pept. Lett.* 13: 235–46.

Lippens, G., J. M. Wieruszeski, A. Leroy, et al. 2004. Proline-directed random-coil chemical shift values as a tool for the NMR assignment of the tau phosphorylation sites. *Chembiochem* 5: 73–8.

Lise, S. and D. T. Jones. 2005. Sequence patterns associated with disordered regions in proteins. *Proteins* 58: 144–50.

Lisse, T., D. Bartels, H. R. Kalbitzer, and R. Jaenicke. 1996. The recombinant dehydrin-like desiccation stress protein from the resurrection plant *Craterostigma plantagineum* displays no defined three-dimensional structure in its native state. *Biol. Chem.* 377: 555–61.

Litingtung, Y., A. M. Lawler, S. M. Sebald, et al. 1999. Growth retardation and neonatal lethality in mice with a homozygous deletion in the C-terminal domain of RNA polymerase II. *Mol. Gen. Genet.* 261: 100–5.

Litvinovich, S. V., S. A. Brew, S. Aota, S. K. Akiyama, C. Haudenschild, and K. C. Ingham. 1998. Formation of amyloid-like fibrils by self-association of a partially unfolded fibronectin type III module. *J. Mol. Biol.* 280: 245–58.

Liu, C. W., M. J. Corboy, G. N. Demartino, and P. J. Thomas. 2003. Endoproteolytic activity of the proteasome. *Science* 299: 408–11.

Liu, J., N. B. Perumal, C. J. Oldfield, E. W. Su, V. N. Uversky, and A. K. Dunker. 2006a. Intrinsic disorder in transcription factors. *Biochemistry* 45: 6873–88.

Liu, J. and B. Rost. 2003. NORSp: Predictions of long regions without regular secondary structure. *Nucleic. Acids Res.* 31: 3833–5.

Liu, J., H. Tan, and B. Rost. 2002. Loopy proteins appear conserved in evolution. *J. Mol. Biol.* 322: 53–64.

Liu, J., Y. Xing, T. R. Hinds, J. Zheng, and W. Xu. 2006b. The third 20 amino acid repeat is the tightest binding site of APC for beta-catenin. *J. Mol. Biol.* 360: 133–44.

Lo Conte, L., C. Chothia, and J. Janin. 1999. The atomic structure of protein–protein recognition sites. *J. Mol. Biol.* 285: 2177–98.

Lobley, A., M. B. Swindells, C. A. Orengo, and D. T. Jones. 2007. Inferring function using patterns of native disorder in proteins. *PLoS Comput. Biol.* 3: e162.

Loftus, S. R., D. Walker, M. J. Mate, et al. 2006. Competitive recruitment of the periplasmic translocation portal TolB by a natively disordered domain of colicin E9. *Proc. Natl. Acad. Sci. USA* 103: 12353–8.

Lohrum, M. A., R. L. Ludwig, M. H. Kubbutat, M. Hanlon, and K. H. Vousden. 2003. Regulation of HDM2 activity by the ribosomal protein L11. *Cancer Cell* 3: 577 – 87.

Longhi, S., V. Receveur-Brechot, D. Karlin, et al. 2003. The C-terminal domain of the measles virus nucleoprotein is intrinsically disordered and folds upon binding to the C-terminal moiety of the phosphoprotein. *J. Biol. Chem.* 278: 18638–48.

Lopez-Garcia, F., R. Zahn, R. Riek, and K. Wuthrich. 2000. NMR structure of the bovine prion protein. *Proc. Natl. Acad. Sci. USA* 97: 8334–9.

Lorsch, J. R. 2002. RNA chaperones exist and DEAD box proteins get a life. *Cell* 109: 797–800.

Love, J. J., X. Li, J. Chung, H. J. Dyson, and P. E. Wright. 2004. The LEF-1 high-mobility group domain undergoes a disorder-to-order transition upon formation of a complex with cognate DNA. *Biochemistry* 43: 8725–34.

Lowry, D. F., A. C. Hausrath, and G. W. Daughdrill. 2008a. A robust approach for analyzing a heterogeneous structural ensemble. *Proteins* 73: 918–28.

Lowry, D. F., A. Stancik, R. M. Shrestha, and G. W. Daughdrill. 2008b. Modeling the accessible conformations of the intrinsically unstructured transactivation domain of p53. *Proteins* 71: 587–98.

Lu, X. and J. C. Hansen. 2004. Identification of specific functional subdomains within the linker histone H10 C-terminal domain. *J. Biol. Chem.* 279: 8701–7.

Lu, Y. and A. Bennick. 1998. Interaction of tannin with human salivary proline-rich proteins. *Arch. Oral Biol.* 43: 717–28.

Luger, K., A. W. Mader, R. K. Richmond, D. F. Sargent, and T. J. Richmond. 1997. Crystal structure of the nucleosome core particle at 2.8 A resolution. *Nature* 389: 251–60.

Luo, Y., J. Hurwitz, and J. Massague. 1995. Cell-cycle inhibition by independent CDK and PCNA binding domains in p21Cip1. *Nature* 375: 159–61.

Lynch, W. P., V. M. Riseman, and A. Bretscher. 1987. Smooth muscle caldesmon is an extended flexible monomeric protein in solution that can readily undergo reversible intra- and intermolecular sulfhydryl cross-linking. A mechanism for caldesmon's F-actin bundling activity. *J. Biol. Chem.* 262: 7429–37.

Ma, H., H. Q. Yang, E. Takano, M. Hatanaka, and M. Maki. 1994. Amino-terminal conserved region in proteinase inhibitor domain of calpastatin potentiates its calpain inhibitory activity by interacting with calmodulin-like domain of the proteinase. *J. Biol. Chem.* 269: 24430–6.

Ma, H., H. Q. Yang, E. Takano, W. J. Lee, M. Hatanaka, and M. Maki. 1993. Requirement of different subdomains of calpastatin for calpain inhibition and for binding to calmodulin-like domains. *J. Biochem. (Tokyo)* 113: 591–9.

Ma, J. 2000. Stimulatory and inhibitory functions of the R domain on CFTR chloride channel. *News Physiol. Sci.* 15: 154–58.

Ma, K., L. Kan, and K. Wang. 2001. Polyproline II helix is a key structural motif of the elastic PEVK segment of titin. *Biochemistry* 40: 3427–38.

Ma, K. and K. Wang. 2003. Malleable conformation of the elastic PEVK segment of titin: non-co-operative interconversion of polyproline II helix, beta-turn and unordered structures. *Biochem J.* 374: 687–95.

Macauley, M. S., W. J. Errington, M. Scharpf, et al. 2006. Beads-on-a-string, characterization of ETS-1 sumoylated within its flexible N-terminal sequence. *J. Biol. Chem.* 281: 4164–72.

Machado, C. and D. J. Andrew. 2000. D-Titin: a giant protein with dual roles in chromosomes and muscles. *J. Cell Biol.* 151: 639–52.

Machida, K., A. Kono-Okada, K. Hongo, T. Mizobata, and Y. Kawata. 2008. Hydrophilic residues 526 KNDAAD 531 in the flexible C-terminal region of the chaperonin GroEL are critical for substrate protein folding within the central cavity. *J. Biol. Chem.* 283: 6886–96.

Magidovich, E., S. J. Fleishman, and O. Yifrach. 2006. Intrinsically disordered C-terminal segments of voltage-activated potassium channels: a possible fishing rod-like mechanism for channel binding to scaffold proteins. *Bioinformatics* 22: 1546–50.

Magidovich, E., I. Orr, D. Fass, U. Abdu, and O. Yifrach. 2007. Intrinsic disorder in the C-terminal domain of the Shaker voltage-activated K+ channel modulates its interaction with scaffold proteins. *Proc. Natl. Acad. Sci. USA* 104: 13022–7.

Makhatadze, G. I. and P. L. Privalov. 1995. Energetics of protein structure. *Adv. Protein Chem.* 47: 307–425.

Makowska, J., S. Rodziewicz-Motowidlo, K. Baginska, et al. 2006. Polyproline II conformation is one of many local conformational states and is not an overall conformation of unfolded peptides and proteins. *Proc. Natl. Acad. Sci. USA* 103: 1744–9.

Malin, E. L., M. H. Alaimo, E. M. Brown, et al. 2001. Solution structures of casein peptides: NMR, FTIR, CD, and molecular modeling studies of alphas1-casein, 1-23. *J Protein Chem* 20: 391–404.

Manalan, A. S. and C. B. Klee. 1983. Activation of calcineurin by limited proteolysis. *Proc. Natl. Acad. Sci. USA* 80: 4291–5.

Mandelkow, E. M., O. Schweers, G. Drewes, et al. 1996. Structure, microtubule interactions, and phosphorylation of tau protein. *Ann. N Y Acad. Sci.* 777: 96–106.

Manning, G. 2005. Genomic overview of protein kinases. *WormBook* 1–19.

Marcotrigiano, J., A. C. Gingras, N. Sonenberg, and S. K. Burley. 1999. Cap-dependent translation initiation in eukaryotes is regulated by a molecular mimic of eIF4G. *Mol. Cell* 3: 707–16.

Marcotte, E. M., M. Pellegrini, T. O. Yeates, and D. Eisenberg. 1999. A census of protein repeats. *J. Mol. Biol.* 293: 151–60.

Mark, W. Y., J. C. Liao, Y. Lu, et al. 2005. Characterization of segments from the central region of BRCA1: an intrinsically disordered scaffold for multiple protein–protein and protein–DNA interactions? *J. Mol. Biol.* 345: 275–87.

Marston, S. B. and C. S. Redwood. 1991. The molecular anatomy of caldesmon. *Biochem J.* 279: 1–16.

Marti, M. J., E. Tolosa, and J. Campdelacreu. 2003. Clinical overview of the synucleinopathies. *Mov. Disord.* 18 Suppl 6: S21–7.

Masino, L., G. Kelly, K. Leonard, Y. Trottier, and A. Pastore. 2002. Solution structure of polyglutamine tracts in GST-polyglutamine fusion proteins. *FEBS Lett.* 513: 267–72.

Massague, J. 1998. TGF-beta signal transduction. *Annu. Rev. Biochem.* 67: 753–91.

McBryant, S. J., C. Krause, and J. C. Hansen. 2006. Domain organization and quaternary structure of the *Saccharomyces cerevisiae* silent information regulator 3 protein, Sir3p. *Biochemistry* 45: 15941–8.

McIntyre, J., E. G. Muller, S. Weitzer, B. E. Snydsman, T. N. Davis, and F. Uhlmann. 2007. In vivo analysis of cohesin architecture using FRET in the budding yeast *Saccharomyces cerevisiae. EMBO J.* 26: 3783–93.

McColl, I. H., E. W. Blanch, L. Hecht, N. R. Kallenbach, and L. D. Barron. 2004. Vibrational Raman optical activity characterization of poly(l-proline) II helix in alanine oligopeptides. *J. Am. Chem. Soc.* 126: 5076–7.

McCubbin, W. D., C. M. Kay, and B. G. Lane. 1985. Hydrodynamic and optical properties of the wheat germ Em protein. *Can. J. Biochem. Cell Biol.* 63: 803–11.

McEwan, I. J., D. Lavery, K. Fischer, and K. Watt. 2007. Natural disordered sequences in the amino terminal domain of nuclear receptors: lessons from the androgen and glucocorticoid receptors. *Nucl. Recept. Signal.* 5: e001.

McMeekin, T. L. 1952. Milk proteins. *J. Food Protect.* 15: 57–63.

McNulty, B. C., G. B. Young, and G. J. Pielak. 2006. Macromolecular crowding in the *Escherichia coli* periplasm maintains alpha-synuclein disorder. *J. Mol. Biol.* 355: 893–7.

McPhie, P., Y. S. Ni, and A. P. Minton. 2006. Macromolecular crowding stabilizes the molten globule form of apomyoglobin with respect to both cold and heat unfolding. *J. Mol. Biol.* 361: 7–10.

Megidish, T., J. H. Xu, and C. W. Xu. 2002. Activation of p53 by protein inhibitor of activated Stat1 (PIAS1). *J. Biol. Chem.* 277: 8255–9.

Meinhart, A. and P. Cramer. 2004. Recognition of RNA polymerase II carboxy-terminal domain by 3′-RNA-processing factors. *Nature* 430: 223–6.

Meininghaus, M., R. D. Chapman, M. Horndasch, and D. Eick. 2000. Conditional expression of RNA polymerase II in mammalian cells. Deletion of the carboxyl-terminal domain of the large subunit affects early steps in transcription. *J. Biol. Chem.* 275: 24375–82.

Melamud, E. and J. Moult. 2003. Evaluation of disorder predictions in CASP5. *Proteins* 53 Suppl 6: 561–5.

Meszaros, B., P. Tompa, I. Simon, and Z. Dosztanyi. 2007. Molecular principles of the interactions of disordered proteins. *J. Mol. Biol.* 372: 549–61.

Michalet, X., S. Weiss, and M. Jager. 2006. Single-molecule fluorescence studies of protein folding and conformational dynamics. *Chem. Rev.* 106: 1785–813.

Miki, Y., J. Swensen, D. Shattuck-Eidens, et al. 1994. A strong candidate for the breast and ovarian cancer susceptibility gene BRCA1. *Science* 266: 66–71.

Minezaki, Y., K. Homma, A. R. Kinjo, and K. Nishikawa. 2006. Human transcription factors contain a high fraction of intrinsically disordered regions essential for transcriptional regulation. *J. Mol. Biol.* 359: 1137–49.

Minezaki, Y., K. Homma, and K. Nishikawa. 2007. Intrinsically disordered regions of human plasma membrane proteins preferentially occur in the cytoplasmic segment. *J. Mol. Biol.* 368: 902–13.

Minor, D. L. Jr. and P. S. Kim. 1996. Context-dependent secondary structure formation of a designed protein sequence. *Nature* 380: 730–4.

Minton, A. P. 2005. Models for excluded volume interaction between an unfolded protein and rigid macromolecular cosolutes: macromolecular crowding and protein stability revisited. *Biophys J.* 88: 971–85.

Mirsky, A. E. and L. Pauling. 1936. On the structure of native, denatured, and coagulated proteins. *Proc. Natl. Acad. Sci. USA* 22: 439–47.

Mittag, T., S. Orlicky, W. Y. Choy, et al. 2008. Dynamic equilibrium engagement of a polyvalent ligand with a single-site receptor. *Proc. Natl. Acad. Sci. USA* 105: 17772–7.

Mohan, A., C. J. Oldfield, P. Radivojac, et al. 2006. Analysis of molecular recognition features (MoRFs). *J. Mol. Biol.* 362: 1043–59.

Mohan, A., W. J. Sullivan Jr., P. Radivojac, A. K. Dunker, and V. N. Uversky. 2008. Intrinsic disorder in pathogenic and non-pathogenic microbes: discovering and analyzing the unfoldomes of early-branching eukaryotes. *Mol. Biosyst.* 4: 328–40.

Moldoveanu, T., K. Gehring, and D. R. Green. 2008. Concerted multi-pronged attack by calpastatin to occlude the catalytic cleft of heterodimeric calpains. *Nature* 456:404–8.

Momma, M., S. Kaneko, K. Haraguchi, and U. Matsukura. 2003. Peptide mapping and assessment of cryoprotective activity of 26/27-kDa dehydrin from soybean seeds. *Biosci. Biotechnol. Biochem.* 67: 1832–5.

Moncoq, K., I. Broutin, C. T. Craescu, P. Vachette, A. Ducruix, and D. Durand. 2004. SAXS study of the PIR domain from the Grb14 molecular adaptor: a natively unfolded protein with a transient structure primer? *Biophys J.* 87: 4056–64.

Moncoq, K., I. Broutin, V. Larue, et al. 2003. The PIR domain of Grb14 is an intrinsically unstructured protein: implication in insulin signaling. *FEBS Lett.* 554: 240–6.

Monsellier, E. and F. Chiti. 2007. Prevention of amyloid-like aggregation as a driving force of protein evolution. *EMBO Rep* 8: 737–42.

Morar, A. S., A. Olteanu, G. B. Young, and G. J. Pielak. 2001. Solvent-induced collapse of alpha-synuclein and acid-denatured cytochrome c. *Protein Sci.* 10: 2195–9.

Morellet, N., N. Jullian, H. De Rocquigny, B. Maigret, J. L. Darlix, and B. P. Roques. 1992. Determination of the structure of the nucleocapsid protein NCp7 from the human immunodeficiency virus type 1 by 1H NMR. *EMBO J.* 11: 3059–65.

Morin, B., J. M. Bourhis, V. Belle, et al. 2006. Assessing induced folding of an intrinsically disordered protein by site-directed spin-labeling electron paramagnetic resonance spectroscopy. *J. Phys. Chem. B* 110: 20596–608.

Mouillon, J. M., P. Gustafsson, and P. Harryson. 2006. Structural investigation of disordered stress proteins. Comparison of full-length dehydrins with isolated peptides of their conserved segments. *Plant Physiol.* 141: 638–50.

Muenzer, J., C. Bildstein, M. Gleason, and D. M. Carlson. 1979. Properties of proline-rich proteins from parotid glands of isoproterenol-treated rats. *J. Biol. Chem.* 254: 5629–34.

Mujtaba, S., Y. He, L. Zeng, et al. 2004. Structural mechanism of the bromodomain of the coactivator CBP in p53 transcriptional activation. *Mol. Cell* 13: 251–63.

Mukhopadhyay, R. and J. H. Hoh. 2001. AFM force measurements on microtubule-associated proteins: the projection domain exerts a long-range repulsive force. *FEBS Lett.* 505: 374–8.

Mukhopadhyay, S., R. Krishnan, E. A. Lemke, S. Lindquist, and A. A. Deniz. 2007. A natively unfolded yeast prion monomer adopts an ensemble of collapsed and rapidly fluctuating structures. *Proc. Natl. Acad. Sci. USA* 104: 2649–54.

Mukrasch, M. D., J. Biernat, M. Von Bergen, C. Griesinger, E. Mandelkow, and M. Zweckstetter. 2005. Sites of tau important for aggregation populate (beta)-structure and bind to microtubules and polyanions. *J. Biol. Chem.* 280: 24978–86.

Mukrasch, M. D., P. Markwick, J. Biernat, et al. 2007a. Highly populated turn conformations in natively unfolded tau protein identified from residual dipolar couplings and molecular simulation. *J. Am. Chem. Soc.* 129: 5235–43.

Mukrasch, M. D., M. Von Bergen, J. Biernat, et al. 2007b. The "jaws" of the tau-microtubule interaction. *J. Biol. Chem.* 282: 12230–9.

Mulder, F. A., L. Bouakaz, A. Lundell, et al. 2004. Conformation and dynamics of ribosomal stalk protein L12 in solution and on the ribosome. *Biochemistry* 43: 5930–6.

Muro-Pastor, M. I., F. N. Barrera, J. C. Reyes, F. J. Florencio, and J. L. Neira. 2003. The inactivating factor of glutamine synthetase, IF7, is a "natively unfolded" protein. *Protein Sci.* 12: 1443–54.

Murray, A. W. 2004. Recycling the cell cycle: cyclins revisited. *Cell* 116: 221–34.

Murzin, A. G., S. E. Brenner, T. Hubbard, and C. Chothia. 1995. SCOP: a structural classification of proteins database for the investigation of sequences and structures. *J. Mol. Biol.* 247: 536–40.

Myers, L. C., C. M. Gustafsson, K. C. Hayashibara, P. O. Brown, and R. D. Kornberg. 1999. Mediator protein mutations that selectively abolish activated transcription. *Proc. Natl. Acad. Sci. USA* 96: 67–72.

Nagao, K., Y. Adachi, and M. Yanagida. 2004. Separase-mediated cleavage of cohesin at interphase is required for DNA repair. *Nature* 430: 1044–8.

Nagao, K. and M. Yanagida. 2006. Securin can have a separase cleavage site by substitution mutations in the domain required for stabilization and inhibition of separase. *Genes Cells* 11: 247–60.

Nallamshetty, S., M. Crook, M. Boehm, T. Yoshimoto, M. Olive, and E. G. Nabel. 2005. The cell cycle regulator p27Kip1 interacts with MCM7, a DNA replication licensing factor, to inhibit initiation of DNA replication. *FEBS Lett.* 579: 6529–36.

Nash, P., X. Tang, S. Orlicky, et al. 2001. Multisite phosphorylation of a CDK inhibitor sets a threshold for the onset of DNA replication. *Nature* 414: 514–21.

Nasir, J., S. B. Floresco, J. R. O'Kusky, et al. 1995. Targeted disruption of the Huntington's disease gene results in embryonic lethality and behavioral and morphological changes in heterozygotes. *Cell* 81: 811–23.

Neduva, V. and R. B. Russell. 2005. Linear motifs: evolutionary interaction switches. *FEBS Lett.* 579: 3342–5.

Neduva, V. and R. B. Russell. 2006. DILIMOT: discovery of linear motifs in proteins. *Nucleic. Acids Res.* 34: W350–5.

Nelson, R., M. R. Sawaya, M. Balbirnie, et al. 2005. Structure of the cross-beta spine of amyloid-like fibrils. *Nature* 435: 773–8.

Neri, D., M. Billeter, G. Wider, and K. Wuthrich. 1992. NMR determination of residual structure in a urea-denatured protein, the 434-repressor. *Science* 257: 1559–63.

Neyroz, P., B. Zambelli, and S. Ciurli. 2006. Intrinsically disordered structure of *Bacillus pasteurii* UreG as revealed by steady-state and time-resolved fluorescence spectroscopy. *Biochemistry* 45: 8918–30.

Ng, K. P., G. Potikyan, R. O. Savene, C. T. Denny, V. N. Uversky, and K. A. Lee. 2007. Multiple aromatic side chains within a disordered structure are critical for transcription and transforming activity of EWS family oncoproteins. *Proc. Natl. Acad. Sci. USA* 104: 479–84.

Nguyen, A. W. and P. S. Daugherty. 2005. Evolutionary optimization of fluorescent proteins for intracellular FRET. *Nat. Biotechnol.* 23: 355–60.

Nicholls, C. D., K. G. Mclure, M. A. Shields, and P. W. Lee. 2002. Biogenesis of p53 involves cotranslational dimerization of monomers and posttranslational dimerization of dimers. Implications on the dominant negative effect. *J. Biol. Chem.* 277: 12937–45.

Nimmo, G. A. and P. Cohen. 1978. The regulation of glycogen metabolism. Purification and characterisation of protein phosphatase inhibitor-1 from rabbit skeletal muscle. *Eur. J. Biochem.* 87: 341–51.

Nishimura, M., T. Yoshida, M. Shirouzu, et al. 2004. Solution structure of ribosomal protein L16 from Thermus thermophilus HB8. *J. Mol. Biol.* 344: 1369–83.

Nonet, M., D. Sweetser, and R. A. Young. 1987. Functional redundancy and structural polymorphism in the large subunit of RNA polymerase II. *Cell* 50: 909–15.

Nooren, I. M. and J. M. Thornton. 2003. Diversity of protein–protein interactions. *EMBO J.* 22: 3486–92.

Nyarko, A., M. Hare, T. S. Hays, and E. Barbar. 2004. The intermediate chain of cytoplasmic dynein is partially disordered and gains structure upon binding to light-chain LC8. *Biochemistry* 43: 15595–603.

Obenauer, J. C., L. C. Cantley, and M. B. Yaffe. 2003. Scansite 2.0: Proteome-wide prediction of cell signaling interactions using short sequence motifs. *Nucleic. Acids Res.* 31: 3635–41.

Obradovic, Z., K. Peng, S. Vucetic, P. Radivojac, C. J. Brown, and A. K. Dunker. 2003. Predicting intrinsic disorder from amino acid sequence. *Proteins* 53 Suppl 6: 566–72.

Obradovic, Z., K. Peng, S. Vucetic, P. Radivojac, and A. K. Dunker. 2005. Exploiting heterogeneous sequence properties improves prediction of protein disorder. *Proteins* 61 Suppl 7: 176–82.

Ohashi, T., S. D. Galiacy, G. Briscoe, and H. P. Erickson. 2007. An experimental study of GFP-based FRET, with application to intrinsically unstructured proteins. *Protein Sci.* 16: 1429–38.

Ohashi, T., C. A. Hale, P. A. De Boer, and H. P. Erickson. 2002. Structural evidence that the P/Q domain of ZipA is an unstructured, flexible tether between the membrane and the C-terminal FtsZ-binding domain. *J. Bacteriol.* 184: 4313–5.

Ohnishi, S., A. L. Lee, M. H. Edgell, and D. Shortle. 2004. Direct demonstration of structural similarity between native and denatured eglin C. *Biochemistry* 43: 4064–70.

Ohno, S. 1984. Repeats of base oligomers as the primordial coding sequences of the primeval earth and their vestiges in modern genes. *J. Mol. Evol.* 20: 313–21.

Ohno, S. 1987. Early genes that were oligomeric repeats generated a number of divergent domains on their own. *Proc. Natl. Acad. Sci. USA* 84: 6486–90.

Ojala, P. M., K. Yamamoto, E. Castanos-Velez, P. Biberfeld, S. J. Korsmeyer, and T. P. Makela. 2000. The apoptotic v-cyclin-CDK6 complex phosphorylates and inactivates Bcl-2. *Nat. Cell Biol.* 2: 819–25.

Oldfield, C. J., Y. Cheng, M. S. Cortese, C. J. Brown, V. N. Uversky, and A. K. Dunker. 2005a. Comparing and combining predictors of mostly disordered proteins. *Biochemistry* 44: 1989–2000.

Oldfield, C. J., Y. Cheng, M. S. Cortese, P. Romero, V. N. Uversky, and A. K. Dunker. 2005b. Coupled folding and binding with alpha-helix-forming molecular recognition elements. *Biochemistry* 44: 12454–70.

Oldfield, C. J., J. Meng, J. Y. Yang, M. Q. Yang, V. N. Uversky, and A. K. Dunker. 2008. Flexible nets: disorder and induced fit in the associations of p53 and 14-3-3 with their partners. *BMC Genomics* 9 (Suppl. 1): S1.

Oldfield, C. J., E. L. Ulrich, Y. Cheng, A. K. Dunker, and J. L. Markley. 2005c. Addressing the intrinsic disorder bottleneck in structural proteomics. *Proteins* 59: 444–53.

Olson, K. E., P. Narayanaswami, P. D. Vise, D. F. Lowry, M. S. Wold, and G. W. Daughdrill. 2005. Secondary structure and dynamics of an intrinsically unstructured linker domain. *J. Biomol. Struct. Dyn.* 23: 113–24.

Orengo, C. A., A. D. Michie, S. Jones, D. T. Jones, M. B. Swindells, and J. M. Thornton. 1997. CATH—a hierarchic classification of protein domain structures. *Structure* 5: 1093–108.

Orphanides, G. and D. Reinberg. 2002. A unified theory of gene expression. *Cell* 108: 439–51.

Ostedgaard, L. S., O. Baldursson, D. W. Vermeer, M. J. Welsh, and A. D. Robertson. 2000. A functional R domain from cystic fibrosis transmembrane conductance regulator is predominantly unstructured in solution. *Proc. Natl. Acad. Sci. USA* 97: 5657–62.

Otzen, D. E., L. S. Itzhaki, N. F. Elmasry, S. E. Jackson, and A. R. Fersht. 1994. Structure of the transition state for the folding/unfolding of the barley chymotrypsin inhibitor 2 and its implications for mechanisms of protein folding. *Proc. Natl. Acad. Sci. USA* 91: 10422–5.

Overall, C. M. and G. S. Butler. 2007. Protease yoga: extreme flexibility of a matrix metalloproteinase. *Structure* 15: 1159–61.

Page, R., W. Peti, I. A. Wilson, R. C. Stevens, and K. Wuthrich. 2005. NMR screening and crystal quality of bacterially expressed prokaryotic and eukaryotic proteins in a structural genomics pipeline. *Proc. Natl. Acad. Sci. USA* 102: 1901–5.

Palmer, M. S. and J. Collinge. 1993. Mutations and polymorphisms in the prion protein gene. *Hum. Mutat.* 2: 168–73.

Pan, H., G. Barany and C. Woodward. 1997. Reduced BPTI is collapsed. A pulsed field gradient NMR study of unfolded and partially folded bovine pancreatic trypsin inhibitor. *Protein Sci.* 6: 1985–92.

Pan, K. M., M. Baldwin, J. Nguyen, et al. 1993. Conversion of alpha-helices into beta-sheets features in the formation of the scrapie prion proteins. *Proc. Natl. Acad. Sci. USA* 90: 10962–6.

Panchal, S. C., D. A. Kaiser, E. Torres, T. D. Pollard, and M. K. Rosen. 2003. A conserved amphipathic helix in WASP/Scar proteins is essential for activation of Arp2/3 complex. *Nat. Struct. Biol.* 10: 591–8.

Panetti, T. S. 2002. Tyrosine phosphorylation of paxillin, FAK, and p130CAS: effects on cell spreading and migration. *Front. Biosci.* 7: d143–50.

Pantazatos, D., J. S. Kim, H. E. Klock, et al. 2004. Rapid refinement of crystallographic protein construct definition employing enhanced hydrogen/deuterium exchange MS. *Proc. Natl. Acad. Sci. USA* 101: 751–6.

Papp, S. and J. M. Vanderkooi. 1989. Tryptophan phosphorescence at room temperature as a tool to study protein structure and dynamics. *Photochem. Photobiol.* 49: 775–84.

Park, I. K. and A. A. DePaoli-Roach. 1994. Domains of phosphatase inhibitor-2 involved in the control of the ATP-Mg-dependent protein phosphatase. *J. Biol. Chem.* 269: 28919–28.

Park, S. M., H. Y. Jung, T. D. Kim, J. H. Park, C. H. Yang, and J. Kim. 2002. Distinct roles of the N-terminal-binding domain and the C-terminal-solubilizing domain of alpha-synuclein, a molecular chaperone. *J. Biol. Chem.* 277: 28512–20.

Parker, D., K. Ferreri, T. Nakajima, et al. 1996. Phosphorylation of CREB at Ser-133 induces complex formation with CREB-binding protein via a direct mechanism. *Mol. Cell Biol.* 16: 694–703.

Parker, D., M. Rivera, T. Zor, et al. 1999. Role of secondary structure in discrimination between constitutive and inducible activators. *Mol. Cell Biol.* 19: 5601–7.

Parrish, J. R., K. D. Gulyas, and R. L. Finley Jr. 2006. Yeast two-hybrid contributions to interactome mapping. *Curr. Opin. Biotechnol.* 17: 387–93.

Pasta, S. Y., B. Raman, T. Ramakrishna, and M. Rao Ch. 2002. Role of the C-terminal extensions of alpha-crystallins. Swapping the C-terminal extension of alpha-crystallin to alphaB-crystallin results in enhanced chaperone activity. *J. Biol. Chem.* 277: 45821–8.

Patel, S. S., B. J. Belmont, J. M. Sante, and M. F. Rexach. 2007. Natively unfolded nucleoporins gate protein diffusion across the nuclear pore complex. *Cell* 129: 83–96.

Patil, A. and H. Nakamura. 2006. Disordered domains and high surface charge confer hubs with the ability to interact with multiple proteins in interaction networks. *FEBS Lett.* 580: 2041–5.

Pattaramanon, N., N. Sangha, and A. Gafni. 2007. The carboxy-terminal domain of heat-shock factor 1 is largely unfolded but can be induced to collapse into a compact, partially structured state. *Biochemistry* 46: 3405–15.

Patthy, L. 1996. Exon shuffling and other ways of module exchange. *Matrix Biol.* 15: 301–10; discussion 11–2.

Patthy, L. 1999. Genome evolution and the evolution of exon-shuffling—a review. *Gene* 238: 103–14.

Patti, J. M., B. L. Allen, M. J. Mcgavin, and M. Hook. 1994. MSCRAMM-mediated adherence of microorganisms to host tissues. *Annu. Rev. Microbiol.* 48: 585–617.

Pauling, L. 1948. The nature of forces between large molecules of biological interest. *Nature* 161: 707–09.

Paulsson, M. and P. Dejmek. 1990. Thermal denaturation of whey proteins in mixtures with caseins studied by DSC. *J. Dairy Sci.* 73: 590–600.

Paunola, E., P. K. Mattila, and P. Lappalainen. 2002. WH2 domain: a small, versatile adapter for actin monomers. *FEBS Lett.* 513: 92–7.

Pawson, T. and P. Nash. 2003. Assembly of cell regulatory systems through protein interaction domains. *Science* 300: 445–52.

Pawson, T. and J. D. Scott. 1997. Signaling through scaffold, anchoring, and adaptor proteins. *Science* 278: 2075–80.

Payens, T. A. J. and H. J. Vreeman. 1982. In: *Solution Behaviour of Surfactants*, 543. New York: Plenum, p. 543.

Paz, A., T. Zeev-Ben-Mordehai, M. Lundqvist, et al. 2008. Biophysical characterization of the unstructured cytoplasmic domain of the human neuronal adhesion protein neuroligin 3. *Biophys J.* 95: 1928–44.

Peng, K., P. Radivojac, S. Vucetic, A. K. Dunker, and Z. Obradovic. 2006. Length-dependent prediction of protein intrinsic disorder. *BMC Bioinformatics* 7: 208.

Penkett, C. J., C. Redfield, I. Dodd, et al. 1997. NMR analysis of main-chain conformational preferences in an unfolded fibronectin-binding protein. *J. Mol. Biol.* 274: 152–9.

Penkett, C. J., C. Redfield, J. A. Jones, et al. 1998. Structural and dynamical characterization of a biologically active unfolded fibronectin-binding protein from *Staphylococcus aureus*. *Biochemistry* 37: 17054–67.

Permyakov, S. E., I. S. Millett, S. Doniach, E. A. Permyakov, and V. N. Uversky. 2003. Natively unfolded C-terminal domain of caldesmon remains substantially unstructured after the effective binding to calmodulin. *Proteins* 53: 855–62.

Persson, M., P. Hammarstrom, M. Lindgren, B. H. Jonsson, M. Svensson and U. Carlsson. 1999. EPR mapping of interactions between spin-labeled variants of human carbonic anhydrase II and GroEL: evidence for increased flexibility of the hydrophobic core by the interaction. *Biochemistry* 38: 432–41.

Perutz, M. F. 1960. Structure of hemoglobin. *Brookhaven Symp. Biol.* 13: 165–83.

Perutz, M. F. 1999. Glutamine repeats and neurodegenerative diseases: molecular aspects. *Trends Biochem. Sci.* 24: 58–63.

Perutz, M. F., R. Staden, L. Moens and I. De Baere. 1993. Polar zippers. *Curr. Biol.* 3: 249–53.

Peters, J. M. 2002. The anaphase-promoting complex: proteolysis in mitosis and beyond. *Mol. Cell* 9: 931–43.

Peti, W., T. Etezady-Esfarjani, T. Herrmann, H. E. Klock, S. A. Lesley, and K. Wuthrich. 2004. NMR for structural proteomics of *Thermotoga maritima*: screening and structure determination. *J. Struct. Funct. Genomics* 5: 205–15.

Petkova, A. T., Y. Ishii, J. J. Balbach, et al. 2002. A structural model for Alzheimer's beta-amyloid fibrils based on experimental constraints from solid state NMR. *Proc. Natl. Acad. Sci. USA* 99: 16742–7.

Phadtare, S., J. Alsina, and M. Inouye. 1999. Cold-shock response and cold-shock proteins. *Curr. Opin. Microbiol.* 2: 175–80.

Pierce, M. M., U. Baxa, A. C. Steven, A. Bax, and R. B. Wickner. 2005. Is the prion domain of soluble Ure2p unstructured? *Biochemistry* 44: 321–8.

Plaxco, K. W. and M. Gross. 1997. Cell biology. The importance of being unfolded. *Nature* 386: 657, 59.

Podlaha, O. and J. Zhang. 2003. Positive selection on protein-length in the evolution of a primate sperm ion channel. *Proc. Natl. Acad. Sci. USA* 100: 12241–6.

Pokholok, D. K., C. T. Harbison, S. Levine, et al. 2005. Genome-wide map of nucleosome acetylation and methylation in yeast. *Cell* 122: 517–27.

Pometun, M. S., E. Y. Chekmenev, and R. J. Wittebort. 2004. Quantitative observation of backbone disorder in native elastin. *J. Biol. Chem.* 279: 7982–7.

Ponting, C. P., J. Schultz, R. R. Copley, M. A. Andrade, and P. Bork. 2000. Evolution of domain families. *Adv. Protein Chem.* 54: 185–244.

Pontius, B. W. 1993. Close encounters: why unstructured, polymeric domains can increase rates of specific macromolecular association. *Trends Biochem. Sci.* 18: 181–6.

Pontius, B. W. and P. Berg. 1990. Renaturation of complementary DNA strands mediated by purified mammalian heterogeneous nuclear ribonucleoprotein A1 protein: implications for a mechanism for rapid molecular assembly. *Proc. Natl. Acad. Sci. USA* 87: 8403–7.

Pontius, B. W. and P. Berg. 1991. Rapid renaturation of complementary DNA strands mediated by cationic detergents: a role for high-probability binding domains in enhancing the kinetics of molecular assembly processes. *Proc. Natl. Acad. Sci. USA* 88: 8237–41.

Popovic, M., M. Coglievina, C. Guarnaccia, et al. 2006. Gene synthesis, expression, purification, and characterization of human Jagged-1 intracellular region. *Protein Expr. Purif.* 47: 398–404.

Poulin, F., A. C. Gingras, H. Olsen, S. Chevalier, and N. Sonenberg. 1998. 4E-BP3, a new member of the eukaryotic initiation factor 4E-binding protein family. *J. Biol. Chem.* 273: 14002–7.

Poy, F., M. Lepourcelet, R. A. Shivdasani, and M. J. Eck. 2001. Structure of a human Tcf4-beta-catenin complex. *Nat. Struct. Biol.* 8: 1053–7.

Prakash, S., L. Tian, K. S. Ratliff, R. E. Lehotzky, and A. Matouschek. 2004. An unstructured initiation site is required for efficient proteasome-mediated degradation. *Nat. Struct. Mol. Biol.* 11: 830–7.

Prasch, S., S. Schwarz, A. Eisenmann, B. M. Wohrl, K. Schweimer, and P. Rosch. 2006. Interaction of the intrinsically unstructured phage lambda N protein with *Escherichia coli* NusA. *Biochemistry* 45: 4542–9.

Price, W. S. 1998. Pulsed-field gradient nuclear magnetic resonance as a tool for studying translational diffusion: part 1. Basic theory. In *Concepts in Magnetic Resonance Part A*, 299–336: Wiley Periodicals, Inc.

Prilusky, J., C. E. Felder, T. Zeev-Ben-Mordehai, et al. 2005. FoldIndex: a simple tool to predict whether a given protein sequence is intrinsically unfolded. *Bioinformatics* 21: 3435–8.

Prince, V. E. and F. B. Pickett. 2002. Splitting pairs: the diverging fates of duplicated genes. *Nat. Rev. Genet.* 3: 827–37.

Privalov, P. L. 1979. Stability of proteins: small globular proteins. *Adv. Protein Chem.* 33: 167–241.

Privalov, P. L. 1982. Stability of proteins. Proteins which do not present a single cooperative system. *Adv. Protein Chem.* 35: 1–104.

Proudfoot, N. J., A. Furger, and M. J. Dye. 2002. Integrating mRNA processing with transcription. *Cell* 108: 501–12.

Prusiner, S. B. 1982. Novel proteinaceous infectious particles cause scrapie. *Science* 216: 136–44.

Prusiner, S. B. 1998. Prions. *Proc. Natl. Acad. Sci. USA* 95: 13363–83.

Prusiner, S. B. 2001. Shattuck lecture—neurodegenerative diseases and prions. *N. Engl. J. Med.* 344: 1516–26.

Ptitsyn, O. B. and V. N. Uversky. 1994. The molten globule is a third thermodynamical state of protein molecules. *FEBS Lett.* 341: 15–8.

Puig, O., F. Caspary, G. Rigaut, et al. 2001. The tandem affinity purification (TAP) method: a general procedure of protein complex purification. *Methods* 24: 218–29.

Punta, M. and B. Rost. 2005. PROFcon: novel prediction of long-range contacts. *Bioinformatics* 21: 2960–8.

Puntervoll, P., R. Linding, C. Gemund, et al. 2003. ELM server: A new resource for investigating short functional sites in modular eukaryotic proteins. *Nucleic. Acids Res.* 31: 3625–30.

Qu, Y. and D. W. Bolen. 2002. Efficacy of macromolecular crowding in forcing proteins to fold. *Biophys. Chem.* 101–2: 155–65.

Rabbitts, T. H. and M. R. Stocks. 2003. Chromosomal translocation products engender new intracellular therapeutic technologies. *Nat. Med.* 9: 383–6.

Radhakrishnan, I., G. C. Perez-Alvarado, H. J. Dyson, and P. E. Wright. 1998. Conformational preferences in the Ser133-phosphorylated and non-phosphorylated forms of the kinase inducible transactivation domain of CREB. *FEBS Lett.* 430: 317–22.

Radhakrishnan, I., G. C. Perez-Alvarado, D. Parker, H. J. Dyson, M. R. Montminy, and P. E. Wright. 1997. Solution structure of the KIX domain of CBP bound to the transactivation domain of CREB: a model for activator:coactivator interactions. *Cell* 91: 741–52.

Radivojac, P., Z. Obradovic, C. J. Brown, and A. K. Dunker. 2002. Improving sequence alignments for intrinsically disordered proteins. *Pac. Symp. Biocomput.* 589–600.

Radivojac, P., S. Vucetic, T. R. O'Connor, V. N. Uversky, Z. Obradovic, and A. K. Dunker. 2006. Calmodulin signaling: analysis and prediction of a disorder-dependent molecular recognition. *Proteins* 63: 398–410.

Ramachandran, G. N., C. Ramakrishnan, and V. Sasisekharan. 1963. Stereochemistry of polypeptide chain configurations. *J. Mol. Biol.* 7: 95–9.

Ramakrishnan, M., P. H. Jensen, and D. Marsh. 2003. Alpha-synuclein association with phosphatidylglycerol probed by lipid spin labels. *Biochemistry* 42: 12919–26.

Rantalainen, K. I., V. N. Uversky, P. Permi, N. Kalkkinen, A. K. Dunker, and K. Makinen. 2008. Potato virus A genome-linked protein VPg is an intrinsically disordered molten globule-like protein with a hydrophobic core. *Virology* 377: 280–8.

Rape, M. and S. Jentsch. 2002. Taking a bite: proteasomal protein processing. *Nat. Cell Biol.* 4: E113–6.

Rauscher, S., S. Baud, M. Miao, F. W. Keeley, and R. Pomes. 2006. Proline and glycine control protein self-organization into elastomeric or amyloid fibrils. *Structure* 14: 1667–76.

Receveur-Brechot, V., J. M. Bourhis, V. N. Uversky, B. Canard, and S. Longhi. 2005. Assessing protein disorder and induced folding. *Proteins* 62: 24–45.

Receveur, V., M. Czjzek, M. Schulein, P. Panine, and B. Henrissat. 2002. Dimension, shape, and conformational flexibility of a two-domain fungal cellulase in solution probed by small angle X-ray scattering. *J. Biol. Chem.* 277: 40887–92.

Rechsteiner, M. and S. W. Rogers. 1996. PEST sequences and regulation by proteolysis. *Trends Biochem. Sci.* 21: 267–71.

Redeker, V., S. Lachkar, S. Siavoshian, et al. 2000. Probing the native structure of stathmin and its interaction domains with tubulin. Combined use of limited proteolysis, size exclusion chromatography, and mass spectrometry. *J. Biol. Chem.* 275: 6841–9.

Redinbo, M. R., L. Stewart, P. Kuhn, J. J. Champoux, and W. G. Hol. 1998. Crystal structures of human topoisomerase I in covalent and noncovalent complexes with DNA. *Science* 279: 1504–13.

Reeves, R. 2001. Molecular biology of HMGA proteins: hubs of nuclear function. *Gene* 277: 63–81.

Reeves, R. and L. Beckerbauer. 2001. HMGI/Y proteins: flexible regulators of transcription and chromatin structure. *Biochim. Biophys. Acta.* 1519: 13–29.

Reinholt, F. P., K. Hultenby, A. Oldberg, and D. Heinegard. 1990. Osteopontin—a possible anchor of osteoclasts to bone. *Proc. Natl. Acad. Sci. USA* 87: 4473–5.

Renault, L., B. Bugyi and M. F. Careier. 2008. Spire and Cordon-bleu: multifunctional regulators of actin dynamics *Trends Cell Biol.* 18: 494–504.

Richards, J. P., H. P. Bachinger, R. H. Goodman, and R. G. Brennan. 1996. Analysis of the structural properties of cAMP-responsive element-binding protein (CREB) and phosphorylated CREB. *J. Biol. Chem.* 271: 13716–23.

Riek, R., S. Hornemann, G. Wider, R. Glockshuber, and K. Wuthrich. 1997. NMR characterization of the full-length recombinant murine prion protein, mPrP(23–231). *FEBS Lett.* 413: 282–8.

Riordan, J. R., J. M. Rommens, B. Kerem, et al. 1989. Identification of the cystic fibrosis gene: cloning and characterization of complementary DNA. *Science* 245: 1066–73.

Ritter, C., M. L. Maddelein, A. B. Siemer, et al. 2005. Correlation of structural elements and infectivity of the HET-s prion. *Nature* 435: 844–8.

Rochet, J. C. and P. T. Lansbury Jr. 2000. Amyloid fibrillogenesis: themes and variations. *Curr. Opin. Struct. Biol.* 10: 60–8.

Rock, R. S., B. Ramamurthy, A. R. Dunn, et al. 2005. A flexible domain is essential for the large step size and processivity of myosin VI. *Mol. Cell* 17: 603–9.

Rodger, A. and B. Nordén. 1997. *Circular Dichroism and Linear Dichroism.* Oxford: Oxford University Press.

Romero, P., Z. Obradovic, and A. K. Dunker. 1999. Folding minimal sequences: the lower bound for sequence complexity of globular proteins. *FEBS Lett.* 462: 363–7.

Romero, P., Z. Obradovic, C. R. Kissinger, J. E. Villafranca and A. K. Dunker. 1997. Identifying disordered regions in proteins from amino acid sequences. *Proc. IEEE Int. Conf. Neural Networks* 1: 90–95.

Romero, P., Z. Obradovic, C. R. Kissinger, et al. 1998. Thousands of proteins likely to have long disordered regions. *Pac. Symp. Biocomputing* 3: 437–48.

Romero, P., Z. Obradovic, X. Li, E. C. Garner, C. J. Brown, and A. K. Dunker. 2001. Sequence complexity of disordered protein. *Proteins* 42: 38–48.

Romero, P. R., S. Zaidi, Y. Y. Fang, et al. 2006. Alternative splicing in concert with protein intrinsic disorder enables increased functional diversity in multicellular organisms. *Proc. Natl. Acad. Sci. USA* 103: 8390–5.

Rose, G. D., P. J. Fleming, J. R. Banavar, and A. Maritan. 2006. A backbone-based theory of protein folding. *Proc. Natl. Acad. Sci. USA* 103: 16623–33.

Rosenblum, G., P. E. Van Den Steen, S. R. Cohen, et al. 2007. Insights into the structure and domain flexibility of full-length pro-matrix metalloproteinase-9/gelatinase B. *Structure* 15: 1227–36.

Ross, E. D., H. K. Edskes, M. J. Terry, and R. B. Wickner. 2005. Primary sequence independence for prion formation. *Proc. Natl. Acad. Sci. USA* 102: 12825–30.

Rousseau, F., J. Schymkowitz, and L. Serrano. 2006. Protein aggregation and amyloidosis: confusion of the kinds? *Curr. Opin. Struct. Biol.* 16: 118–26.

Rout, M. P., J. D. Aitchison, A. Suprapto, K. Hjertaas, Y. Zhao, and B. T. Chait. 2000. The yeast nuclear pore complex: composition, architecture, and transport mechanism. *J. Cell Biol.* 148: 635–51.

Russo, A. A., P. D. Jeffrey, A. K. Patten, J. Massague, and N. P. Pavletich. 1996. Crystal structure of the p27Kip1 cyclin-dependent-kinase inhibitor bound to the cyclin A-Cdk2 complex. *Nature* 382: 325–31.

Ruthenburg, A. J., C. D. Allis, and J. Wysocka. 2007. Methylation of lysine 4 on histone H3: intricacy of writing and reading a single epigenetic mark. *Mol. Cell* 25: 15–30.

Ruvolo, V., E. Wang, S. Boyle, and S. Swaminathan. 1998. The Epstein–Barr virus nuclear protein SM is both a post-transcriptional inhibitor and activator of gene expression. *Proc. Natl. Acad. Sci. USA* 95: 8852–7.

Rye, H. S., S. G. Burston, W. A. Fenton, et al. 1997. Distinct actions of cis and trans ATP within the double ring of the chaperonin GroEL. *Nature* 388: 792–8.

Salamino, F., R. De Tullio, M. Michetti, P. Mengotti, E. Melloni, and S. Pontremoli. 1994. Modulation of calpastatin specificity in rat tissues by reversible phosphorylation and dephosphorylation. *Biochem. Biophys. Res. Commun.* 199: 1326–32.

Sanchez-Puig, N., D. B. Veprintsev, and A. R. Fersht. 2005. Human full-length securin is a natively unfolded protein. *Protein Sci.* 14: 1410–8.

Sanchez, C., J. Diaz-Nido, and J. Avila. 2000. Phosphorylation of microtubule-associated protein 2 (MAP2) and its relevance for the regulation of the neuronal cytoskeleton function. *Prog. Neurobiol.* 61: 133–68.

Sandal, M., F. Valle, I. Tessari, et al. 2008. Conformational Equilibria in monomeric alpha-synuclein at the single-molecule level. *PLoS Biol.* 6: e6.

Sands, W. A. and T. M. Palmer. 2008. Regulating gene transcription in response to cyclic AMP elevation. *Cell Signal.* 20: 460–6.

Sarkar, T., A. Mitra, S. Gupta et al. 2004. MAP2 prevents protein aggregation and facilitates reactivation of unfolded enzymes. *Eur. J. Biochem.* 271: 1488–96.

Saxena, A. M., J. B. Udgaonkar, and G. Krishnamoorthy. 2006. Characterization of Intramolecular distances and site-specific dynamics in chemically unfolded barstar: evidence for denaturant-dependent non-random structure. *J. Mol. Biol.* 359: 174–89.

Schaffar, G., P. Breuer, R. Boteva, et al. 2004. Cellular toxicity of polyglutamine expansion proteins: mechanism of transcription factor deactivation. *Mol. Cell* 15: 95–105.

Scheele, U., J. Alves, R. Frank, M. Duwel, C. Kalthoff, and E. Ungewickell. 2003. Molecular and functional characterization of clathrin- and AP-2-binding determinants within a disordered domain of auxilin. *J. Biol. Chem.* 278: 25357–68.

Scheibel, T. and S. L. Lindquist. 2001. The role of conformational flexibility in prion propagation and maintenance for Sup35p. *Nat. Struct. Biol.* 8: 958–62.

Schlessinger, A., J. Liu, and B. Rost. 2007a. Natively unstructured loops differ from other loops. *PLoS Comput. Biol.* 3: e140.

Schlessinger, A., M. Punta, and B. Rost. 2007b. Natively unstructured regions in proteins identified from contact predictions. *Bioinformatics*

Schmid, F. X. 1989. Spectral probes of conformation. In: *Protein Structure: A practical Approach*, 251–85. Oxford: IRL Press, Oxford University Press.

Schmidt, E. E. and C. J. Davies. 2007. The origins of polypeptide domains. *Bioessays* 29: 262–70.

Schneider, B. L., Q. H. Yang, and A. B. Futcher. 1996. Linkage of replication to start by the Cdk inhibitor Sic1. *Science* 272: 560–2.

Schwarz-Linek, U., E. S. Pilka, A. R. Pickford, et al. 2004. High affinity streptococcal binding to human fibronectin requires specific recognition of sequential f1 modules. *J. Biol. Chem.* 279: 39017–25.

Schwarz-Linek, U., J. M. Werner, A. R. Pickford, et al. 2003. Pathogenic bacteria attach to human fibronectin through a tandem beta-zipper. *Nature* 423: 177–81.

Schwarzinger, S., G. J. Kroon, T. R. Foss, J. Chung, P. E. Wright, and H. J. Dyson. 2001. Sequence-dependent correction of random coil NMR chemical shifts. *J. Am. Chem. Soc.* 123: 2970–8.

Schwarzinger, S., G. J. Kroon, T. R. Foss, P. E. Wright, and H. J. Dyson. 2000. Random coil chemical shifts in acidic 8-M urea: implementation of random coil shift data in NMRView. *J. Biomol. NMR* 18: 43–8.

Schweers, O., E. Schonbrunn-Hanebeck, A. Marx, and E. Mandelkow. 1994. Structural studies of tau protein and Alzheimer paired helical filaments show no evidence for beta-structure. *J. Biol. Chem.* 269: 24290–7.

Schwob, E., T. Bohm, M. D. Mendenhall, and K. Nasmyth. 1994. The B-type cyclin kinase inhibitor p40SIC1 controls the G1 to S transition in *S. cerevisiae*. *Cell* 79: 233–44.

Sebolt-Leopold, J. S. and J. M. English. 2006. Mechanisms of drug inhibition of signalling molecules. *Nature* 441: 457–62.

Sedzik, J. and D. A. Kirschner. 1992. Is myelin basic protein crystallizable? *Neurochem. Res.* 17: 157–66.

Seet, B. T., I. Dikic, M. M. Zhou, and T. Pawson. 2006. Reading protein modifications with interaction domains. *Nat. Rev. Mol. Cell Biol.* 7: 473–83.

Selenko, P., G. Gregorovic, R. Sprangers, et al. 2003. Structural basis for the molecular recognition between human splicing factors U2AF65 and SF1/mBBP. *Mol. Cell* 11: 965–76.

Selenko, P., Z. Serber, B. Gadea, J. Ruderman, and G. Wagner. 2006. From the cover: Quantitative NMR analysis of the protein G B1 domain in *Xenopus laevis* egg extracts and intact oocytes. *Proc. Natl. Acad. Sci. USA* 103: 11904–9.

Selenko, P. and G. Wagner. 2007. Looking into live cells with in-cell NMR spectroscopy. *J. Struct. Biol.* 158: 244–53.

Selkoe, D. J. 2003. Folding proteins in fatal ways. *Nature* 426: 900–4.

Semrad, K., R. Green, and R. Schroeder. 2004. RNA chaperone activity of large ribosomal subunit proteins from *Escherichia coli*. *RNA* 10: 1855–60.

Serber, Z. and V. Dotsch. 2001. In-cell NMR spectroscopy. *Biochemistry* 40: 14317–23.

Serpell, L. C., M. Sunde, M. D. Benson, G. A. Tennent, M. B. Pepys, and P. E. Fraser. 2000. The protofilament substructure of amyloid fibrils. *J. Mol. Biol.* 300: 1033–9.

Shannon, C. E. 1948. A mathematical theory of communication. *Bell Syst. Tech. J.* 27: 379–423, 623–56.

Sheaff, R. J., J. D. Singer, J. Swanger, M. Smitherman, J. M. Roberts, and B. E. Clurman. 2000. Proteasomal turnover of p21Cip1 does not require p21Cip1 ubiquitination. *Mol. Cell* 5: 403–10.

Sheng, M. and E. Kim. 2000. The Shank family of scaffold proteins. *J. Cell Sci.* 113: 1851–6.

Sherr, C. J. and J. M. Roberts. 1999. CDK inhibitors: positive and negative regulators of G1-phase progression. *Genes Dev.* 13: 1501–12.

Shi, Z., C. A. Olson, G. D. Rose, R. L. Baldwin, and N. R. Kallenbach. 2002. Polyproline II structure in a sequence of seven alanine residues. *Proc. Natl. Acad. Sci. USA* 99: 9190–5.

Shieh, S. Y., Y. Taya, and C. Prives. 1999. DNA damage-inducible phosphorylation of p53 at N-terminal sites including a novel site, Ser20, requires tetramerization. *EMBO J.* 18: 1815–23.

Shilatifard, A. 2006. Chromatin modifications by methylation and ubiquitination: implications in the regulation of gene expression. *Annu. Rev. Biochem.* 75: 243–69.

Shimura, H., N. Hattori, S. Kubo, et al. 2000. Familial Parkinson disease gene product, parkin, is a ubiquitin-protein ligase. *Nat. Genet.* 25: 302–5.

Shoemaker, B. A., J. J. Portman, and P. G. Wolynes. 2000. Speeding molecular recognition by using the folding funnel: the fly-casting mechanism. *Proc. Natl. Acad. Sci. USA* 97: 8868–73.

Shogren-Knaak, M., H. Ishii, J. M. Sun, M. J. Pazin, J. R. Davie, and C. L. Peterson. 2006. Histone H4–K16 acetylation controls chromatin structure and protein interactions. *Science* 311: 844–7.

Shortle, D. 1996. The denatured state (the other half of the folding equation) and its role in protein stability. *FASEB J.* 10: 27–34.

Shortle, D. and M. S. Ackerman. 2001. Persistence of native-like topology in a denatured protein in 8-M urea. *Science* 293: 487–9.

Si, K., M. Giustetto, A. Etkin, et al. 2003a. A neuronal isoform of CPEB regulates local protein synthesis and stabilizes synapse-specific long-term facilitation in aplysia. *Cell* 115: 893–904.

Si, K., S. Lindquist and E. R. Kandel. 2003b. A neuronal isoform of the aplysia CPEB has prion-like properties. *Cell* 115: 879–91.

Sickmeier, M., J. A. Hamilton, T. Legall, et al. 2007. DisProt: the database of disordered proteins. *Nucleic. Acids Res.* 35: D786–93.

Sigalov, A., D. Aivazian, and L. Stern. 2004. Homooligomerization of the cytoplasmic domain of the T cell receptor zeta chain and of other proteins containing the immunoreceptor tyrosine-based activation motif. *Biochemistry* 43: 2049–61.

Sigalov, A. B. 2004. Multichain immune recognition receptor signaling: different players, same game? *Trends Immunol.* 25: 583–89.

Sigalov, A. B., D. A. Aivazian, V. N. Uversky, and L. J. Stern. 2006. Lipid-binding activity of intrinsically unstructured cytoplasmic domains of multichain immune recognition receptor signaling subunits. *Biochemistry* 45: 15731–39.

Sigler, P. B. 1988. Transcriptional activation. Acid blobs and negative noodles. *Nature* 333: 210–2.

Simmons, L. K., P. C. May, K. J. Tomaselli, et al. 1994. Secondary structure of amyloid beta peptide correlates with neurotoxic activity in vitro. *Mol. Pharmacol.* 45: 373–9.

Simon, S. M., F. J. Sousa, R. Mohana-Borges, and G. C. Walker. 2008. Regulation of *Escherichia coli* SOS mutagenesis by dimeric intrinsically disordered umuD gene products. *Proc. Natl. Acad. Sci. USA* 105: 1152–7.

Singh, G. P., M. Ganapathi, and D. Dash. 2006. Role of intrinsic disorder in transient interactions of hub proteins. *Proteins* 66: 761–65.

Singh, K., M. W. Devouge, and B. B. Mukherjee. 1990. Physiological properties and differential glycosylation of phosphorylated and nonphosphorylated forms of osteopontin secreted by normal rat kidney cells. *J. Biol. Chem.* 265: 18696–701.

Singh, V. K., I. Pacheco, V. N. Uversky, S. P. Smith, R. J. Macleod, and Z. Jia. 2008. Intrinsically disordered human C/EBP homologous protein regulates biological activity of colon cancer cells during calcium stress. *J. Mol. Biol.* 380: 313–26.

Sivakolundu, S. G., D. Bashford, and R. W. Kriwacki. 2005. Disordered p27Kip1 exhibits intrinsic structure resembling the Cdk2/cyclin A-bound conformation. *J. Mol. Biol.* 353: 1118–28.

Slijper, M., R. Boelens, A. L. Davis, et al. 1997. Backbone and side chain dynamics of lac repressor headpiece (1–56) and its complex with DNA. *Biochemistry* 36: 249–54.

Smet, C., A. Leroy, A. Sillen, J. M. Wieruszeski, I. Landrieu, and G. Lippens. 2004. Accepting its random coil nature allows a partial NMR assignment of the neuronal tau protein. *Chembiochem* 5: 1639–46.

Smith, C. L., R. Horowitz-Scherer, J. F. Flanagan, C. L. Woodcock, and C. L. Peterson. 2003. Structural analysis of the yeast SWI/SNF chromatin remodeling complex. *Nat. Struct. Biol.* 10: 141–5.

Smith, J. L. E. A. 2004. Kinetic profiles of p300 occupancy in vivo predict common features of promoter structure and coactivator recruitment. *Proc. Natl. Acad. Sci. USA* 101: 11554–59.

Smith, M. J., R. A. Crowther, and M. Goedert. 2000. The natural osmolyte trimethylamine-N-oxide (TMAO) restores the ability of mutant tau to promote microtubule assembly. *FEBS Lett.* 484: 265–70.

Smulders, R., J. A. Carver, R. A. Lindner, M. A. Van Boekel, H. Bloemendal, and W. W. De Jong. 1996. Immobilization of the C-terminal extension of bovine alphaA-crystallin reduces chaperone-like activity. *J. Biol. Chem.* 271: 29060–6.

Smyth, E., C. D. Syme, E. W. Blanch, L. Hecht, M. Vasak, and L. D. Barron. 2001. Solution structure of native proteins with irregular folds from Raman optical activity. *Biopolymers* 58: 138–51.

Soding, J. and A. N. Lupas. 2003. More than the sum of their parts: on the evolution of proteins from peptides. *Bioessays* 25: 837–46.

Solt, I., C. Magyar, I. Simon, P. Tompa, and M. Fuxreiter. 2006. Phosphorylation-induced transient intrinsic structure in the kinase-inducible domain of CREB facilitates its recognition by the KIX domain of CBP. *Proteins* 64: 749–57.

Sorek, R. 2007. The birth of new exons: mechanisms and evolutionary consequences. *RNA* 13: 1603–8.

Sorenson, M. K., S. S. Ray, and S. A. Darst. 2004. Crystal structure of the flagellar sigma/anti-sigma complex sigma(28)/FlgM reveals an intact sigma factor in an inactive conformation. *Mol. Cell* 14: 127–38.

Soulages, J. L., K. Kim, E. L. Arrese, C. Walters, and J. C. Cushman. 2003. Conformation of a group 2 late embryogenesis abundant protein from soybean. Evidence of poly (L-proline)-type II structure. *Plant Physiol.* 131: 963–75.

Soulages, J. L., K. Kim, C. Walters, and J. C. Cushman. 2002. Temperature-induced extended helix/random coil transitions in a group 1 late embryogenesis-abundant protein from soybean. *Plant Physiol.* 128: 822–32.

Sparrow, J. C. 1999. *Actin*. Oxford: Oxford University Press.

Spera, S., M. Ikura, and A. Bax. 1991. Measurement of the exchange rates of rapidly exchanging amide protons: application to the study of calmodulin and its complex with a myosin light chain kinase fragment. *J. Biomol. NMR* 1: 155–65.

Spolar, R. S. and M. T. Record Jr. 1994. Coupling of local folding to site-specific binding of proteins to DNA. *Science* 263: 777–84.

Srisodsuk, M., T. Reinikainen, M. Penttila, and T. T. Teeri. 1993. Role of the interdomain linker peptide of *Trichoderma reesei* cellobiohydrolase I in its interaction with crystalline cellulose. *J. Biol. Chem.* 268: 20756–61.

Stahl, N., M. A. Baldwin, D. B. Teplow, et al. 1993. Structural studies of the scrapie prion protein using mass spectrometry and amino acid sequencing. *Biochemistry* 32: 1991–2002.

Steiner, B., E. M. Mandelkow, J. Biernat, et al. 1990. Phosphorylation of microtubule-associated protein tau: identification of the site for Ca2(+)-calmodulin dependent kinase and relationship with tau phosphorylation in Alzheimer tangles. *EMBO J.* 9: 3539–44.

Stryer, L. 1978. Fluorescence energy transfer as a spectroscopic ruler. *Annu. Rev. Biochem.* 47: 819–46.

Stryer, L. 1995. *Biochemistry*. New York: W.H. Freeman and Co.

Sugase, K., H. J. Dyson, and P. E. Wright. 2007. Mechanism of coupled folding and binding of an intrinsically disordered protein. *Nature* 447: 1021–5.

Sunde, M. and C. Blake. 1997. The structure of amyloid fibrils by electron microscopy and X-ray diffraction. *Adv. Protein Chem.* 50: 123–59.

Suzuki, K., S. Hata, Y. Kawabata, and H. Sorimachi. 2004. Structure, activation, and biology of calpain. *Diabetes* 53 Suppl 1: S12–8.

Svergun, D. I. and M. H. Koch. 2002. Advances in structure analysis using small-angle scattering in solution. *Curr. Opin. Struct. Biol.* 12: 654–60.

Svergun, D. I. and M. H. J. Koch. 2003. Small-angle scattering studies of biological macromolecules in solution. *Rep. Prog. Phys.* 66: 1735–82.

Sweet, R. M. and D. Eisenberg. 1983. Correlation of sequence hydrophobicities measures similarity in three-dimensional protein structure. *J. Mol. Biol.* 171: 479–88.

Syme, C. D., E. W. Blanch, C. Holt, et al. 2002. A Raman optical activity study of rheomorphism in caseins, synucleins and tau. *Eur. J. Biochem.* 269: 148–56.

Szilagyi, A., D. Gyorffy, and P. Zavodszky. 2008. The twilight zone between protein order and disorder. *Biophys J.* 95: 1612–26.

Szollosi, E., M. Bokor, A. Bodor, et al. 2008. Intrinsic structural disorder of DF31, a *Drosophila* protein of chromatin decondensation and remodeling activities. *J. Proteome. Res.* 7: 2291–9.

Taatjes, D. J., A. M. Naar, F. Andel III, E. Nogales, and R. Tjian. 2002. Structure, function, and activator-induced conformations of the CRSP coactivator. *Science* 295: 1058–62.

Taatjes, D. J., T. Schneider-Poetsch, and R. Tjian. 2004. Distinct conformational states of nuclear receptor-bound CRSP-Med complexes. *Nat. Struct. Mol. Biol.* 11: 664–71.

Tabuchi, K., T. Biederer, S. Butz, and T. C. Sudhof. 2002. CASK participates in alternative tripartite complexes in which Mint 1 competes for binding with caskin 1, a novel CASK-binding protein. *J. Neurosci.* 22: 4264–73.

Takagi, Y., G. Calero, H. Komori, et al. 2006. Head module control of mediator interactions. *Mol. Cell* 23: 355–64.

Takahashi, M., M. Itakura, and M. Kataoka. 2003. New aspects of neurotransmitter release and exocytosis: regulation of neurotransmitter release by phosphorylation. *J. Pharmacol. Sci.* 93: 41–5.

Takano, E., H. Ma, H. Q. Yang, M. Maki, and M. Hatanaka. 1995. Preference of calcium-dependent interactions between calmodulin-like domains of calpain and calpastatin subdomains. *FEBS Lett.* 362: 93–7.

Takano, E., M. Maki, H. Mori, et al. 1988. Pig heart calpastatin: identification of repetitive domain structures and anomalous behavior in polyacrylamide gel electrophoresis. *Biochemistry* 27: 1964–72.

Tanford, C. 1968. Protein denaturation. *Adv. Protein Chem.* 23: 121–282.

Tanford, C., K. Kawahara, and S. Lapanje. 1966. Proteins in 6-M guanidine hydrochloride. Demonstration of random coil behavior. *J. Biol. Chem.* 241: 1921–23.

Tao, W. and A. J. Levine. 1999. P19(ARF) stabilizes p53 by blocking nucleocytoplasmic shuttling of Mdm2. *Proc. Natl. Acad. Sci. USA* 96: 6937–41.

Tfelt-Hansen, J., D. Kanuparthi, and N. Chattopadhyay. 2006. The emerging role of pituitary tumor transforming gene in tumorigenesis. *Clin. Med. Res.* 4: 130–7.

Thapar, R., G. A. Mueller, and W. F. Marzluff. 2004. The N-terminal domain of the *Drosophila* histone mRNA binding protein, SLBP, is intrinsically disordered with nascent helical structure. *Biochemistry* 43: 9390–400.

Thirone, A. C., C. Huang, and A. Klip. 2006. Tissue-specific roles of IRS proteins in insulin signaling and glucose transport. *Trends Endocrinol. Metab.* 17: 72–8.

Thomas, B. and M. F. Beal. 2007. Parkinson's disease. *Hum Mol Genet* 16 Spec No. 2: R183–94.

Thomas, J., S. M. Van Patten, P. Howard, et al. 1991. Expression in *Escherichia coli* and characterization of the heat-stable inhibitor of the cAMP-dependent protein kinase. *J. Biol. Chem.* 266: 10906–11.

Thomas, P. D. and K. A. Dill. 1996. An iterative method for extracting energy-like quantities from protein structures. *Proc. Natl. Acad. Sci. USA* 93: 11628–33.

Thomas, W. H., U. Weser, and K. Hempel. 1977. Conformational changes induced by ionic strength and pH in two bovine myelin basic proteins. *Hoppe Seylers Z. Physiol. Chem.* 358: 1345–52.

Thompson, J. D., D. G. Higgins, and T. J. Gibson. 1994. CLUSTAL W: improving the sensitivity of progressive multiple sequence alignment through sequence weighting, position-specific gap penalties and weight matrix choice. *Nucleic. Acids Res.* 22: 4673–80.

Thorn, D. C., S. Meehan, M. Sunde, et al. 2005. Amyloid fibril formation by bovine milk kappa-casein and its inhibition by the molecular chaperones alpha(S)- and beta-casein. *Biochemistry* 44: 17027–36.

Tisne, C., B. P. Roques, and F. Dardel. 2001. Heteronuclear NMR studies of the interaction of tRNA(Lys)3 with HIV-1 nucleocapsid protein. *J. Mol. Biol.* 306: 443–54.

Todd, M. J., G. H. Lorimer, and D. Thirumalai. 1996. Chaperonin-facilitated protein folding: optimization of rate and yield by an iterative annealing mechanism. *Proc. Natl. Acad. Sci. USA* 93: 4030–5.

Tofaris, G. K., R. Layfield, and M. G. Spillantini. 2001. alpha-synuclein metabolism and aggregation is linked to ubiquitin-independent degradation by the proteasome. *FEBS Lett.* 509: 22–6.

Tokuriki, N., M. Kinjo, S. Negi, et al. 2004. Protein folding by the effects of macromolecular crowding. *Protein Sci.* 13: 125–33.

Tompa, P. 2002. Intrinsically unstructured proteins. *Trends Biochem. Sci.* 27: 527–33.

Tompa, P. 2003a. The functional benefits of protein disorder. *J. Mol. Struct. THEOCHEM* 666–667: 361–71.

Tompa, P. 2003b. Intrinsically unstructured proteins evolve by repeat expansion. *BioEssays* 25: 847–55.

Tompa, P. 2005. The interplay between structure and function in intrinsically unstructured proteins. *FEBS Lett.* 579: 3346–54.

Tompa, P., P. Banki, M. Bokor, et al. 2006a. Protein-water and protein-buffer interactions in the aqueous solution of an intrinsically unstructured plant dehydrin: NMR intensity and DSC aspects. *Biophys J.* 91: 2243–9.

Tompa, P., P. Buzder-Lantos, A. Tantos, et al. 2004. On the sequential determinants of calpain cleavage. *J. Biol. Chem.* 279: 20775–85.

Tompa, P. and P. Csermely. 2004. The role of structural disorder in the function of RNA and protein chaperones. *FASEB J* 18: 1169–75.

Tompa, P., Z. Dosztanyi, and I. Simon. 2006b. Prevalent structural disorder in *E. coli* and *S. cerevisiae* proteomes. *J. Proteome. Res.* 5: 1996–2000.

Tompa, P. and M. Fuxreiter. 2008. Fuzzy complexes: polymorphism and structural disorder in protein–protein interactions. *Trends Biochem. Sci.* 33: 2–8.

Tompa, P., M. Fuxreiter, C. J. Oldfield, I. Simon, A. K. Dunker, and V. N. Uversky. 2009. Close encounters of the third kind: disordered domains and the interactions of proteins. *BioEssays* 31: 328–35.

Tompa, P., J. Prilusky, I. Silman, and J. L. Sussman. 2008. Structural disorder serves as a weak signal for intracellular protein degradation. *Proteins* 71: 903–9.

Tompa, P., C. Szasz, and L. Buday. 2005. Structural disorder throws new light on moonlighting. *Trends Biochem. Sci.* 30: 484–9.

Tong, K. I., Y. Katoh, H. Kusunoki, K. Itoh, T. Tanaka, and M. Yamamoto. 2006. Keap1 recruits Neh2 through binding to ETGE and DLG motifs: characterization of the two-site molecular recognition model. *Mol. Cell Biol.* 26: 2887–900.

Torok, M., S. Milton, R. Kayed, et al. 2002. Structural and dynamic features of Alzheimer's Abeta peptide in amyloid fibrils studied by site-directed spin labeling. *J. Biol. Chem.* 277: 40810–5.

Toth-Petroczy, A., C. J. Oldfield, I. Simon, et al. 2008. Malleable machines in transcription regulation: the mediator complex. *PLoS Comput. Biol.* 4: e1000243.

Tozawa, K., C. J. Macdonald, C. N. Penfold, et al. 2005. Clusters in an intrinsically disordered protein create a protein-binding site: the TolB-binding region of colicin E9. *Biochemistry* 44: 11496–507.

Triezenberg, S. J. 1995. Structure and function of transcriptional activation domains. *Curr. Opin. Genet. Dev.* 5: 190–6.

Trombitas, K., M. Greaser, S. Labeit, et al. 1998. Titin extensibility in situ: entropic elasticity of permanently folded and permanently unfolded molecular segments. *J. Cell Biol.* 140: 853–9.

Tsien, R. Y. 1998. The green fluorescent protein. *Annu. Rev. Biochem.* 67: 509–44.

Tsukazaki, T., T. A. Chiang, A. F. Davison, L. Attisano, and J. L. Wrana. 1998. SARA, a FYVE domain protein that recruits Smad2 to the TGFbeta receptor. *Cell* 95: 779–91.

Tsvetkov, P., G. Asher, A. Paz, et al. 2008. Operational definition of intrinsically unstructured protein sequences based on susceptibility to the 20S proteasome. *Proteins* 70: 1357–66.

Tucker, M. M., J. B. Robinson Jr., and E. Stellwagen. 1981. The effect of proteolysis on the calmodulin activation of cyclic nucleotide phosphodiesterase. *J. Biol. Chem.* 256: 9051–8.

Tucker, P. K. and B. L. Lundrigan. 1993. Rapid evolution of the sex determining locus in Old World mice and rats. *Nature* 364: 715–7.

Tung, H. Y., W. Wang, and C. S. Chan. 1995. Regulation of chromosome segregation by Glc8p, a structural homolog of mammalian inhibitor 2 that functions as both an activator and an inhibitor of yeast protein phosphatase 1. *Mol. Cell Biol.* 15: 6064–74.

Tunnacliffe, A. and M. J. Wise. 2007. The continuing conundrum of the LEA proteins. *Naturwissenschaften* 94: 791–812.

Turner, C. F. and P. B. Moore. 2004. The solution structure of ribosomal protein L18 from *Bacillus stearothermophilus*. *J. Mol. Biol.* 335: 679–84.

Tyukhtenko, S., L. Deshmukh, V. Kumar, et al. 2008. Characterization of the neuron-specific L1-CAM cytoplasmic tail: naturally disordered in solution it exercises different binding modes for different adaptor proteins. *Biochemistry* 47: 4160–8.

Ueda, K., H. Fukushima, E. Masliah, et al. 1993. Molecular cloning of cDNA encoding an unrecognized component of amyloid in Alzheimer's disease. *Proc. Natl. Acad. Sci. USA* 90: 11282–6.

Uesugi, M., O. Nyanguile, H. Lu, A. J. Levine, and G. L. Verdine. 1997. Induced alpha helix in the VP16 activation domain upon binding to a human TAF. *Science* 277: 1310–3.

Ulfers, A. L., J. L. Mcmurry, D. A. Kendall, and D. F. Mierke. 2002. Structure of the third intracellular loop of the human cannabinoid 1 receptor. *Biochemistry* 41: 11344–50.

Uversky, V. N. 1993. Use of fast protein size-exclusion liquid chromatography to study the unfolding of proteins which denature through the molten globule. *Biochemistry* 32: 13288–98.

Uversky, V. N. 2002a. Natively unfolded proteins: A point where biology waits for physics. *Protein Sci.* 11: 739–56.

Uversky, V. N. 2002b. What does it mean to be natively unfolded? *Eur. J. Biochem.* 269: 2–12.

Uversky, V. N. 2003. A protein-chameleon: conformational plasticity of alpha-synuclein, a disordered protein involved in neurodegenerative disorders. *J. Biomol. Struct. Dyn.* 21: 211–34.

Uversky, V. N. 2007. Neuropathology, biochemistry, and biophysics of alpha-synuclein aggregation. *J. Neurochem.* 103: 17–37.

Uversky, V. N. and A. L. Fink. 2002. The chicken–egg scenario of protein folding revisited. *FEBS Lett.* 515: 79–83.

Uversky, V. N. and A. L. Fink. 2004. Conformational constraints for amyloid fibrillation: the importance of being unfolded. *Biochim. Biophys. Acta.* 1698: 131–53.

Uversky, V. N., J. R. Gillespie, and A. L. Fink. 2000a. Why are "natively unfolded" proteins unstructured under physiologic conditions? *Proteins* 41: 415–27.

Uversky, V. N., J. R. Gillespie, I. S. Millett, et al. 2000b. Zn(2+)-mediated structure formation and compaction of the "natively unfolded" human prothymosin alpha. *Biochem. Biophys. Res. Commun.* 267: 663–8.

Uversky, V. N., J. R. Gillespie, I. S. Millett, et al. 1999. Natively unfolded human prothymosin alpha adopts partially folded collapsed conformation at acidic pH. *Biochemistry* 38: 15009–16.

Uversky, V. N., M. D. Kirkitadze, N. V. Narizhneva, S. A. Potekhin and A. Tomashevski. 1995. Structural properties of alpha-fetoprotein from human cord serum: the protein molecule at low pH possesses all the properties of the molten globule. *FEBS Lett.* 364: 165–7.

Uversky, V. N., H. J. Lee, J. Li, A. L. Fink, and S. J. Lee. 2001a. Stabilization of partially folded conformation during alpha-synuclein oligomerization in both purified and cytosolic preparations. *J. Biol. Chem.* 276: 43495–8.

Uversky, V. N., J. Li, and A. L. Fink. 2001b. Evidence for a partially folded intermediate in alpha-synuclein fibril formation. *J. Biol. Chem.* 276: 10737–44.

Uversky, V. N., J. Li, and A. L. Fink. 2001c. Metal-triggered structural transformations, aggregation, and fibrillation of human alpha-synuclein. A possible molecular NK between Parkinson's disease and heavy metal exposure. *J. Biol. Chem.* 276: 44284–96.

Uversky, V. N., J. Li, and A. L. Fink. 2001d. Trimethylamine-N-oxide-induced folding of alpha-synuclein. *FEBS Lett.* 509: 31–5.

Uversky, V. N. and O.B. Ptytsin, 1994. "Partly folded" state, a new equilibrium state of protein molecules: four-state guanidinium chloride-induced unfolding of beta-lactamase at low temperature. *Biochemistry* 33: 2782–91.

Uversky, V. N., C. J. Oldfield, and A. K. Dunker. 2005. Showing your ID: intrinsic disorder as an ID for recognition, regulation and cell signaling. *J. Mol. Recognit.* 18: 343–84.

Uversky, V. N., C. J. Oldfield, and A. K. Dunker. 2008. Intrinsically disordered proteins in human diseases: introducing the D2 concept. *Annu. Rev. Biophys.* 37: 215–46.

Uversky, V. N., A. Roman, C. J. Oldfield, and A. K. Dunker. 2006. Protein intrinsic disorder and human papillomaviruses: increased amount of disorder in E6 and E7 oncoproteins from high risk HPVs. *J. Proteome. Res.* 5: 1829–42.

Uversky, V. N., S. Winter, O. V. Galzitskaya, L. Kittler, and G. Lober. 1998. Hyperphosphorylation induces structural modification of tau protein. *FEBS Lett.* 439: 21–5.

Vacic, V., C. J. Oldfield, A. Mohan, et al. 2007. Characterization of molecular recognition features, MoRFs, and their binding partners. *J. Proteome. Res.* 6: 2351–66.

Vamvaca, K., B. Vogeli, P. Kast, K. Pervushin, and D. Hilvert. 2004. An enzymatic molten globule: efficient coupling of folding and catalysis. *Proc. Natl. Acad. Sci. USA* 101: 12860–4.

Van Gilst, M. R., W. A. Rees, A. Das, and P. H. Von Hippel. 1997. Complexes of N antitermination protein of phage lambda with specific and nonspecific RNA target sites on the nascent transcript. *Biochemistry* 36: 1514–24.

Van Leeuwen, H. C., M. J. Strating, M. Rensen, W. De Laat, and P. C. Van Der Vliet. 1997. Linker length and composition influence the flexibility of Oct-1 DNA binding. *EMBO J.* 16: 2043–53.

Van Montfort, R. L., E. Basha, K. L. Friedrich, C. Slingsby, and E. Vierling. 2001. Crystal structure and assembly of a eukaryotic small heat shock protein. *Nat. Struct. Biol.* 8: 1025–30.

Vassilev, L. T., B. T. Vu, B. Graves, et al. 2004. In vivo activation of the p53 pathway by small-molecule antagonists of MDM2. *Science* 303: 844–8.

Vaynberg, J., T. Fukuda, K. Chen, et al. 2005. Structure of an ultraweak protein–protein complex and its crucial role in regulation of cell morphology and motility. *Mol. Cell* 17: 513–23.

Venkatraman, P., R. Wetzel, M. Tanaka, N. Nukina, and A. L. Goldberg. 2004. Eukaryotic proteasomes cannot digest polyglutamine sequences and release them during degradation of polyglutamine-containing proteins. *Mol. Cell* 14: 95–104.

Venkitaraman, A. R. 2002. Cancer susceptibility and the functions of BRCA1 and BRCA2. *Cell* 108: 171–82.

Veprintsev, D. B., S. M. Freund, A. Andreeva, et al. 2006. Core domain interactions in full-length p53 in solution. *Proc. Natl. Acad. Sci. USA* 103: 2115–9.

Vergnaud, G. and F. Denoeud. 2000. Minisatellites: mutability and genome architecture. *Genome Res* 10: 899–907.

Verkhivker, G. M. 2004. Protein conformational transitions coupled to binding in molecular recognition of unstructured proteins: hierarchy of structural loss from all-atom Monte Carlo simulations of p27Kip1 unfolding-unbinding and structural determinants of the binding mechanism. *Biopolymers* 75: 420–33.

Verkhivker, G. M. 2005. Protein conformational transitions coupled to binding in molecular recognition of unstructured proteins: deciphering the effect of intermolecular interactions on computational structure prediction of the p27Kip1 protein bound to the cyclin A-cyclin-dependent kinase 2 complex. *Proteins* 58: 706–16.

Verkhivker, G. M., D. Bouzida, D. K. Gehlhaar, P. A. Rejto, S. T. Freer, and P. W. Rose. 2003. Simulating disorder-order transitions in molecular recognition of unstructured proteins: where folding meets binding. *Proc. Natl. Acad. Sci. USA* 100: 5148–53.

Vihinen, M., E. Torkkila, and P. Riikonen. 1994. Accuracy of protein flexibility predictions. *Proteins* 19: 141–9.

Vise, P., B. Baral, A. Stancik, D. F. Lowry, and G. W. Daughdrill. 2007. Identifying long-range structure in the intrinsically unstructured transactivation domain of p53. *Proteins* 67: 526–30.

Vise, P. D., B. Baral, A. J. Latos, and G. W. Daughdrill. 2005. NMR chemical shift and relaxation measurements provide evidence for the coupled folding and binding of the p53 transactivation domain. *Nucleic. Acids Res.* 33: 2061–77.

Vitalis, A., X. Wang, and R. V. Pappu. 2007. Quantitative characterization of intrinsic disorder in polyglutamine: insights from analysis based on polymer theories. *Biophys J.* 93: 1923–37.

Vogel, C., M. Bashton, N. D. Kerrison, C. Chothia, and S. A. Teichmann. 2004. Structure, function and evolution of multidomain proteins. *Curr. Opin. Struct. Biol.* 14: 208–16.

Voges, D., P. Zwickl, and W. Baumeister. 1999. The 26S proteasome: a molecular machine designed for controlled proteolysis. *Annu. Rev. Biochem.* 68: 1015–68.

von Bergen, M., P. Friedhoff, J. Biernat, J. Heberle, E. M. Mandelkow, and E. Mandelkow. 2000. Assembly of tau protein into Alzheimer paired helical filaments depends on a local sequence motif ((306)VQIVYK(311)) forming beta structure. *Proc. Natl. Acad. Sci. USA* 97: 5129–34.

von der Haar, T., Y. Oku, M. Ptushkina, et al. 2006. Folding transitions during assembly of the eukaryotic mRNA cap-binding complex. *J. Mol. Biol.* 356: 982–92.

von Mering, C., L. J. Jensen, B. Snel, et al. 2005. STRING: known and predicted protein–protein associations, integrated and transferred across organisms. *Nucleic. Acids Res.* 33: D433–D37.

von Ossowski, I., J. T. Eaton, M. Czjzek, et al. 2005. Protein disorder: conformational distribution of the flexible linker in a chimeric double cellulase. *Biophys J.* 88: 2823–32.

Vrhovski, B. and A. S. Weiss. 1998. Biochemistry of tropoelastin. *Eur. J. Biochem.* 258: 1–18.

Vucetic, S., C. J. Brown, A. K. Dunker, and Z. Obradovic. 2003. Flavors of protein disorder. *Proteins* 52: 573–84.

Vucetic, S., Z. Obradovic, V. Vacic, et al. 2005. DisProt: a database of protein disorder. *Bioinformatics* 21: 137–40.

Vullo, A., O. Bortolami, G. Pollastri, and S. C. Tosatto. 2006. Spritz: a server for the prediction of intrinsically disordered regions in protein sequences using kernel machines. *Nucleic. Acids Res.* 34: W164–8.

Waizenegger, I., J. F. Gimenez-Abian, D. Wernic, and J. M. Peters. 2002. Regulation of human separase by securin binding and autocleavage. *Curr. Biol.* 12: 1368–78.

Waldsich, C., R. Grossberger, and R. Schroeder. 2002. RNA chaperone StpA loosens interactions of the tertiary structure in the td group I intron in vivo. *Genes Dev.* 16: 2300–12.

Walker, F. O. 2007. Huntington's disease. *Lancet* 369: 218–28.

Wall, J., M. Schell, C. Murphy, R. Hrncic, F. J. Stevens, and A. Solomon. 1999. Thermodynamic instability of human lambda 6 light chains: correlation with fibrillogenicity. *Biochemistry* 38: 14101–8.

Wallon, G., J. Rappsilber, M. Mann, and L. Serrano. 2000. Model for stathmin/OP18 binding to tubulin. *EMBO J.* 19: 213–22.

Wang, J. Q., A. Arora, L. Yang, et al. 2005. Phosphorylation of AMPA receptors: mechanisms and synaptic plasticity. *Mol. Neurobiol.* 32: 237–49.

Wang, S., W. R. Trumble, H. Liao, C. R. Wesson, A. K. Dunker, and C. H. Kang. 1998. Crystal structure of calsequestrin from rabbit skeletal muscle sarcoplasmic reticulum. *Nat. Struct. Biol.* 5: 476–83.

Ward, J. J., J. S. Sodhi, L. J. Mcguffin, B. F. Buxton, and D. T. Jones. 2004. Prediction and functional analysis of native disorder in proteins from the three kingdoms of life. *J. Mol. Biol.* 337: 635–45.

Watanabe, K., P. Nair, D. Labeit, et al. 2002. Molecular mechanics of cardiac titin's PEVK and N2B spring elements. *J. Biol. Chem.* 277: 11549–58.

Watt, I. M. 1997. *The Principles & Practice of Electron Microscopy*. Cambridge: Cambridge University Press.

Watts, J. D., P. D. Cary, P. Sautiere, and C. Crane-Robinson. 1990. Thymosins: both nuclear and cytoplasmic proteins. *Eur. J. Biochem.* 192: 643–51.

Weathers, E. A., M. E. Paulaitis, T. B. Woolf, and J. H. Hoh. 2004. Reduced amino acid alphabet is sufficient to accurately recognize intrinsically disordered protein. *FEBS Lett.* 576: 348–52.

Weikl, T., K. Abelmann, and J. Buchner. 1999. An unstructured C-terminal region of the Hsp90 co-chaperone p23 is important for its chaperone function. *J. Mol. Biol.* 293: 685–91.

Weinreb, P. H., W. Zhen, A. W. Poon, K. A. Conway, and P. T. Lansbury Jr. 1996. NACP, a protein implicated in Alzheimer's disease and learning, is natively unfolded. *Biochemistry* 35: 13709–15.

Weiss, M. A., T. Ellenberger, C. R. Wobbe, J. P. Lee, S. C. Harrison, and K. Struhl. 1990. Folding transition in the DNA-binding domain of GCN4 on specific binding to DNA. *Nature* 347: 575–8.

Wells, M., H. Tidow, T. J. Rutherford, et al. 2008. Structure of tumor suppressor p53 and its intrinsically disordered N-terminal transactivation domain. *Proc. Natl. Acad. Sci. USA* 105: 5762–7.

Wells, R. D. 1996. Molecular basis of genetic instability of triplet repeats. *J. Biol. Chem.* 271: 2875–8.

Wendt, A., V. F. Thompson, and D. E. Goll. 2004. Interaction of calpastatin with calpain: a review. *Biol. Chem.* 385: 465–72.

Westermark, P., S. Araki, M. D. Benson, et al. 1999. Nomenclature of amyloid fibril proteins. Report from the meeting of the International Nomenclature Committee on Amyloidosis, August 8–9, 1998. Part 1. *Amyloid* 6: 63–6.

Westhof, E., D. Altschuh, D. Moras, et al. 1984. Correlation between segmental mobility and the location of antigenic determinants in proteins. *Nature* 311: 123–6.

Wetlaufer, D. B. 1973. Nucleation, rapid folding, and globular intrachain regions in proteins. *Proc. Natl. Acad. Sci. USA* 70: 697–701.

Whitfield, L. S., R. Lovell-Badge, and P. N. Goodfellow. 1993. Rapid sequence evolution of the mammalian sex-determining gene SRY. *Nature* 364: 713–5.

Wickner, R. B. 2005. Scrapie in ancient China? *Science* 309: 874.

Wickner, R. B., H. K. Edskes, M. L. Maddelein, K. L. Taylor, and H. Moriyama. 1999. Prions of yeast and fungi. Proteins as genetic material. *J. Biol. Chem.* 274: 555–8.

Wickner, R. B., H. K. Edskes, B. T. Roberts, et al. 2004. Prions: proteins as genes and infectious entities. *Genes Dev.* 18: 470–85.

Wickner, R. B., K. L. Taylor, H. K. Edskes, and M. L. Maddelein. 2000. Prions: Portable prion domains. *Curr. Biol.* 10: R335–7.

Wilkins, D. K., S. B. Grimshaw, V. Receveur, C. M. Dobson, J. A. Jones, and L. J. Smith. 1999. Hydrodynamic radii of native and denatured proteins measured by pulsed field gradient NMR techniques. *Biochemistry* 38: 16424–31.

Wille, H., E. M. Mandelkow, J. Dingus, R. B. Vallee, L. I. Binder, and E. Mandelkow. 1992a. Domain structure and antiparallel dimers of microtubule-associated protein 2 (MAP2). *J. Struct. Biol.* 108: 49–61.

Wille, H., E. M. Mandelkow, and E. Mandelkow. 1992b. The juvenile microtubule-associated protein MAP2c is a rod-like molecule that forms antiparallel dimers. *J. Biol. Chem.* 267: 10737–42.

Williams, R. J. 1989. NMR studies of mobility within protein structure. *Eur. J. Biochem.* 183: 479–97.

Williams, R. M., Z. Obradovic, V. Mathura, et al. 2001. The protein non-folding problem: amino acid determinants of intrinsic order and disorder. *Pac. Symp. Biocomput.* 6: 89–100.

Williamson, M. P. 1994. The structure and function of proline-rich regions in proteins. *Biochem J.* 297: 249–60.

Winzeler, E. A., D. D. Shoemaker, A. Astromoff, et al. 1999. Functional characterization of the *S. cerevisiae* genome by gene deletion and parallel analysis. *Science* 285: 901–6.

Wise, M. J. and A. Tunnacliffe. 2004. POPP the question: what do LEA proteins do? *Trends Plant. Sci.* 9: 13–7.

Wishart, D. S., C. G. Bigam, A. Holm, R. S. Hodges, and B. D. Sykes. 1995. 1H, 13C and 15N random coil NMR chemical shifts of the common amino acids. I. Investigations of nearest-neighbor effects. *J. Biomol. NMR* 5: 67–81.

Wisniewskia, M., R. Webba, R. Balsamob, T. Closec, X. Yud, and M. Griffithd. 1999. Purification, immunolocalization, cryoprotective, and antifreeze activity of PCA60: a dehydrin from peach (*Prunus persica*). *Physol. Plant* 105: 600–08.

Wissmann, R., T. Baukrowitz, H. Kalbacher, et al. 1999. NMR structure and functional characteristics of the hydrophilic N terminus of the potassium channel beta-subunit Kvbeta1.1. *J. Biol. Chem.* 274: 35521–5.

Wittmann, T., G. M. Bokoch, and C. M. Waterman-Storer. 2004. Regulation of microtubule destabilizing activity of Op18/stathmin downstream of Rac1. *J. Biol. Chem.* 279: 6196–203.

Wool, I. G. 1996. Extraribosomal functions of ribosomal proteins. *Trends Biochem. Sci.* 21: 164–5.

Wootton, J. C. 1994a. Non-globular domains in protein sequences: automated segmentation using complexity measures. *Computers Chem.* 18: 269–85.

Wootton, J. C. 1994b. Sequences with "unusual" amino acid compositions. *Curr. Opin. Struct. Biol.* 4: 413–21.

Wootton, J. C. and M. H. Drummond. 1989. The Q-linker: a class of interdomain sequences found in bacterial multidomain regulatory proteins. *Protein Eng* 2: 535–43.

Wopfner, F., G. Weidenhofer, R. Schneider, et al. 1999. Analysis of 27 mammalian and 9 avian PrPs reveals high conservation of flexible regions of the prion protein. *J. Mol. Biol.* 289: 1163–78.

Wright, P. E. and H. J. Dyson. 1999. Intrinsically unstructured proteins: re-assessing the protein structure-function paradigm. *J. Mol. Biol.* 293: 321–31.

Wu, G., Y. G. Chen, B. Ozdamar, et al. 2000. Structural basis of Smad2 recognition by the Smad anchor for receptor activation. *Science* 287: 92–7.

Wu, K., M. E. Bottazzi, C. De La Fuente, et al. 2004. Protein profile of tax-associated complexes. *J. Biol. Chem.* 279: 495–508.

Wutrich, K. 1986. *NMR of Proteins and Nucleic Acids*. New York: Wiley Interscience.

Xiao, H., R. Sandaltzopoulos, H. M. Wang, et al. 2001. Dual functions of largest NURF subunit NURF301 in nucleosome sliding and transcription factor interactions. *Mol. Cell* 8: 531–43.

Xie, H., S. Vucetic, L. M. Iakoucheva, et al. 2007. Functional anthology of intrinsic disorder. 1. Biological processes and functions of proteins with long disordered regions. *J. Proteome. Res.* 6: 1882–98.

Xie, Z. and L. H. Tsai. 2004. Cdk5 phosphorylation of FAK regulates centrosome-associated miocrotubules and neuronal migration. *Cell Cycle* 3: 108–10.

Yamamoto, A., V. Guacci, and D. Koshland. 1996. Pds1p is required for faithful execution of anaphase in the yeast, *Saccharomyces cerevisiae. J. Cell Biol.* 133: 85–97.

Yamamoto, T., S. Izumi, and K. Gekko. 2004. Mass spectrometry on segment-specific hydrogen exchange of dihydrofolate reductase. *J. Biochem. (Tokyo)* 135: 17–24.

Yang, J., T. D. Hurley, and A. A. Depaoli-Roach. 2000. Interaction of inhibitor-2 with the catalytic subunit of type 1 protein phosphatase. Identification of a sequence analogous to the consensus type 1 protein phosphatase-binding motif. *J. Biol. Chem.* 275: 22635–44.

Yang, W. Z., T. P. Ko, L. Corselli, R. C. Johnson, and H. S. Yuan. 1998. Conversion of a beta-strand to an alpha-helix induced by a single-site mutation observed in the crystal structure of Fis mutant Pro26Ala. *Protein Sci.* 7: 1875–83.

Yang, X. J. 2004a. The diverse superfamily of lysine acetyltransferases and their roles in leukemia and other diseases. *Nucleic. Acids Res.* 32: 959–76.

Yang, X. J. 2004b. Lysine acetylation and the bromodomain: a new partnership for signaling. *BioEssays* 26: 1076–87.

Yang, X. J. 2005. Multisite protein modification and intramolecular signaling. *Oncogene* 24: 1653–62.

Yang, Z. R., R. Thomson, P. Mcneil, and R. M. Esnouf. 2005. RONN: the bio-basis function neural network technique applied to the detection of natively disordered regions in proteins. *Bioinformatics* 21: 3369–76.

Yap, K. L., J. Kim, K. Truong, M. Sherman, T. Yuan, and M. Ikura. 2000. Calmodulin target database. *J. Struct. Funct. Genomics* 1: 8–14.

Yong, C., H. Mitsuyasu, Z. Chun, S. Oshiro, N. Hamasaki, and S. Kitajima. 1998. Structure of the human transcription factor TFIIF revealed by limited proteolysis with trypsin. *FEBS Lett.* 435: 191–4.

Young, R. A. 1991. RNA polymerase II. *Annu. Rev. Biochem.* 60: 689–715.

Yu, H., J. K. Chen, S. Feng, D. C. Dalgarno, A. W. Brauer, and S. L. Schreiber. 1994. Structural basis for the binding of proline-rich peptides to SH3 domains. *Cell* 76: 933–45.

Zagotta, W. N., T. Hoshi, and R. W. Aldrich. 1990. Restoration of inactivation in mutants of Shaker potassium channels by a peptide derived from ShB. *Science* 250: 568–71.

Zahn, R., A. Liu, T. Luhrs, et al. 2000. NMR solution structure of the human prion protein. *Proc. Natl. Acad. Sci. USA* 97: 145–50.

Zambelli, B., M. Stola, F. Musiani, et al. 2005. UreG, a chaperone in the urease assembly process, is an intrinsically unstructured GTPase that specifically binds Zn2+. *J. Biol. Chem.* 280: 4684–95.

Zeev-Ben-Mordehai, T., E. H. Rydberg, A. Solomon, et al. 2003. The intracellular domain of the *Drosophila* cholinesterase-like neural adhesion protein, gliotactin, is natively unfolded. *Proteins* 53: 758–67.

Zhang, J. and J. L. Corden. 1991. Phosphorylation causes a conformational change in the carboxyl-terminal domain of the mouse RNA polymerase II largest subunit. *J. Biol. Chem.* 266: 2297–302.

Zhang, M. and P. Coffino. 2004. Repeat sequence of Epstein–Barr virus-encoded nuclear antigen 1 protein interrupts proteasome substrate processing. *J. Biol. Chem.* 279: 8635–41.

Zhang, X., M. A. Perugini, S. Yao, et al. 2008. Solution conformation, backbone dynamics and lipid interactions of the intrinsically unstructured malaria surface protein MSP2. *J. Mol. Biol.* 379: 105–21.

Zhang, Y., Y. Kim, N. Genoud, et al. 2006. Determinants for dephosphorylation of the RNA polymerase II C-terminal domain by Scp1. *Mol. Cell* 24: 759–70.

Zhang, Y., B. Stec, and A. Godzik. 2007. Between order and disorder in protein structures: analysis of "dual personality" fragments in proteins. *Structure* 15: 1141–7.

Zheng-Fischhofer, Q., J. Biernat, E. M. Mandelkow, S. Illenberger, R. Godemann, and E. Mandelkow. 1998. Sequential phosphorylation of tau by glycogen synthase kinase-3beta and protein kinase A at Thr212 and Ser214 generates the Alzheimer-specific epitope of antibody AT100 and requires a paired-helical-filament-like conformation. *Eur. J. Biochem.* 252: 542–52.

Zhu, F., J. Kapitan, G. E. Tranter, et al. 2007. Residual structure in disordered peptides and unfolded proteins from multivariate analysis and ab initio simulation of Raman optical activity data. *Proteins* 70: 823–33.

Zhuang, S., K. Mabuchi, and C. A. Wang. 1996. Heat treatment could affect the biochemical properties of caldesmon. *J. Biol. Chem.* 271: 30242–8.

Zitzewitz, J. A., B. Ibarra-Molero, D. R. Fishel, K. L. Terry, and C. R. Matthews. 2000. Preformed secondary structure drives the association reaction of GCN4-p1, a model coiled-coil system. *J. Mol. Biol.* 296: 1105–16.

Zor, T., B. M. Mayr, H. J. Dyson, M. R. Montminy, and P. E. Wright. 2002. Roles of phosphorylation and helix propensity in the binding of the KIX domain of CREB-binding protein by constitutive (c-Myb) and inducible (CREB) activators. *J. Biol. Chem.* 277: 42241–8.

Zou, H., T. J. Mcgarry, T. Bernal, and M. W. Kirschner. 1999. Identification of a vertebrate sister-chromatid separation inhibitor involved in transformation and tumorigenesis. *Science* 285: 418–22.

Index

C

D

E

F

G

H

I

N

Q

R

S

T

U